# Coatings and Surface Engineering

Edited by **Guy Lennon**

**C**WILLFORD PRESS

New York

Published by Willford Press,
118-35 Queens Blvd., Suite 400,
Forest Hills, NY 11375, USA
www.willfordpress.com

**Coatings and Surface Engineering**
Edited by Guy Lennon

International Standard Book Number: 978-1-68285-142-5 (Hardback)

# Contents

# Preface

This book has been a concerted effort by a group of academicians, researchers and scientists, who have contributed their research works for the realization of the book. This book has materialized in the wake of emerging advancements and innovations in this field. Therefore, the need of the hour was to compile all the required researches and disseminate the knowledge to a broad spectrum of people comprising of students, researchers and specialists of the field.

Coatings and surface engineering is a very significant discipline as it has applications ranging in diverse branches like chemistry, mechanical engineering, and electrical engineering. With state-of-the-art inputs by acclaimed experts of this field, this book targets students, professionals and academicians of various engineering branches. It covers the applications of coatings in different industries like optics and food industry. It also discusses the methods and techniques for ceramic coatings, magnetic coatings, multilayer coatings, etc. Coherent flow of topics, student-friendly language and extensive use of examples make this book an invaluable source of knowledge.

At the end of the preface, I would like to thank the authors for their brilliant chapters and the publisher for guiding us all-through the making of the book till its final stage. Also, I would like to thank my family for providing the support and encouragement throughout my academic career and research projects.

**Editor**

# Microstructures and Photovoltaic Properties of Zn(Al)O/Cu₂O-Based Solar Cells Prepared by Spin-Coating and Electrodeposition

**Takeo Oku \*, Tetsuya Yamada, Kazuya Fujimoto and Tsuyoshi Akiyama**

Department of Materials Science, The University of Shiga Prefecture, 2500 Hassaka, Hikone, Shiga 522-8533, Japan; E-Mails: zn21tyamada@yahoo.co.jp (T.Y.); ze61kfujimoto@ec.usp.ac.jp (K.F.); akiyama.t@mat.usp.ac.jp (T.A.)

\* Author to whom correspondence should be addressed; E-Mail: oku@mat.usp.ac.jp

**Abstract:** Copper oxide (Cu₂O)-based heterojunction solar cells were fabricated by spin-coating and electrodeposition methods, and photovoltaic properties and microstructures were investigated. Zinc oxide (ZnO) and Cu₂O were used as n- and p-type semiconductors, respectively, to fabricate photovoltaic devices based on In-doped tin oxide/ZnO/Cu₂O/Au heterojunction structures. Short-circuit current and fill factor increased by aluminum (Al) doping in the ZnO layer, which resulted in the increase of the conversion efficiency. The efficiency was improved further by growing ZnO and Cu₂O layers with larger crystallite sizes, and by optimizing the Al-doping by spin coating.

**Keywords:** Cu₂O; ZnO; photovoltaic cell

## 1. Introduction

Semiconductor oxides are a promising alternative to silicon based solar cells as they possess high optical absorption, nontoxicity, and have low production costs. Copper oxides (Cu₂O) are known to be p-type semiconductor oxides, and the crystal structure of Cu₂O is cubic system with a space group of Pn3m [1], as shown in Figure 1a. The Cu₂O is a suitable material for high efficiency solar cells because of its direct bandgap of 2.1 eV.

**Figure 1.** Crystal structures of (**a**) $Cu_2O$ and (**b**) ZnO; (**c**) Device structure of $ZnO/Cu_2O$ heterojunction solar cells.

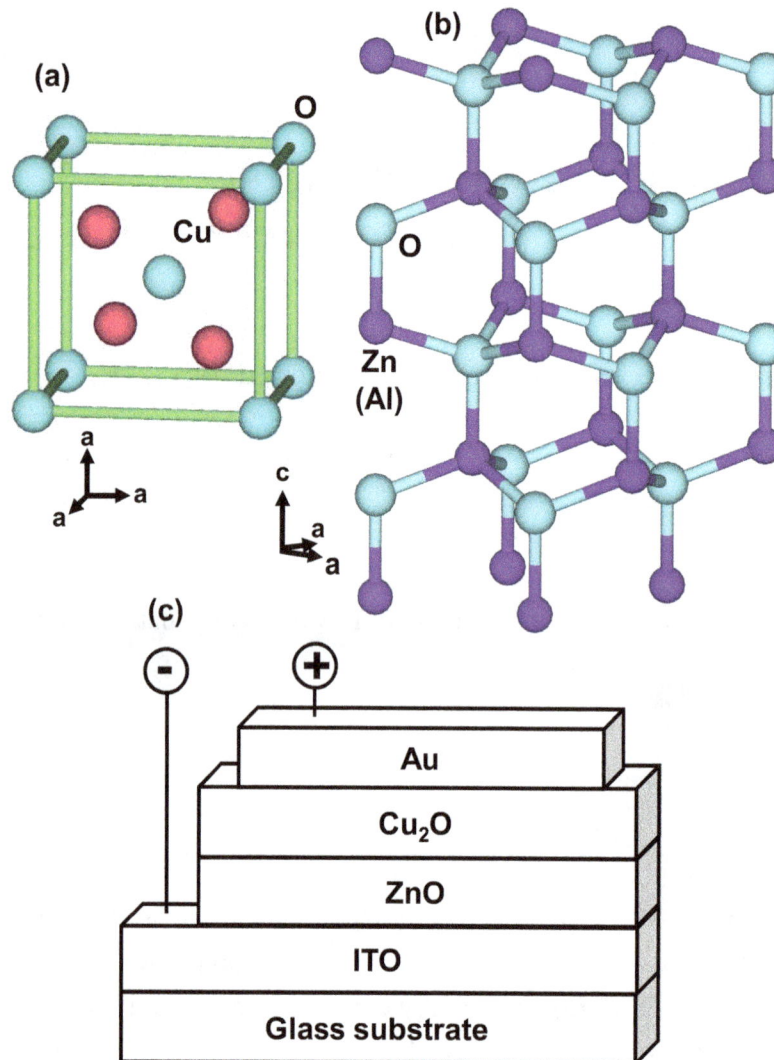

A maximum efficiency of 5.38% has been obtained for $Cu_2O$ solar cells fabricated by high-temperature annealing and pulsed laser deposition [2]. $Cu_2O$-based solar cells fabricated by electrodeposition and photochemical deposition have been reported [3–12], and zinc oxide (ZnO)/$Cu_2O$ thin film solar cells prepared by electrodeposition have also been reported [13–22]. Electrodeposition is a low-temperature method for solar cell fabrication with a low process cost. Izaki *et al.* [3] reported high conversion efficiency of 1.28% for electrodeposited $ZnO/Cu_2O$ solar cells using an electrolyte containing KOH for $Cu_2O$ deposition. Conversion efficiencies of 1.06 and 0.88% were also reported for p-$Cu_2O$/n-$Cu_2O$ [12] and $ZnO/Cu_2O$ [11] solar cells prepared by electrodeposition. Progress of the conversion efficiency up to 1.43% for electrodeposited $ZnO/Cu_2O$ solar cells was also reported by using an electrolyte containing LiOH for $Cu_2O$ deposition [23].

The purpose of the present work was to fabricate $ZnO/Cu_2O$ thin-film solar cells by spin-coating and electrodeposition and to investigate the effect of doping element into the ZnO layers. ZnO is a good electron acceptor with a hexagonal crystal system (space group P63mc) [24] as shown in Figure 1b, and has been used as an n-type semiconductor active layer for inorganic thin-film solar cells [25–30]. To improve the carrier transport in the ZnO structure, aluminum (Al) was selected in the present work for doping element at the Zn sites in the ZnO structure. The Al-doped ZnO (AZO) materials have also been reported for solar cell materials [31–35]. Spin coating is a low-cost method, and electrodeposition is a method for homogeneous thin film formation, which is essential for the mass production of any solar cells. The low cost methods have been already reported [3,36,37], and the spin coating of Al-doped ZnO in the present work would be also one of the useful and easy fabrication methods. In the present work, The ZnO/Cu$_2$O solar cells prepared in the present study were investigated by structural analysis, optical absorption, and photovoltaic measurements.

## 2. Experimental Procedures

A thin layer of zinc oxide (ZnO) was prepared by a sol-gel method [38,39]. A 2-methoxyethanol (0.97 mL, Wako, Osaka, Japan) solution containing zinc acetate dihydrate (0.1098 g/mL, Wako, 99%) and monoethanolamine (0.03 mL, Nacalai Tesque, Kyoto, Japan) was spin-coated at 2000 rpm on a pre-cleaned indium-doped tin oxide (ITO, Xin Yan Technology, Kowloon, Hong Kong, ~10 Ω/□) glass substrate. Aluminum nitrate enneahydrate (0.105 g, Wako, 98.0%) was also added into the above solution to form AZO layers with a composition of Al$_3$Zn$_{97}$. To form ZnO layers by a dipping method (ZnO-D), the spin-coated and annealed sample was also dipped in a solution (50 mL) of zinc nitrate hexahydrate (0.185 g, Wako, 99%) and hexamethylenetetramine (0.0876 g, Nacalai Tesque, Kyoto, Japan) at 90 °C for 60 min, and rinsed by distilled water and dried in air. Then, the ZnO layers were annealed at 300 °C for 60 min. The device process is schematically illustrated in Figure 2. The film thicknesses were measured by an atomic force microscope (AFM, SPA400-AFM, Hitachi High-Technologies, Tokyo, Japan), and the thickness of the ZnO and AZO was ~200 nm.

**Figure 2.** Device fabrication process ZnO/Cu$_2$O heterojunction solar cells.

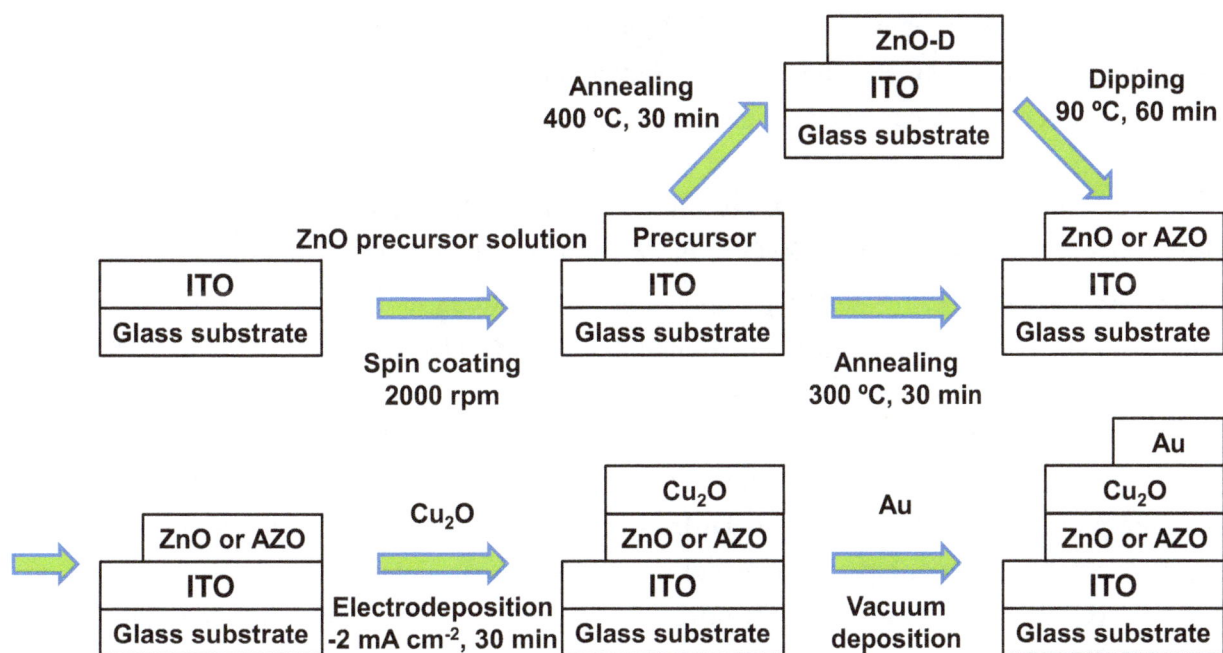

A Cu$_2$O layer was prepared on the ZnO layer by electrodeposition method using a platinum counter electrode. Copper (II) sulfate (CuSO$_4$, 5.107 g, Wako, 97.5%) and L-lactic acid (80 mL, Wako) were dissolved in distilled water. The pH of the electrolyte was adjusted to 12.5 by the addition of NaOH (Wako). The electrolyte temperature was kept at 65 °C during electrodeposition. The electrodeposition of Cu$_2$O layer was carried out at a current density of 2 mA cm$^{-2}$ using a potentio/galvanostat (Model 1110, Husou, Kanagawa, Japan). The film thickness of the Cu$_2$O layer was measured to be ~1.5 μm. Gold (Au) metal contacts were deposited as top electrodes. The structure of the heterojunction solar cells is thus ITO/ZnO/Cu$_2$O/Au, which is shown in Figure 1c as a schematic illustration.

Current density–voltage (J-V) characteristics of the solar cells were measured (HSV-110, Hokuto Denko, Tokyo, Japan) in the dark and under illumination at 100 mW·cm$^{-2}$ using an AM 1.5 solar simulator (XES-301S, San-ei Electric, Osaka, Japan). The solar cells were illuminated through the side of the FTO substrates and the illuminated area was 0.16 cm$^2$. Optical absorption of the thin films was investigated by UV–visible spectroscopy (V-670, Jasco, Tokyo, Japan) using the reflection spectra, and the thin films were illuminated through the side of the ITO substrates. The microstructures of the solar cells were investigated by X-ray diffractometry (X'Pert-MPD System, Philips, Amsterdam, The Netherlands) with CuKα radiation operating at 40 kV and 40 mA

## 3. Results and Discussion

X-ray diffraction (XRD) patterns of ZnO/Cu$_2$O-based solar cells are shown in Figure 3a. Diffraction peaks due to Cu$_2$O, ZnO, Au electrode and ITO substrate are observed in the XRD pattern. A strong diffraction peak of ZnO 002 is observed only for the ZnO-D/Cu$_2$O cell, which indicates the highly-oriented c-axis of ZnO perpendicular to the ITO substrate. No sharp peak due to ZnO was observed for the ZnO/Cu$_2$O and AZO/Cu$_2$O solar cells.

To investigate the microstructure of ZnO, XRD patterns of spin-coated ZnO and AZO thin films prepared on the glass plates were also measured as shown in Figure 3b, which indicates 002 diffraction peaks of ZnO and AZO. Lattice constants and crystallite sizes of ZnO are summarized as listed in Table 1, and the crystallite sizes were estimated using Scherrer's equation: $D = 0.9\ \lambda/B\cos\theta$, where $\lambda$, $B$, and $\theta$ represent the wavelengths of the X-ray source, the full width at half maximum, and the Bragg angle, respectively. A crystallite size of ZnO in the ZnO-D sample is the largest, which indicates the crystal growth of ZnO during annealing and dipping the substrate. A lattice constant c of ZnO in the AZO sample is the smallest, which suggests Al doping at the Zn sites in the ZnO crystal because the ionic radius of Al$^{3+}$ (0.054 nm) is smaller compared with that of Zn$^{2+}$ (0.074 nm).

Table 2 shows lattice constants and crystallite sizes of Cu$_2$O determined from the XRD measurements. All cells have almost the same lattice constants as the reported Cu$_2$O structure [1], which indicates the high crystallinity of the present Cu$_2$O layers. Diffraction intensity ratios I$_{111}$/I$_{200}$ of Cu$_2$O on the ZnO for all the devices were larger compared to the normal randomly oriented powder samples. As listed in Table 2, Cu$_2$O crystals on the AZO are highly oriented along the [111] direction. The crystallite size of the Cu$_2$O on the ZnO-D was found to be the largest compared with those of other cells.

**Figure 3.** X-ray diffraction (XRD) patterns of (**a**) $Cu_2O$-based solar cells and (**b**) ZnO and Al-doped ZnO (AZO) thin films.

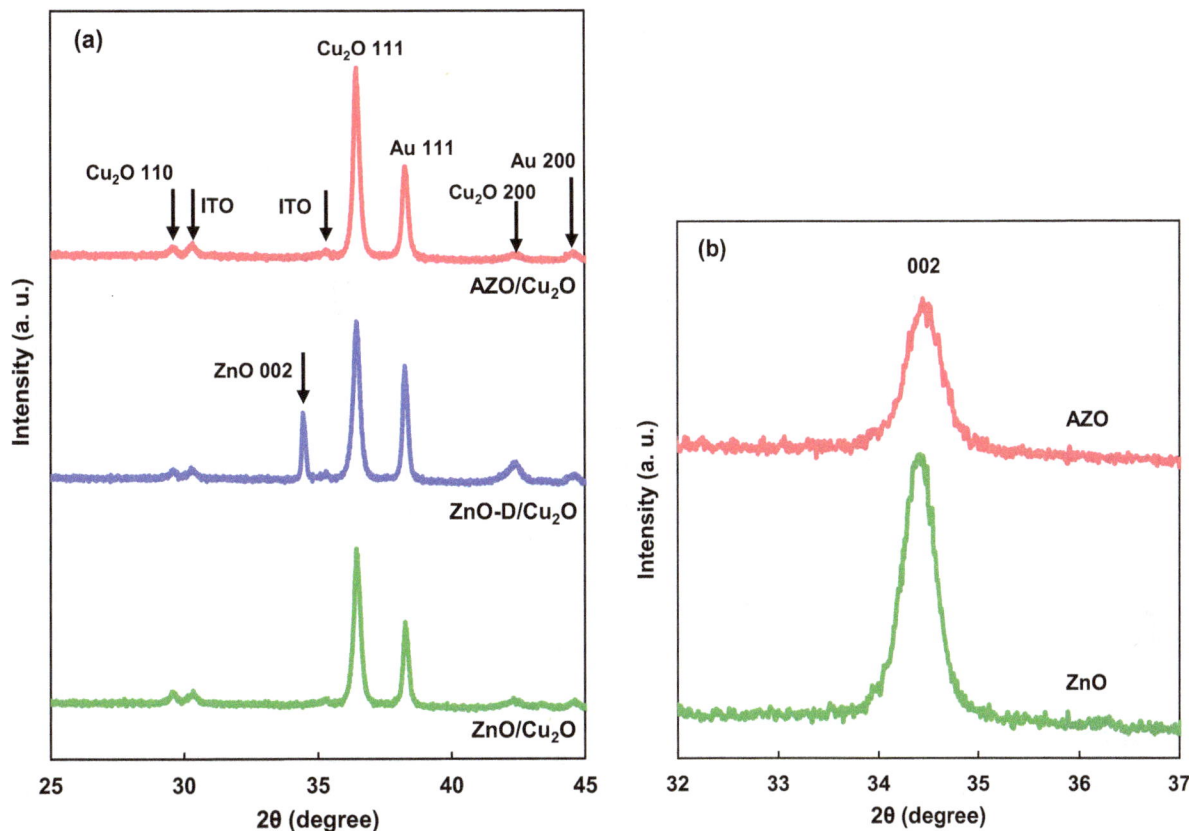

**Table 1.** Lattice constants and crystallite sizes of ZnO.

| ZnO formation | Lattice constant $c$ (nm) | Crystallite size (nm) |
|---|---|---|
| ZnO | 0.5211 | 25 |
| ZnO-D | 0.5208 | 104 |
| AZO | 0.5203 | 22 |
| ZnO [24] | 0.52066 | – |

**Table 2.** Lattice constants and crystallite sizes of $Cu_2O$.

| $Cu_2O$ formation | Lattice constant $a$ (nm) | Crystallite size (nm) | $I_{111}/I_{200}$ |
|---|---|---|---|
| $Cu_2O$ on ZnO | 0.4268 | 52 | 19.9 |
| $Cu_2O$ on ZnO-D | 0.4268 | 189 | 8.63 |
| $Cu_2O$ on AZO | 0.4268 | 86 | 35.7 |
| $Cu_2O$ [1] | 0.42696 | – | 2.7 |

Figure 4a shows optical absorption of the $Cu_2O$ and ZnO thin films. Absorption edges at ~630 nm and ~390 nm in Figure 4a are due to $Cu_2O$ and ZnO structure, which correspond to energy gaps of ~2.0 and ~3.2 eV, respectively. Figure 4b shows the optical absorption of the solar cells. The ZnO/$Cu_2O$ structures show high absorption in the range of 400–600 nm, which are due to $Cu_2O$ crystals. The origin of the absorption background observed above the wavelength of ~650 nm might be light scattering by $Cu_2O$ grains with the large grain size. The reason for the variation in its dependence on types of ZnO would be due to the film thickness. The absorbance difference of the three types of solar cells would be related to the $I_{111}$ orientations of $Cu_2O$ grains. As the $I_{111}$ orientations of $Cu_2O$ layers increased, the absorptions also increased, which would be due to the light confinement by the arrangement of the $Cu_2O$ grain with preferred orientations.

**Figure 4.** Optical absorption of (**a**) $Cu_2O$, ZnO, ZnO-D and AZO and (**b**) ZnO/$Cu_2O$-based solar cells.

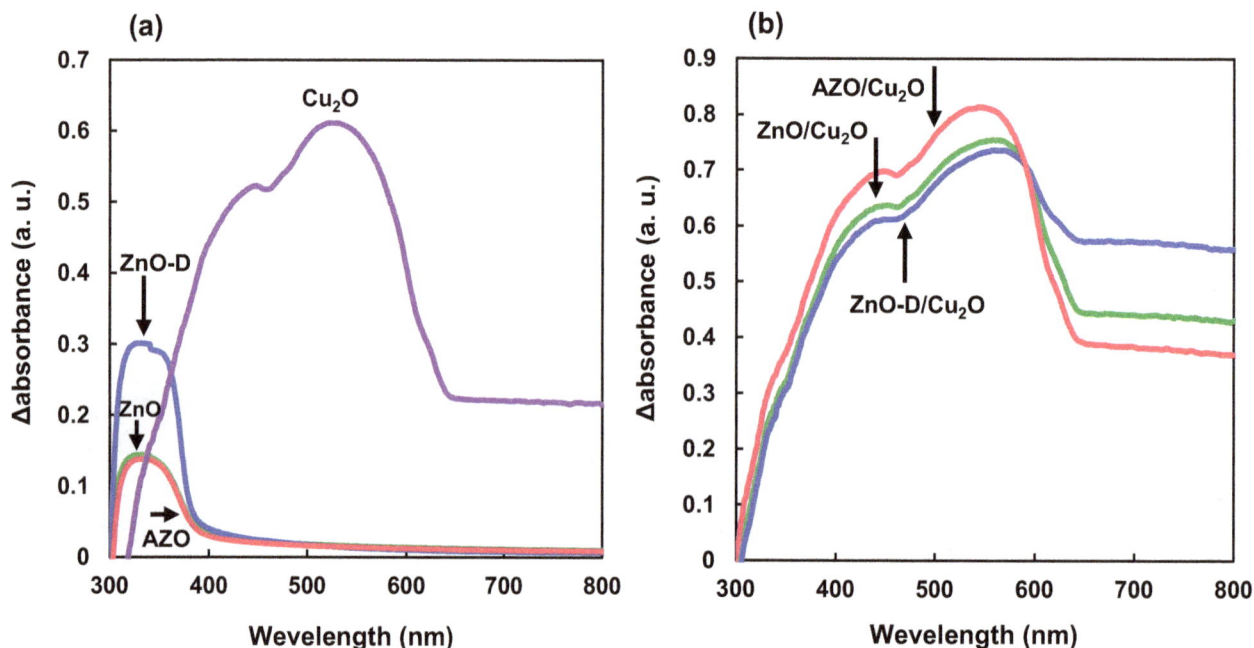

The J-V characteristics of the ZnO/$Cu_2O$ structures under illumination at 100 mW·cm$^{-2}$ obtained using an AM 1.5 solar simulator are shown in Figure 5. The photocurrent was observed under illumination, and the ZnO/$Cu_2O$ structure showed characteristic curves with regard to the short-circuit current and open-circuit voltage. A solar cell with the ITO/ZnO-D/$Cu_2O$/Au structure provided the highest power conversion efficiency ($\eta$) of 0.30%, a fill factor (FF) of 0.37, a short-circuit current density ($J_{SC}$) of 3.6 mA·cm$^{-2}$, and an open-circuit voltage ($V_{OC}$) of 0.24 V. The measured parameters of these ZnO/$Cu_2O$-based solar cells are summarized in Table 3.

An energy level diagram of the present ZnO/$Cu_2O$-based solar cells is summarized as shown in Figure 6. Measured and previously reported values were used for the energy levels [6,7]. Light was irradiated from the ITO substrate side, and was absorbed in the $Cu_2O$ layer. Charges were exited in the $Cu_2O$ layer, and were separated at the ZnO/$Cu_2O$ interface. Electrons are transported to the ITO substrate, and holes are transported to the Au electrode. It has been reported that $V_{OC}$ is nearly proportional to the band gap of the semiconductors [22], and control of the energy levels is important to increase efficiency.

In the present work, Al-doping to ZnO and dipping method for ZnO after spin-coating ZnO were found to be effective in improving the conversion efficiency of the ZnO/$Cu_2O$ solar cells. The Al-doping at the Zn sites in the ZnO structure improve the carrier transport in the ZnO structure, which would result in the increase of short-circuit current as listed in Table 3. In addition, the $Cu_2O$ prepared on the AZO provided the highest $I_{111}/I_{200}$, which indicates the formation of the highest {111}-oriented $Cu_2O$ crystallites.

**Figure 5.** J-V characteristic of ZnO/Cu$_2$O-based solar cells.

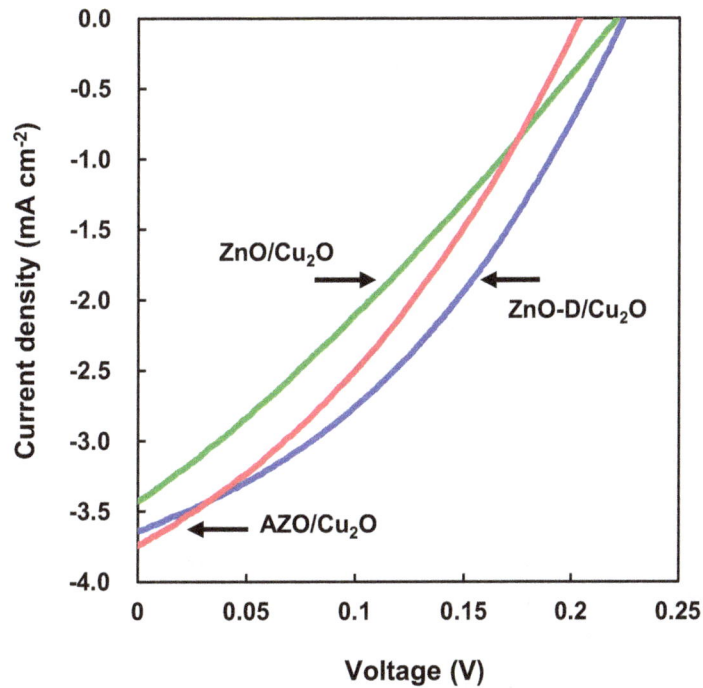

**Table 3.** Measured parameters of the present ZnO/Cu$_2$O-based solar cells.

| Devices | $J_{SC}$ (mA·cm$^{-2}$) | $V_{OC}$ (V) | FF | $\eta$ (%) | $R_s$ ($\Omega$·cm$^2$) | $R_{sh}$ ($\Omega$·cm$^2$) |
|---|---|---|---|---|---|---|
| ZnO/Cu$_2$O | 3.4 | 0.22 | 0.29 | 0.22 | 53 | 94 |
| ZnO-D/Cu$_2$O | 3.6 | 0.22 | 0.37 | 0.30 | 32 | 150 |
| AZO/Cu$_2$O | 3.7 | 0.20 | 0.34 | 0.26 | 32 | 100 |

**Figure 6.** Energy level diagram of ZnO/Cu$_2$O-based solar cells.

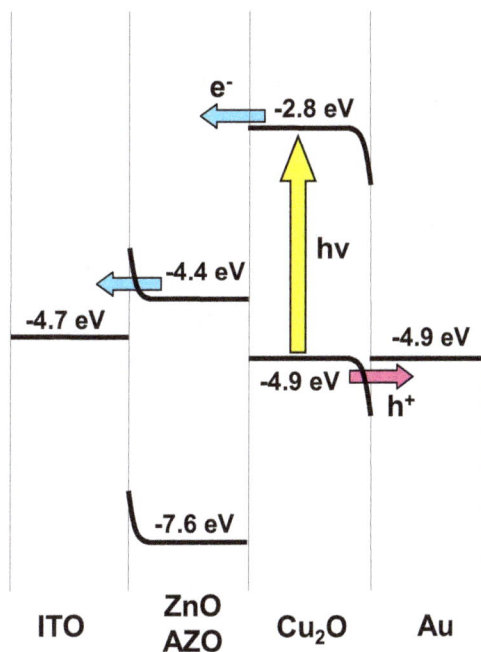

The ZnO layer prepared by a dipping method after spin-coating provided the largest crystallite size of ZnO and high 002-orientation, which indicates the crystal growth of ZnO during the dipping. A nanorod structure of ZnO might be formed by the dipping method [39], which could be also explained from optical absorption in Figure 3a. The Cu$_2$O prepared on the ZnO-D also provided the largest crystallite size, which would result in an increase in the shunt resistance ($R_{sh}$) and a decrease in the series resistance ($R_s$) of the ZnO-D/Cu$_2$O solar cell, as listed in Table 3. The increase in the $R_{sh}$ and the decrease in the $R_s$ indicate increase of carrier separation and decrease of inner electrical resistance. The ZnO-D/Cu$_2$O solar cell provided the highest fill factor in the present work, which indicates suppression of carrier recombination at the ZnO/Cu$_2$O interface. A decrease in the leakage current and an increase in $R_{sh}$ resulted in the improved photoelectric parameters.

Thickness effects were discussed in the previous works [36,37,40], and the effect of thickness reduction or decomposition of ZnO during electrodeposition of Cu$_2$O layer, which would be dependent on the several electrodeposition conditions. Optimization of thickness of active layers in the present solar cells could increase the efficiency of the solar cells.

In the present work, a simple spin-coating technique was applied for thin film coatings of ZnO layer. Al doping in the ZnO layer by simply adding aluminum nitrate enneahydrate to zinc acetate dehydrate solution resulted in the increase of the conversion efficiency. In addition, a simple dipping method in a zinc nitrate hexahydrate solution was applied to form ZnO layers on the spin-coated ZnO layer, which also resulted in the efficiency increase. These simple fabrication methods are useful for the fabrication of preliminary solar cells with a new structure.

## 4. Conclusions

ITO/ZnO(Al)/Cu$_2$O/Au-based heterojunction solar cells were fabricated by spin-coating and electrodeposition, and the photovoltaic properties and microstructures were investigated. The conversion efficiency was improved by aluminum doping in the ZnO layer, which was attributed to the increase of short-circuit current and fill factor. The efficiencies were improved further by growing the ZnO and Cu$_2$O layers with larger crystallite sizes and high [002]-orientation of ZnO crystal, which resulted in the increase of fill factor. The present method of spin-coating using aluminum doping and electrodeposition is expected for the simple fabrication method for Cu-based solar cells.

## Author Contributions

Takeo Oku wrote the manuscript and summarized the project. Tetsuya Yamada fabricated and characterized the solar cells, and summarized the results. Kazuya Fujimoto developed the deposition method for Cu$_2$O layers by the electrochemical deposition. Tsuyoshi Akiyama also developed the Cu$_2$O deposition method, and supported the project.

## Conflicts of Interest

The authors declare no conflict of interest.

# References

1. Swanson, H.E.; Fuyat, R.K. *Standard X-ray Diffraction Powder Patterns*; U.S. Dept. of Commerce, National Bureau of Standards: Washington, DC, USA, 1953.

2. Minami, T.; Nishi, Y.; Miyata, T. High-efficiency $Cu_2O$-based heterojunction solar cells fabricated using a $Ga_2O_3$ thin film as n-type layer. *Appl. Phys. Express* **2013**, *6*, doi:10.7567/APEX.6.044101.

3. Izaki, M.; Shinagawa, T.; Mizuno, K.; Ida, Y.; Inaba, M.; Tasaka, A. Electrochemically constructed p-$Cu_2O$/n-ZnO heterojunction diode for photovoltaic device. *J. Phys. D* **2007**, *40*, 3326–3329.

4. Duan, Z.; Pasquier, A.D.; Lu, Y.; Xu, Y.; Garfunkel, E. Effects of Mg composition on open circuit voltage of $Cu_2O$–$Mg_xZn_{1-x}O$ heterojunction solar cells. *Sol. Energy Mater. Sol. Cells* **2012**, *96*, 292–297.

5. Shao, F.; Sun, J.; Gao, L.; Luo, J.; Liu, Y.; Yang, S. High efficiency semiconductor-liquid junction solar cells based on $Cu/Cu_2O$. *Adv. Funct. Mater.* **2012**, *22*, 3907–3913.

6. Oku, T.; Takeda, A.; Nagata, A.; Kidowaki, H.; Kumada, K.; Fujimoto, K.; Suzuki, A.; Akiyama, T.; Yamasaki, Y.; Ōsawa, E. Microstructures and photovoltaic properties of $C_{60}$ based solar cells with copper oxides, $CuInS_2$, phthalocyanines, porphyrin, PVK, nanodiamond, germanium and exciton diffusion blocking layers. *Mater. Technol.* **2013**, *28*, 21–39.

7. Oku, T.; Motoyoshi, R.; Fujimoto, K.; Akiyama, T.; Jeyadevan, B.; Cuya, J. Structures and photovoltaic properties of copper oxides/fullerene solar cells. *J. Phys. Chem. Solids* **2011**, *72*, 1206–1211.

8. Jeong, S.S.; Mittiga, A.; Salza, E.; Masci, A.; Passerini, S. Electrodeposited $ZnO/Cu_2O$ heterojunction solar cells. *Electrochim. Acta* **2008**, *53*, 2226–2231.

9. Izaki, M.; Mizuno, K.; Shinagawa, T.; Inaba, M.; Tasaka, A. Photochemical construction of photovoltaic device composed of p-Copper(I) Oxide and n-Zinc Oxide. *J. Electrochem. Soc.* **2006**, *153*, C668–C672.

10. Anandan, S.; Wen, X.; Yang, S. Room temperature growth of CuO nanorod arrays on copper and their application as a cathode in dye-sensitized solar cells. *Mater. Chem. Phys.* **2005**, *93*, 35–40.

11. Cui, J.; Gibson, U.J. A simple two-step electrodeposition of $Cu_2O/ZnO$ nanopillar solar cells. *J. Phys. Chem. C* **2010**, *114*, 6408–6412.

12. McShane, C.M.; Choi, K.-S. Junction studies on electrochemically fabricated p-n $Cu_2O$ homojunction solar cells for efficiency enhancement. *Phys. Chem. Chem. Phys.* **2012**, *14*, 6112–6118.

13. Nakaoka, K.; Ueyama, J.; Ogura, K. Photoelectrochemical behavior of electrodeposited CuO and $Cu_2O$ thin films on conducting substrates. *J. Electrochem. Soc.* **2004**, *151*, C661–C665.

14. Leopold, S.; Herranen, M.; Carlsson, J.-O. Spontaneous potential oscillations in the Cu(II)/Tartrate and lactate systems, aspects of mechanisms and film deposition. *J. Electrochem. Soc.* **2001**, *148*, C513–C517.

15. Izaki, M.; Omi, T. Electrolyte optimization for cathodic growth of zinc oxide films. *J. Electrochem. Soc.* **1996**, *143*, L53–L55.

16. Golden, T.D.; Shumsky, M.G.; Zhou, Y.; VanderWerf, R.A.; Van Leeuwen, R.A.; Switzer, J.A. Electrochemical deposition of copper(I) oxide films. *Chem. Mater.* **1996**, *8*, 2499–2504.

17. Mizuno, K.; Izaki, M.; Murase, K.; Shinagawa, T.; Chigane, M.; Inaba, M.; Tasaka, A.; Awakura, Y. Structural and electrical characterizations of electrodeposited p-type semiconductor $Cu_2O$ films. *J. Electrochem. Soc.* **2005**, *152*, C179–C182.

18. Mahalingam, T.; John, V.S.; Hsu, L.S. Microstructural analysis of electrodeposited zinc oxide thin films. *J. New Mater. Electrochem. Syst.* **2007**, *10*, 9–14.

19. Georgieva, V.; Ristov, M. Electrodeposited cuprous oxide on indium tin oxide for solar applications. *Sol. Energy Mater. Sol. Cells* **2002**, *73*, 67–73.

20. Caballero-Briones, F.; Artés, J.M.; Díez-Pérez, I.; Gorostiza, P.; Sanz, F. Direct observation of the valence band edge by *in situ* ECSTM-ECTS in p-type $Cu_2O$ layers prepared by copper anodization. *J. Phys. Chem. C* **2009**, *113*, 1028–1036.

21. Chu, C.W.; Shrotriya, V.; Li, G.; Yang, Y. Tuning acceptor energy level for efficient charge collection in copper-phthalocyanine-based organic solar cells. *Appl. Phys. Lett.* **2006**, *88*, doi:10.1063/1.2194207.

22. Green, M.A.; Emery, K.; King, D.L.; Hishikawa, Y.; Warta, W. Solar cell efficiency tables (version 28). *Prog. Photovolt.* **2006**, *14*, 455–461.

23. Fujimoto, K.; Oku, T.; Akiyama, T. Fabrication and characterization of $ZnO/Cu_2O$ solar cells prepared by electrodeposition. *Appl. Phys. Express* **2013**, *6*, doi:10.7567/APEX.6.086503.

24. McMurdie, H.; Morris, M.; Evans, E.; Paretzkin, B.; Wong-Ng, W.; Ettlinger, L.; Hubbard, C. Standard X-ray diffraction powder patterns from the JCPDS research associateship. *Powder Diffr.* **1986**, *1*, 64–77.

25. Minami, T.; Miyata, T.; Ihara, K.; Minamino, Y.; Tsukada, S. Effect of ZnO film deposition methods on the photovoltaic properties of $ZnO–Cu_2O$ heterojunction devices. *Thin Solid Films* **2006**, *494*, 47–52.

26. Wei, H.; Gong, H.; Wang, Y.; Hu, X.; Chen, L.; Xu, H.; Liub, P.; Cao, B. Three kinds of $Cu_2O/ZnO$ heterostructure solar cells fabricated with electrochemical deposition and their structure-related photovoltaic properties. *CrystEngComm* **2011**, *13*, 6065–6070.

27. Fujimoto, K.; Oku, T.; Akiyama, T.; Suzuki, A. Fabrication and characterization of copper oxide-zinc oxide solar cells prepared by electrodeposition. *J. Phys. Conf. Ser.* **2013**, *433*, doi:10.1088/1742-6596/433/1/012024.

28. Mittiga, A.; Salza, E.; Sarto, F.; Tucci, M.; Vasanthi, R. Heterojunction solar cell with 2% efficiency based on a $Cu_2O$ substrate. *Appl. Phys. Lett.* **2006**, *88*, doi:10.1063/1.2194315.

29. Raksa, P.; Nilphai, S.; Gardchareon, A.; Choopun, S. Copper oxide thin film and nanowire as a barrier in ZnO dye-sensitized solar cells. *Thin Solid Films* **2009**, *517*, 4741–4744.

30. Minami, T.; Nishi, Y.; Miyata, T.; Nomoto, J. High-efficiency oxide solar cells with $ZnO/Cu_2O$ heterojunction fabricated on thermally oxidized $Cu_2O$ sheets. *Appl. Phys. Express* **2011**, *4*, doi:10.1143/APEX.4.062301.

31. Minami, T.; Tanaka, H.; Shimakawa, T.; Miyata, T.; Sato, H. High-efficiency oxide heterojunction solar cells using $Cu_2O$ sheets. *Jpn. J. Appl. Phys.* **2004**, *43*, L917–L919.

32. Fan, X.; Fang, G.; Guo, S.; Liu, N.; Gao, H.; Qin, P.; Li, S.; Long, H.; Zheng, Q.; Zhao, X. Controllable synthesis of flake-like Al-doped ZnO nanostructures and its application in inverted organic solar cells. *Nanoscale Res. Lett.* **2011**, *6*, 546:1–546:6.

33. Septina, W.; Ikeda, S.; Khan, M.A.; Hirai, T.; Harada, T.; Matsumura, M.; Peter, L.M. Potentiostatic electrodeposition of cuprous oxide thin films for photovoltaic applications. *Electrochim. Acta* **2011**, *56*, 4882–4888.

34. Kang, D.W.; Kuk, S.H.; Ji, K.S.; Lee, H.M.; Han, M.K. Effects of ITO precursor thickness on transparent conductive Al doped ZnO film for solar cell applications. *Sol. Energy Mater. Sol. Cells* **2011**, *95*, 138–141.

35. Minami, T.; Nishi, Y.; Miyata, T.; Abe, S. Photovoltaic properties in Al-doped ZnO/non-doped $Zn_{1-x}Mg_xO$/Cu$_2$O heterojunction solar cells. *ECS Trans.* **2013**, *50*, 59–68.

36. Liu, Y.; Turley, H.K.; Tumbleston, J.R.; Samulski, E.T.; Lopez, R. Minority carrier transport length of electrodeposited in heterojunction solar cells. *Appl. Phys. Lett.* **2011**, *98*, doi:10.1063/1.3579259.

37. Musselman, K.P.; Wisnet, A.; Iza, D.C.; Hesse, H.C.; Scheu, C.; MacManus-Driscoll, J.L.; Schmidt-Mende, L. Strong efficiency improvements in ultra-low-cost inorganic nanowire solar cells. *Adv. Mater.* **2010**, *22*, E254–E258.

38. Yi, S.H.; Choi, S.K.; Jang, J.M.; Kim, J.A.; Jung, W.G. Properties of aluminum doped zinc oxide thin film by sol-gel process. *Proc. SPIE* **2008**, *6831*, doi:10.1117/12.757447.

39. Xu, T.; Venkatesan, S.; Galipeau, D.; Qiao, Q. Study of polymer/ZnO nanostructure interfaces by Kelvin probe force microscopy. *Sol. Energy Mater. Sol. Cells* **2013**, *108*, 246–251.

40. Marin, A.T.; Muñoz-Rojas, D.; Iza, D.C.; Gershon, T.; Musselman, K.P.; MacManus-Driscoll, J.L. Novel atmospheric growth technique to improve both light absorption and charge collection in ZnO/Cu$_2$O thin film solar cells. *Adv. Funct. Mater.* **2013**, *23*, 3413–3419.

# 2

# Photocatalytic TiO$_2$ and Doped TiO$_2$ Coatings to Improve the Hygiene of Surfaces Used in Food and Beverage Processing—A Study of the Physical and Chemical Resistance of the Coatings

Parnia Navabpour [1,*], Soheyla Ostovarpour [2], Carin Tattershall [3], Kevin Cooke [1], Peter Kelly [2], Joanna Verran [2], Kathryn Whitehead [2], Claire Hill [3], Mari Raulio [4] and Outi Priha [4]

[1]  Teer Coatings Ltd., Miba Coating Group, West Stone House, Berry Hill Industrial Estate, Droitwich WR9 9AS, UK; E-Mail: kevin.cooke@miba.com

[2]  Faculty of Science and Engineering, Manchester Metropolitan University, Chester Street, Manchester M1 5GD, UK; E-Mails: soheyla.ostovarpour@stu.mmu.ac.uk (S.O.); peter.kelly@mmu.ac.uk (P.K.); j.verran@mmu.ac.uk (J.V.); k.a.whitehead@mmu.ac.uk (K.W.)

[3]  Cristal Pigment UK Ltd., P.O. Box 26, Grimsby, North East Lincolnshire, DN41 8DP, UK; E-Mails: carin.tattershall@cristal.com (C.T.); claire.hill@cristal.com (C.H.)

[4]  VTT Technical Research Centre of Finland, P.O. Box 1000, FI-02044 VTT Espoo, Finland; E-Mails: mari.raulio@tikkurila.com (M.R.); outi.priha@vtt.fi (O.P.)

*  Author to whom correspondence should be addressed; E-Mail: parnia.navabpour@miba.com

---

**Abstract:** TiO$_2$ coatings deposited using reactive magnetron sputtering and spray coating methods, as well as Ag- and Mo-doped TiO$_2$ coatings were investigated as self-cleaning surfaces for beverage processing. The mechanical resistance and retention of the photocatalytic properties of the coatings were investigated over a three-month period in three separate breweries. TiO$_2$ coatings deposited using reactive magnetron sputtering showed better mechanical durability than the spray coated surfaces, whilst the spray-deposited coating showed enhanced retention of photocatalytic properties. The presence of Ag and Mo dopants improved the photocatalytic properties of TiO$_2$ as well as the retention of these properties. The spray-coated TiO$_2$ was the only coating which showed light-induced hydrophilicity, which was retained in the coatings surviving the process conditions.

**Keywords:** photocatalytic; $TiO_2$; magnetron sputtering; spray coating; beverage processing

## 1. Introduction

In aquatic environments, microorganisms have a tendency to attach to surfaces along with organic and inorganic soil. For example in breweries, microorganisms have been shown to accumulate on sterile stainless steel surfaces within hours after the start of production [1].

Consumer demand is driving the development of a new group of more sensitive beverages with less alcohol, hop substances, and preservatives; however, these products are more prone to spoilage than are traditional drinks [2]. There are numerous operations involved in making beer. Each stage has a level of cleanliness that needs to be achieved and fouling is encountered at each stage [3]. Attachment of primary colonizers to stainless steel has been shown to be increased by sugars and sweeteners [1]. Thus removal of these deposits is essential since conditioning of a surface may be followed by biofilm formation. Biofilms on bottling plant surfaces are considered as serious sources for potential product spoiling microorganisms in the brewing industry [4]. Further, Fornalik [5] noted that minor fouling organisms resistant to cleaning in place (CIP) may become more resistant with time. Rheological studies indicated that increasing the temperature of the deposit generated a more elastic deposit which may decrease cleanability [3]. Thus, regular daily cleaning is needed. The following media are usually used in the cleaning process in brewing industry: water and steam, peroxide and alcohol based disinfectants, alkaline and acidic detergents and organic solvents [6,7]. There are however numerous drivers for a revision of CIP operations including the need to minimise utility usage (energy and water) and production downtime, minimisation of waste and greenhouse gas (GHG) emissions, and the need for product safety and quality [3].

One way to reduce cleaning costs and to improve process hygiene could be to use self-cleaning and antimicrobial coatings which can prevent the attachment of microorganisms and soil, or facilitate their efficient removal in the cleaning process.

$TiO_2$ is a widely used semiconductor. It has many different applications in optics [8], the environment [9], photovoltaics and solar cells [10,11], self-cleaning [12,13] and antimicrobial coatings [14]. In the self-cleaning and antimicrobial applications, the intended mechanism of action is often photocatalytic; in which the action of light on the $TiO_2$ coating generates active species that may be detrimental to microbes. For these applications, thin $TiO_2$ films with submicron thicknesses are usually employed. Several studies have been carried out to investigate the effect of crystal structure on the photocatalytic performance of $TiO_2$. Whilst some studies have found a higher activity of the anatase form [15,16], others have reported the mixed phase anatase/rutile to show a better photocatalytic performance [17]. Comparative studies of single phase anatase and rutile $TiO_2$ have concluded that the photocatalytic activity is dependent on the reaction being studied and different kinetics and intermediaries may be produced in each case [18,19]. As the surfaces used in the food and beverage industries are exposed to adverse environments (contact with water and beverages, cleaning solutions, abrasive wear during cleaning), scratch and corrosion resistance play important roles in their mechanical durability and chemical stability. Hence, it is important to satisfy several requirements, including good adhesion to the substrate, the retention of high activity and resistance to chemicals.

The adhesion of any film to its substrate is one of the most important properties of a thin film. The level of adhesion depends on the force required to separate atoms or molecules at the interface between film and substrate. The adhesion of a film to the substrate is strongly dependent on the chemical nature, cleanliness, and microscopic topography of the substrate surface [20]. The presence of contaminants on the substrate surface may increase or decrease the adhesion depending on whether the adsorption energy is increased or decreased, respectively. Also the adhesion of a film can be improved by providing more nucleation sites on the substrate, for instance, by using a fine-grained substrate or a substrate pre-coated with suitable materials. Of the deposition processes available, magnetron sputtering has been shown to produce well adhered and uniform coatings over wide areas [11]. In this process, the adhesion of the film to the substrate can be improved by ion-cleaning of the substrate prior to the coating deposition as well as additional ion bombardment during coating deposition which improves adhesion by providing intermixing on an atomic scale [21].

It has been shown throughout the literature that the chemical and structural properties of the active film have a profound impact on the overall photocatalytic performance. Photocatalytic performance is influenced by film characteristics including; composition, bulk and surface structure and nanostructure, atomic to nanoscale roughness, hydroxyl concentration, and impurity concentration (e.g., Fe and Cr) [22–25].

The work described in this paper investigates the chemical and mechanical durability, wettability and the retention of photocatalytic activity of selected coatings after being placed in different brewery process environments, in this case bottle/can filling lines in three Finnish breweries.

## 2. Experimental Section

### 2.1. Preparation of Coated Surfaces

The substrate material for all coatings was stainless steel AISI 304 2B ($75 \times 25 \times 1.6$ mm$^3$). Coatings were produced using either closed field unbalanced magnetron sputtering (CFUBMS) [21] or by spray-coating with a TiO$_2$ sol. Table 1 shows the coatings produced.

**Table 1.** Preparation method of coated surfaces.

| Code | Coating | Deposition Method |
|------|---------|-------------------|
| T1 | TiO$_2$ | Reactive magnetron sputtering |
| T2 | TiO$_2$-Ag (low) | – |
| T3 | TiO$_2$-Ag (high) | – |
| U1 | TiO$_2$ | Reactive magnetron sputtering + heat treatment |
| U2 | TiO$_2$-Mo | – |
| MC | TiO$_2$ | Spray-coated with TiO$_2$ sol |

Coatings T1–T3 were deposited using reactive magnetron sputtering in a Teer Coatings UDP 450 coating system. One titanium target (99.5% purity) was used for the deposition of TiO$_2$. Argon (99.998% purity) was used as the working gas and oxygen (99.5% purity) as the reactive gas. The working pressure was 1 mbar. Ag (99.95% purity) was used as the dopant. Advanced Energy Pinnacle Plus pulsed DC power supplies were used to power the titanium magnetrons and bias the substrates. An Advanced Energy DC power supply was used to power the silver target. 10–30 substrates were ultrasonically cleaned in acetone prior to loading into the chamber in order to remove surface contaminants. The substrates were aligned on a flat plate parallel to the surface of the metal targets at a distance of 150 mm from the target plane. A high rotational speed of 10 rpm was applied to the substrates to ensure enhanced mixing of silver and titanium within the coatings rather than the preferential formation of multilayer coatings. The substrates were ion-cleaned for a period of 20 min prior to the coating deposition using a bias voltage of −400 V and a low current of 0.2–0.35 A on the targets. The coatings were deposited at a bias voltage of −40 V. A thin layer of Ti was initially deposited as the adhesion layer prior to the introduction of oxygen to the deposition chamber. The amount of oxygen was controlled using an optical emission monitor, using conditions known to produce stoichiometric $TiO_2$ [26]. A pulsed-DC power of 2.5 kW was used on the Ti target at frequency 50 kHz and a duty of 97.75% (in synchronous mode). A continuous DC power of 70 W in the case of T2 and 150 W in the case of T3 was applied to the Ag target to vary the dopant content. The deposition rate was 17–22 nm/min depending on Ag content and coatings with thickness of 0.8–1 µm were produced. No additional heating was used during the coating process and the temperature did not exceed 200 °C during the process.

U1 and U2 coatings were deposited using reactive magnetron sputtering in a Teer Coatings UDP 450 coating system as described above. Two opposing magnetrons were fitted with titanium targets and one with the Mo dopant metal target (99.5% purity). The magnetrons with the titanium targets were in the closed field configuration and driven in pulsed DC sputtering mode using a dual channel Advanced Energy Pinnacle Plus supply at a frequency of 100 kHz and a duty of 50% (in synchronous mode). The Mo metal target was driven in a continuous DC mode (Advanced Energy MDX). The Ti targets were operated at a constant time-averaged power of 1 kW and the dopant target was operated at 180 W. Stainless steel samples were mounted on a substrate holder, which was rotated between the magnetrons at 4 rpm during deposition. The target to substrate separation was 8 cm. The titanium and Mo targets were cleaned by pre-sputtering in a pure argon atmosphere for 10 min. Deposition times were adapted to obtain a film thickness of 0.8–1 µm (deposition rate was 7.5 nm/min). The sputtered films were post deposition annealed at 600 °C for 30 min. in air.

Coating MC was prepared by spray-coating with a proprietary water-based $TiO_2$ sol using the following method. This transparent, neutral sol contained 2% $TiO_2$ (as anatase). Degreased stainless steel coupons were fixed to aluminium panels (approximately $150 \times 100$ mm$^2$). The panels with attached coupons were accurately weighed. The $TiO_2$ sol (0.2–0.3 g) was sprayed onto the aluminium panel with the attached coupons in a slow, steady motion, sweeping the panel in horizontal stripes from top to bottom, using a Badger Airbrush 200-3 model spray kit (Badger Air-Brush Co., Franklin Park, IL, USA). After air-drying for at least 15 min, the spraying procedure was repeated until 0.8–1.0 g/m$^2$ of $TiO_2$ sol was delivered to the surface. After air-drying overnight, the aluminium panel with the attached stainless steel coupons was re-weighed to give an accurate measurement of the weight per area of the coating.

## 2.2. Wettability

Water contact angle measurement is a practical tool to determine the wettability of a surface. Contact angle values were measured using a Digidrop instrument. At least two drops were measured for each surface and the measurements averaged. Measurements were conducted after exposure in light, either SUNTEST CPS+ (xenon arc, filtered with special window glass, 550 W/m$^2$ across the irradiance range 320–800 nm) or UVA light (Philips blacklight, 10–12 W/m$^2$ across the irradiance range 350–400 nm), or after storage in the dark.

## 2.3. Adhesion of Coatings

The scratch and wear resistance of the coatings were assessed using a Teer ST3001 scratch–wear tester (Teer Coatings Ltd, Droitwich, UK) [27]. The coated surfaces were evaluated using a Rockwell diamond tip (radius 200 μm). A load rate of 100 N·min$^{-1}$ and a constant sliding speed of 10.0 mm·min$^{-1}$ were used with the load increasing from 10 to 40 N. The scratch tracks were examined using a Cambridge Stereoscan 200 scanning electron microscope (Cambridge Instruments, Cambridge, UK) in order to detect any flaking.

## 2.4. Photocatalytic Characterization of Coatings

The photocatalytic activities of the coatings were analyzed using the methylene blue (MB) degradation assay under UV and fluorescent light sources. In brief, MB solutions were made up to an initial concentration of 0.0105 mMol·L$^{-1}$. Photocatalytic surfaces were placed in 10 mL of the MB solution and irradiated at an integrated power flux of 40 W/m$^2$ with two 15 W UV lamps (365 nm wavelength). Tests were also carried out using two 15 W fluorescent tubes in place of the UV tubes to simulate typical lighting environments. The integrated power flux to the coatings with the fluorescent tubes was 64 W/m$^2$, of which the UV component (300–400 nm) was 13 W/m$^2$. A 10 cm distance between the light source and MB solution was used. Samples of the MB solution were taken before testing and at 1 hour intervals up to a total of 5–8 h. and analyzed using a UV-Vis spectrophotometer (Perkin Elmer, Waltham, MA, USA). Spectra were taken in the range of 650–668 nm and the height of the absorption peak in this region was monitored.

A graph of peak height absorbance against irradiation time, which has an exponential form was generated. An index of photocatalytic activity (*Pa*) was defined by comparing the degradation rate of the MB solution in contact with the coated surfaces to the rate for an irradiated MB solution with no coating present. The equation below was used to calculate the photocatalytic activity of each of the films. Two parameters were defined: *Pa*$_{UV}$ for UV irradiation and *Pa*$_{FL}$ for fluorescent light irradiation [28].

$$Pa = 1 - C_0 \left[ e^{-mx} \Big/ e^{-cx} \right] \qquad (1)$$

where $C_0$ = peak height at time = 0; $C_0 e^{-mx}$ = decay rate of methylene blue; $C_0 e^{-cx}$ = decay rate of methylene blue in contact with photocatalytic coating.

## 2.5. Process Tests

Coated stainless steel pieces were placed on process surfaces within three breweries for a period of three months. Figure 1 shows an example of samples in location. Details of the location of test pieces in each brewery are given in Table 2. There was no special provision of lighting for the photocatalytic coatings; the process test took place under the usual brewery conditions of lighting, with coupons receiving varying amounts of light depending on their position in each machine. Furthermore, all samples underwent the normal process conditions and cleaning regimes used in each brewery which included acid and alkaline cleaning chemicals such as acetic acid and sodium hydroxide, ethanol, steam and mechanical brushing. For each coating, two replicates were used in each of the breweries. Additionally, two replicates were retained as controls and were kept in the dark for the same period. After three months, each replicate was cut into six sections. The mechanical durability, photocatalytic activity and wettability were evaluated each on two of these sections.

**Figure 1.** Samples in location at Brewery B.

**Table 2.** Coatings evaluated in process tests (for a period of three months).

| Coating | Control | Brewery A [1] | Brewery B [2] | Brewery C [3] |
|---------|---------|-----------|-----------|-----------|
| $TiO_2$ (T1) | T1-R1 T1-R2 | T1-1 T1-2 | T1-3 T1-4 | T1-5 T1-6 |
| $TiO_2$-Ag (low) (T2) | T2-R1 T2-R2 | T2-1 T2-2 | T2-3 T2-4 | T2-5 T2-6 |
| $TiO_2$-Ag (high) (T3) | T3-R1 T3-R2 | T3-1 T3-2 | T3-3 T3-4 | T3-5 T3-6 |
| $TiO_2$ (U1) | U1-R1 U1-R2 | U1-1 U1-2 | U1-3 U1-4 | U1-5 U1-6 |
| $TiO_2$-Mo (U2) | U2-R1 U2-R2 | U2-1 U2-2 | U2-3 U2-4 | U2-5 U2-6 |
| $TiO_2$ (MC) | MC-R1 MC-R2 | MC-1 MC-2 | MC-3 MC-4 | MC-5 MC-6 |

[1] Filler table of beer canning machine; [2] Seamer of beer canning machine; [3] Filler table of a water and soft drinks PET line, inclined 10°.

## 3. Results and Discussion

This work compared three $TiO_2$ surfaces: as-deposited and heat-treated coatings deposited by reactive magnetron sputtering (T1 and U1, respectively), and a spray-coated $TiO_2$ (MC). Two dopants (Ag and Mo) were also investigated. Ag was used as it is a well-known antimicrobial material which could impart additional antimicrobial functionality to the coating. Mo was used as a dopant to reduce the band gap of $TiO_2$ in order to improve the visible light activity of $TiO_2$. Mo-$TiO_2$ has been reported to shift the band gap of $TiO_2$ by $-0.20$ eV [28]. The photoactivity and mechanical properties of the surfaces were studied for the as-prepared coatings and those having undergone process conditions. The effect of the process conditions on the properties of the coatings was investigated.

### 3.1. As Prepared Coatings

SEM and EDX were used to analyze the topography and dopant concentration (as atomic percent of total metals) in the as-prepared doped coatings. Ag-$TiO_2$ and Mo-$TiO_2$ surfaces showed small submicron sized particles which were characterized by EDX as silver rich phases, suggesting that the dopant separated from the matrix $TiO_2$. The silver content was $0.50 \pm 0.05$ at% in T2 and $30.0 \pm 3.1$ at% in T3. The Mo content in U2 was $7.0 \pm 0.8$ at %. The structure of coatings was analysed using XRD (Figure 2). The as-deposited $TiO_2$ coating (T1), showed an anatase structure. Ag-$TiO_2$ coatings showed strong silver peaks. The heat treated $TiO_2$ and Mo-$TiO_2$ (U1 and U2) showed anatase and rutile peaks as well as monoclinic $\beta$-$TiO_2$ which were very strong in the case of the doped coating.

**Figure 2.** Microstructure of coatings as evaluated using XRD, (**a**) as deposited $TiO_2$ and Ag-$TiO_2$ coatings (T1–T3 ); and (**b**) $TiO_2$ and Mo-$TiO_2$ coatings after heat treatment (U1 and U2) (S—substrate, An—anatase, Ru—rutile).

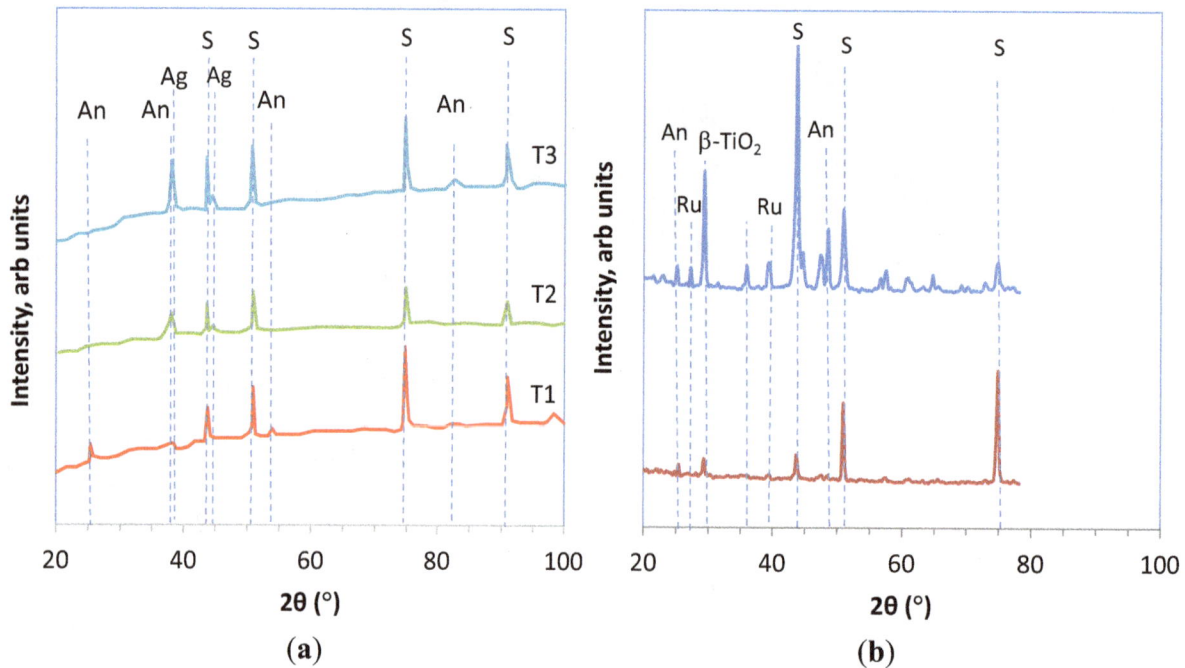

(a)                                                        (b)

Figure 3 shows the photocatalytic activity for the as-prepared coatings and compares these values with those obtained for Pilkington Activ™ as a standard commercial product. As can be seen, all coatings showed high photocatalytic activity. In the case of T3, a change was also observed in the colour of the solution. This was thought to have been caused by leaching of silver from the surface. SEM analysis of the coating was performed before and after immersion in water for 2 h and showed the presence of microparticles on the surface which EDX confirmed to be silver (Figure 4). The silver microparticles in the as deposited coating were embedded in the matrix. Immersion in water resulted in the silver particles to protrude from the surface and EDX showed a reduction in the silver content, confirming that silver was indeed diffusing out of the coating.

**Figure 3.** Photocatalytic activity of the as-deposited coatings and comparison with a commercially available photocatalytic surface (Pilkington Activ™).

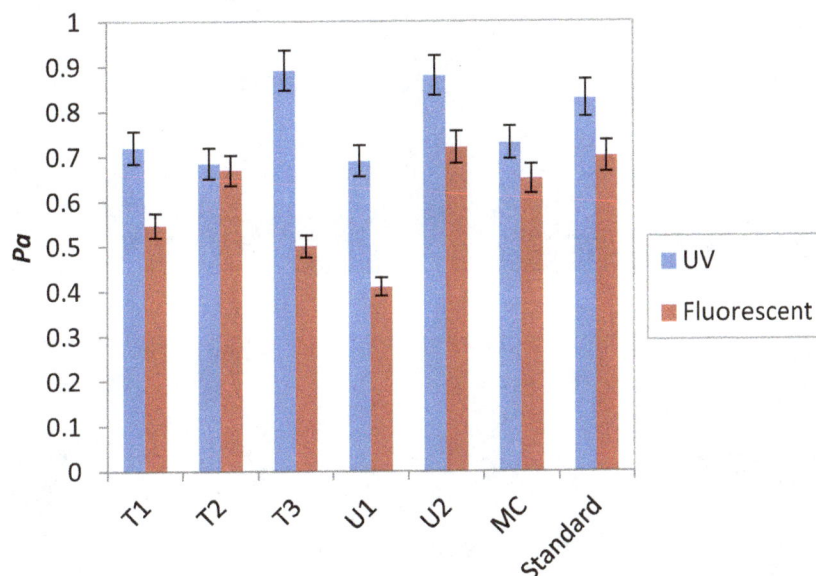

**Figure 4.** SEM micrographs of T3, (**a**) as deposited coating; and (**b**) after being under water for 2 h.

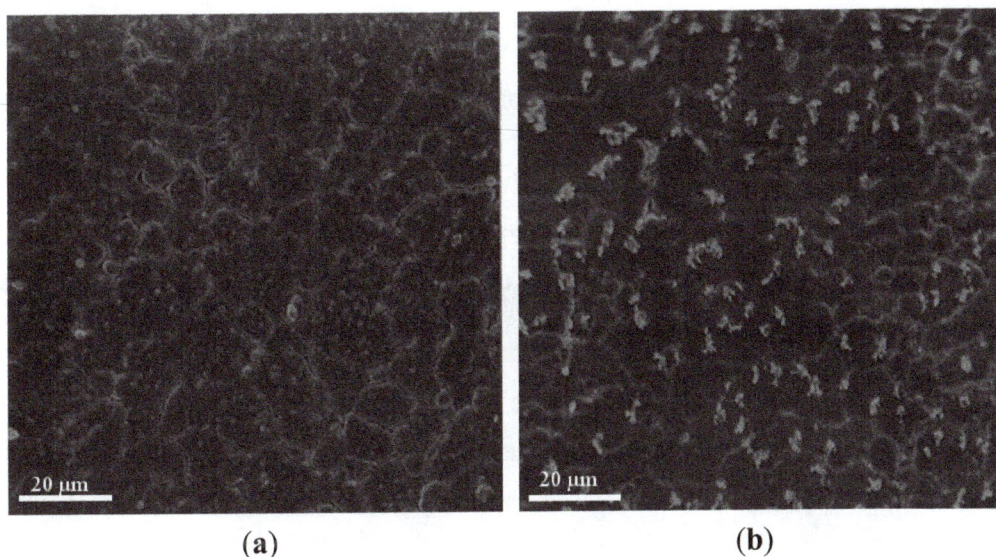

(a)                                (b)

Mechanical resistance of the coatings was analyzed using scratch testing. Figure 5 shows the scratch tracks of the coatings after production, as observed using the SEM. Coatings T1–T3 and MC showed excellent adhesion to the stainless steel substrate and no flaking was observed around the scratch tracks. Slight flaking was observed in U1 and U2, which was localized to the area immediately next to the scratch track. This may have been caused by the lack of a Ti adhesion layer in these coatings or due to the stresses applied to the coating during annealing. Given the destructive nature of the scratch test and the high load levels used in this test, all coatings were deemed to show sufficient mechanical resistance for use on food and drinks processing surfaces.

**Figure 5.** Progressive load scratch tracks of **(a)** T1; **(b)** T2; **(c)** T3; **(d)** U1; **(e)** U2 and **(f)** MC.

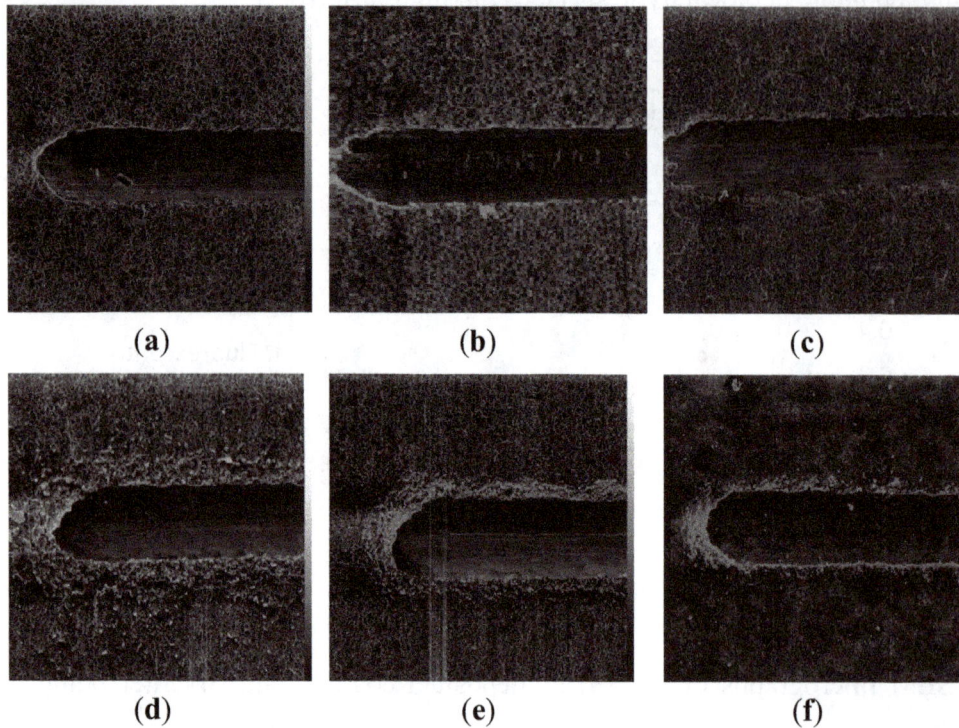

|       |       |       |
|:-----:|:-----:|:-----:|
| **(a)** | **(b)** | **(c)** |
| **(d)** | **(e)** | **(f)** |

## 3.2. Properties of the Surfaces after Process Tests

Visual inspection of the coatings after the three months process trial and their comparison with the control surfaces showed that all coatings prepared by magnetron sputtering (T1, T2, T3, U1 and U2) were physically present, although some color changes were apparent (Figure 6). The $TiO_2$ sol coating (MC) appeared to be still present after the process test at Brewery C but was at least partially removed at the other two breweries. It was noticeable that many of the surfaces were heavily soiled, particularly those that had been on trial at Breweries A and B.

**Figure 6.** Images of coupons after the three month brewery trial. See Table 2 for sample descriptions.

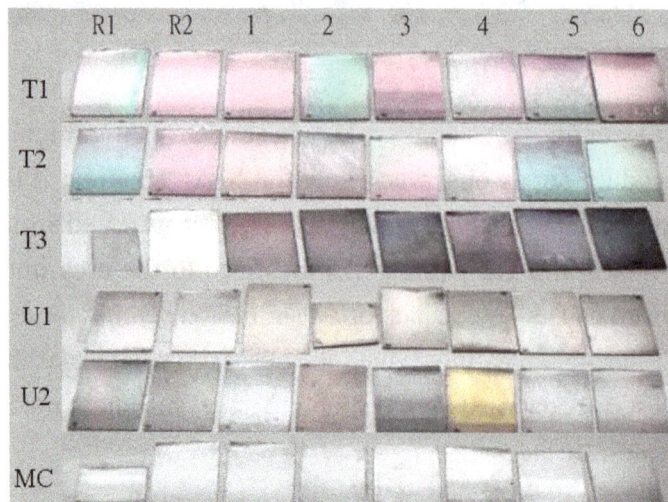

## 3.3. Mechanical Durability of the Coatings

The results of scratch adhesion tests performed on samples after the process trial confirmed the observations made on the appearance of coatings. Representative results are shown in Figure 7. T1–T3 coatings showed good adhesion with no flaking after the process studies. U1–U2 coatings showed some flaking, which in most cases was confined to the area immediately next to the scratch track. Some of the samples, however, showed a more widespread flaking. This was most likely caused by the lack of a Ti base layer, which can enhance the adhesion of $TiO_2$ to the stainless steel substrate or alternatively could be a result of the heat treatment. The MC coating from Breweries A and C showed some flaking near the scratch track. Samples removed from Brewery B showed no flaking. EDX analysis of these samples showed a Ti peak which had been greatly reduced compared to that of the control samples, suggesting that the coating had been heavily worn. This could be due to the different cleaning regimes, e.g., chemicals and scrubbing methods used in the different breweries, with some conditions exceeding the chemical and mechanical resistance of the coating.

**Figure 7.** SEM micrographs showing the scratch tracks of coatings before and after process tests at Brewery C.

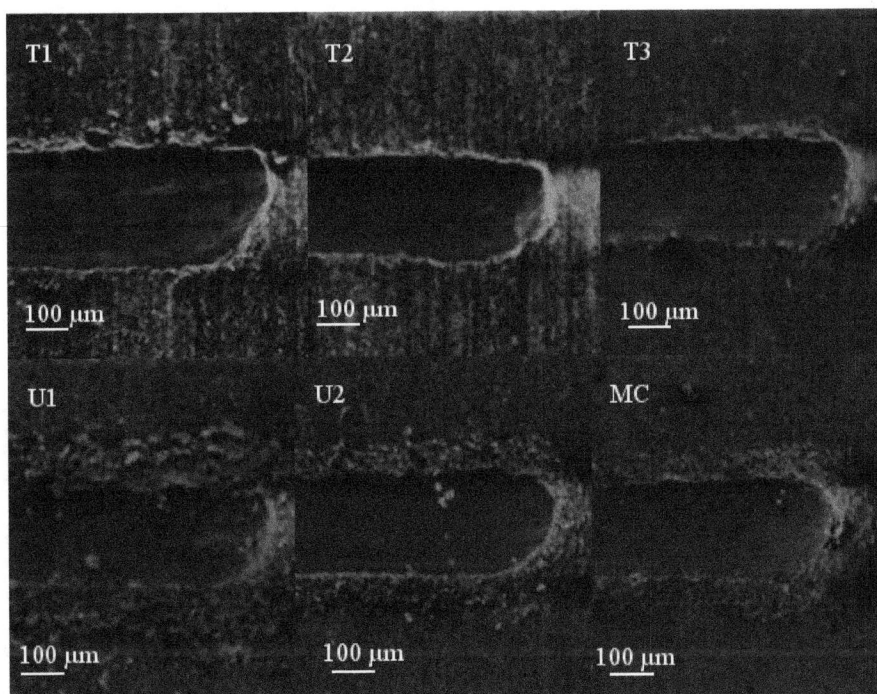

## 3.4. Composition of the Coatings

EDX results showed that the Ag content in T2 remained fairly constant. T3 showed a high level of Ag leaching possibly caused due to the poor dispersion and segregation of Ag within the coating as was seen from the SEM image of this coating (Figure 3). U2 showed a fairly constant concentration of Mo, except that in the areas where coating had been partially removed, it was not possible to measure the relative concentration of Mo in the coatings due to the weak signal and overlapping of the emission lines from the coating with those from the substrate (Table 3).

**Table 3.** Concentration of dopant as analysed using EDX (error in the measurements was ±10%).

| Coating | As Deposited | Control | | Brewery A | | Brewery B | | Brewery C | |
|---|---|---|---|---|---|---|---|---|---|
| | | R1 | R2 | 1 | 2 | 3 | 4 | 5 | 6 |
| TiO$_2$-Ag (low Ag) (T2) | 0.5 | 0.5 | | 0.2 | 0.1 | 0.3 | 0.6 | 0.7 | 0.1 |
| TiO$_2$-Ag (high Ag) (T3) | 30.0 | 32.0 | 34.0 | 1.8 | 1.6 | 9.2 | 1.1 | 3.6 | 10.3 |
| TiO$_2$-Mo (U2) | 7.0 | 7.3 | 7.2 | – | – | – | 8 | 8.1 | 8.0 |

## 3.5. Photocatalytic Properties

Figure 8 shows the photocatalytic activity of the coatings under fluorescent and UV irradiation.

**Figure 8.** Photoactivity of TiO$_2$ and doped TiO$_2$ coatings under UV (blue bars) and fluorescent light (red bars) irradiation. (**a**) T1; (**b**) T2; (**c**) T3; (**d**) U1; (**e**) U2; (**f**) MC.

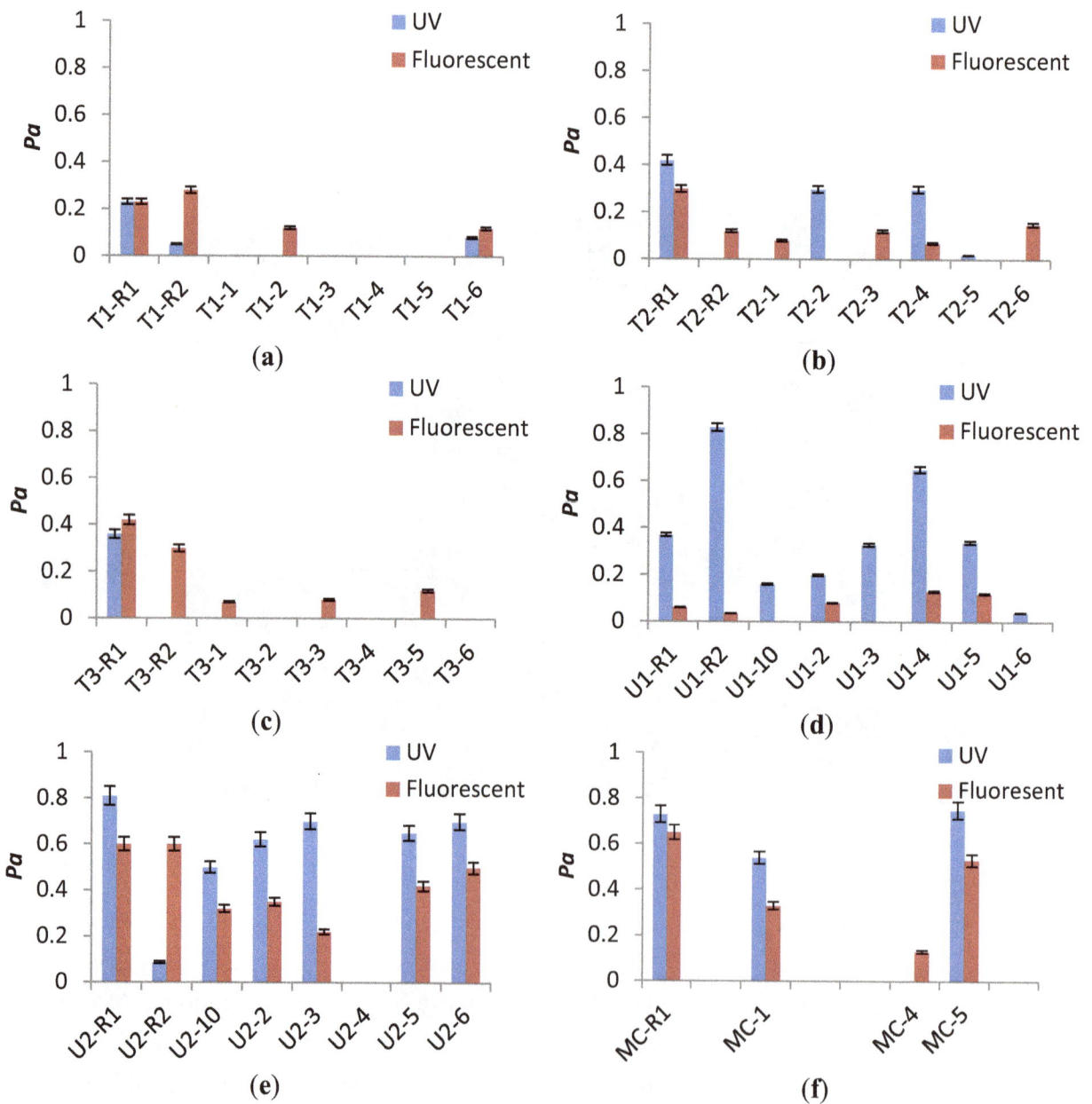

A loss of activity for T1–T3 coatings under UV light following the brewery trials was seen to varying degrees. The lower content $TiO_2$-Ag surface (T2) retained the most activity with the exception of samples received from Brewery C. A greater loss of photocatalytic properties of the higher doped Ag coatings was seen, possibly due to the leaching of silver during the process studies. The controls also lost activity following three months storage in the dark compared to the as-deposited samples (UV light). Similar results were seen when photocatalytic activity was assessed under fluorescent light. Comparison of the photocatalytic properties of U1 and U2, showed that the addition of Mo to the heat-treated $TiO_2$ surface increased its photocatalytic activity under UV and fluorescent light and this remained the case following the process studies. Photoactivity was largely retained for Mo-doped surfaces from all breweries with the exception of one of the two samples received from Brewery B. $TiO_2$ alone retained some of its photoactivity to varying degrees when irradiated with UV, although values between the duplicate samples differ. Less activity was shown under fluorescent light exposure, as expected and controls also showed lower photocatalytic activity compared to the as-deposited samples. Compared to the controls stored in the dark, the MC $TiO_2$ surfaces retained much of their photocatalytic activity, with the exception of samples received from Brewery B (under UV), where scratch test and EDX results had shown very little coating had been left on the substrate surface after the trial. As a small area of the substrate remained uncoated during the spray coating process, duplicate samples were not available in the case of MC surfaces.

The differences in photocatalytic activities of the surfaces received from the breweries could be due to the position of the samples and the cleaning regimes used. Work by others has shown that canning machines were markedly less prone to accumulation of microorganisms than bottling machines which use recycled glass bottles [1]. Further, it has been suggested that horizontal surfaces were prone to microbial accumulation and should be avoided in constructions as much as possible. Biofilm formation has also been shown to occur on certain surfaces despite daily cleaning and disinfection [1]. Thus, deposits formed by reaction processes or microbes usually cannot be wholly removed with water from stainless steel [29]. Various cleaners may have different success. In a surface test without soil a hypochlorite-based disinfectant was shown to be effective after an exposure of 10 min against all the microbes tested whereas an isopropanol-based cleaning agent was effective against all the vegetative cells tested [30]. In the presence of soil, hypochlorite was effective against *Listeria monocytogenes* and *Pseudomonas aeruginosa* [30]. The nature of clean may also affect efficacy. At 30 and 50 °C water rinsing at the flow velocities investigated could remove up to 85% of a yeast deposit. At a water rinsing temperature of 70 °C, less yeast deposit could be removed overall [3]. If surfaces were soiled with chemical residue and not cleaned sufficiently, it is possible that this may have an effect on photocatalytic activity. Conversely over aggressive cleaners might damage the surface, as noted previously.

## 3.6. Wettability

Photo-induced hydrophilicity is often associated with photocatalytic $TiO_2$ coatings [31]. Large differences in the wettability of $TiO_2$ coatings after irradiation by light or after storage in the dark are believed to be due to the generation of hydrophilic radicals on the $TiO_2$ surface by the action of light. Measurement of contact angle had been found to be an effective and easy method of detecting the presence of the $TiO_2$ sol coating, MC. In addition, it is expected that the contact of contaminants with the surface is enhanced in the case of hydrophilic surfaces, resulting in an increase in the effect of the photocatalyst. Thus water contact angle measurements were made on each test coupon listed in Table 3 to help determine the presence and activity of each coating.

Contact angles were firstly measured for the coupons immediately on unpacking (dark), and then after 20 hours irradiation. It was noticeable that many of the coupons were heavily soiled so a portion of each sample was cleaned by wiping with 2-propanol on a soft cloth and then with water. Contact angles were re-measured after 20 h. under UVA light, and again after 6–7 days in the dark. The results are shown in Figures 9–11.

Figures 9–11 show that for most coatings, the effect of light on the wettability was more pronounced in the case of the reference surfaces than those having undergone the processing conditions. This may indicate changes in the coating activity resulting from the exposure to the cleaning chemicals *etc.* used during the processing.

**Figure 9.** Water contact angle measurements for (**a**) $TiO_2$ (T1); and (**b**) $TiO_2$-Ag (low) (T2) coupons after three-month Brewery Trial.

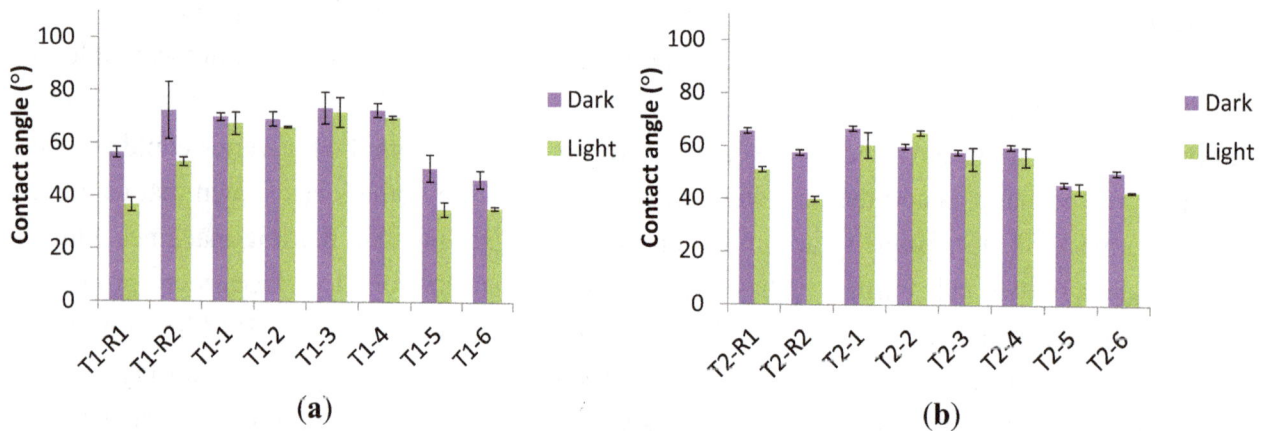

(**a**)                                                                                      (**b**)

**Figure 10.** Water contact angle measurements for (**a**) $TiO_2$-Ag (high) (T3) and (**b**) $TiO_2$ (U1) coupons after three month Brewery Trial.

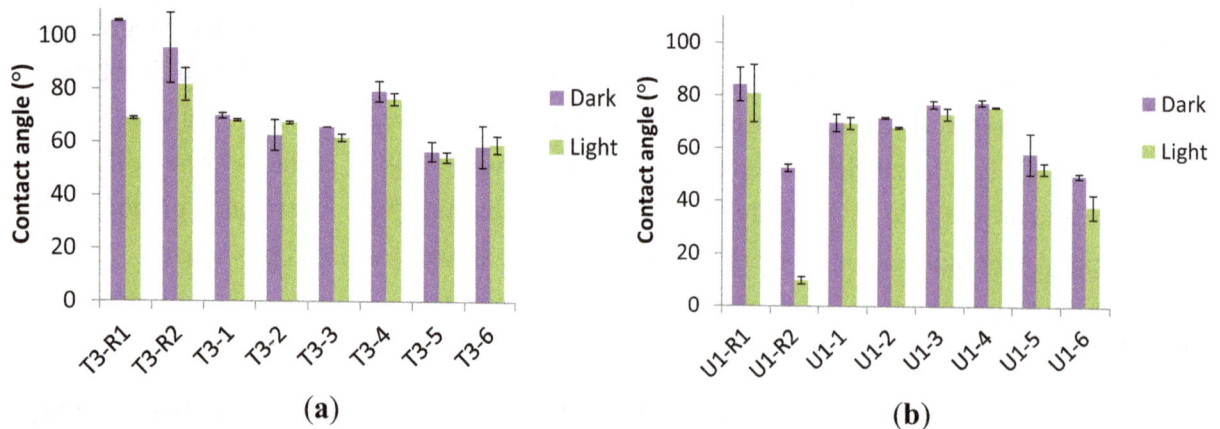

(**a**)                                                                                      (**b**)

For the $TiO_2$ sol coating, MC, the two coupons sited at the Brewery C showed similar wettability to the control sample (MC-R2), after cleaning, both in the dark and the light. Visual inspection of the coupons sited at the other two breweries showed that the coating was wholly or partly removed from these coupons, and the contact angle measurements reflect this loss of coating (Figure 11b). Contact angle values on blank stainless steel surfaces after cleaning were 70°–80°.

**Figure 11.** Water contact angle measurements for (**a**) $TiO_2$-Mo (U2); and (**b**) $TiO_2$ (MC) coupons after three month Brewery Trial (Coated area of MC-R1 was too small to test).

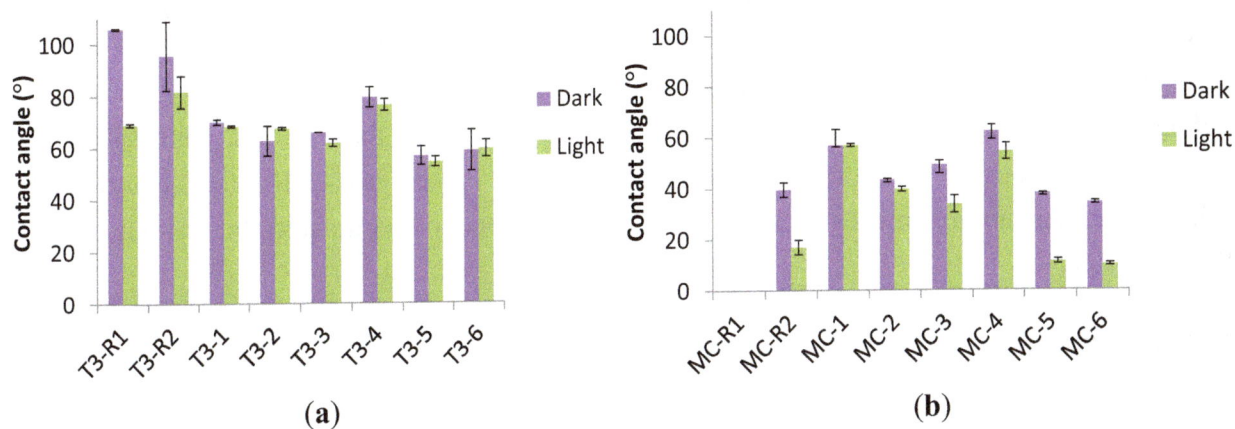

(**a**)                                    (**b**)

## 4. Conclusions

$TiO_2$ coatings were deposited either using reactive magnetron sputtering, both with and without subsequent heat treatment, or prepared by spray coating. Photocatalytic activity, determined by methylene blue degradation, was high under UV irradiation. The coatings were also active under fluorescent irradiation. Doping of magnetron sputtered $TiO_2$ with Ag- (0.5 at%) and Mo- (7 at%) increased the activity under fluorescent light. High Ag loading (~30%) had a detrimental effect on the fluorescent light induced photoactivity, possibly due to the replacement of Ti atoms in the $TiO_2$ matrix with Ag. The coatings were placed in three different breweries for three months. The magnetron sputtered $TiO_2$ surfaces which had not undergone heat treatment showed the best mechanical resistance, whilst the spray coated $TiO_2$ and Mo-$TiO_2$ showed the best retention of photoactivity. Irradiation of the coatings resulted in an increase in wettability, but the spray-coated $TiO_2$ was the only coating showing light-induced hydrophilicity after the process trial.

This work presented the potential of magnetron sputtered $TiO_2$ and doped $TiO_2$ coatings for surfaces used in food and beverage processing where there is a requirement for robust coatings. Selection of the optimum deposition parameters and dopants can lead to coatings which retain photoactivity and are durable in harsh processing conditions. The use of spray coatings is preferred on surfaces which do not experience severe mechanical wear and abrasion.

## Acknowledgments

This work was part of the MATERA+ Project "Disconnecting", refer MFM-1855, Project No. 620015. Funding from the Technology Strategy Board, the UK's innovation agency and Finnish Funding Agency for Technology and Innovation (Tekes) is gratefully acknowledged. The authors would also like to thank Oy Panimolaboratorio - Bryggerilaboratorium Ab (PBL Brewing Laboratory) for their collaboration in this work.

## Author Contributions

Parnia Navabpour, Kevin Cooke, Soheyla Ostovarpour, Peter Kelly, Joanna Verran and Kathryn Whitehead provided the planning and experimental work on magnetron sputtered coatings and the sections of the paper on these coatings. Teer Coatings authors also carried out the scratch tests and SEM work on all coatings and were responsible for writing the sections of the manuscript on these results. Manchester Metropolitan University authors were responsible for the photocatalytic tests, as well as the manuscript sections on these results. Carin Tattershall and Claire Hill carried out the work on MC coating and contact angle analysis and wrote the article sections on these results. Outi Priha and Mari Raulio were responsible for coordinating and performing the process tests. All authors contributed in the discussion and improvement of the paper. Parnia Navabpour coordinated the writing of the overall manuscript.

## Conflicts of Interest

The authors declare no conflict of interest.

## References

1.  Storgårds, E.; Tapani, K.; Hartwall, P.; Saleva, R.; Suihko, M. Microbial attachment and biofilm formation in brewery bottling plants. *J. Am. Soc. Brew. Chem.* **2006**, *64*, 8–15.
2.  Priha, O.; Laakso, J.; Levänen, E.; Kolari, M.; Mäntylä, T.; Storgårds, E. Effect of photocatalytic and hydrophobic coatings on brewery surface microorganisms. *J. Food Prot.* **2011**, *11*, 1788–1989.
3.  Goode, K.R.; Asteriadou, K.; Fryer, P.J.; Picksley, M.; Robbins, P.T. Characterising the cleanign mechanisms of yeast and the implicaitons for Cleaning In Place (CIP). *Food Bioprod. Proc.* **2010**, *88*, 365–374.
4.  Timke, M.; Wang-Lieu, N.Q.; Altendorf, K.; Lipski, A. Identity, beer spoiling and biofilm forming potential of yeasts from beer bottling plant associated biofilms. *Antonie Van Leeuwenhoek* **2008**, *93*, 151–161.
5.  Fornalik, M. Biofouling and process cleaning: A practical approach to understanding what is happening on the walls of your pipes. *Master Brew. Assoc. Am. Tech. Q.* **2008**, *45*, 340–344.
6.  Storgårds, E. Process Hygiene Control in Beer Production and Dispensing. Available online: http://www.vtt.fi/inf/pdf/publications/2000/P410.pdf (accessed on 8 July 2014).
7.  Rezić, T.; Rezić, I.; Blaženović, I.; Šantek, B. Optimization of corrosion processes of stainless steel during cleaning in steel brewery tanks. *Mater. Corros.* **2013**, *64*, 321–327.
8.  Farahani, N.; Kelly, P.J.; West, G.; Ratova, M.; Hill, C.; Vishnyakov, V. Photocatalytic activity of reactively sputtered and directly sputtered titania coatings. *Thin Solid Films* **2011**, *520*, 1464–1469.
9.  Caballero, L.; Whitehead, K.A.; Allen, N.S.; Verran, J. Inactivation of E.coli on immobilized $TiO_2$ using fluorescent light. *J. Photochem. Photobiol. A* **2009**, *202*, 92–98.
10. Bandaranayake, K.M.P.; Indika Senevirathna, M.K.; Prasad Weligamuwa, P.M.G.M.; Tennakone, K. Dye-sensitized solar cells made from nanocrystalline $TiO_2$ films coated with outer layers of different oxide materials. *Coord. Chem. Rev.* **2004**, *248*, 1277–1281.

11. Sung, Y.M.; Kim, H.J. Sputter deposition and surface treatment of $TiO_2$ films for dye-sensitized solar cells using reactive RF plasma. *Thin Solid Films* **2007**, *515*, 4996–4999.

12. Pakdel, E.; Daoud, W.A.; Wang, X. Self-cleaning and superhydrophilic wool by $TiO_2/SiO_2$ nanocomposite. *Appl. Surf. Sci.* **2013**, *275*, 397–402.

13. Samal, S.S.; Jeyaraman, P.; Vishwakarma, V. Sonochemical Coating of Ag-$TiO_2$ Nanoparticles on Textile Fabrics for Stain Repellency and Self-Cleaning- The Indian Scenario: A Review. *J. Miner. Mater. Charact. Eng.* **2010**, *9*, 519–525.

14. Hájková, P.; Špatenka, P.; Krumeich, J.; Exnar, P.; Kolouch, A.; Matoušek J.; Koči, P. Antibacterial effect of silver modified $TiO_2$/PECVD films. *J. Eur. Phy. D* **2009**, *54*, 189–193.

15. Miao, L.; Tanemura, S.; Kondo, Y.; Iwata, M.; Toh, S.; Kaneko, K. Microstructure and bactericidal ability of photocatalytic $TiO_2$ thin films prepared by rf helicon magnetron sputtering. *Appl. Surf. Sci.* **2004**, *238*, 125–131.

16. Tanemura, S.; Miao, L.; Wunderlich, W.; Tanemura, M.; Mori, Y.; Toh, S.; Kaneko, K. Fabrication and characterization of anatase/rutile-$TiO_2$ thin films by magnetron sputtering: A review. *Sci. Technol. Adv. Mater.* **2005**, *6*, 11–17.

17. Jiang, D.; Zhang, S.; Zhao, H. Photocatalytic Degradation Characteristics of Different Organic Compounds at $TiO_2$ Nanoporous Film Electrodes with Mixed Anatase/Rutile Phases. *Environ. Sci. Technol.* **2007**, *41*, 303–308.

18. Andersson, M.; Österlund, L.; Ljungström, S.; Palmqvist, A. Preparation of nanosize anatase and rutile $TiO_2$ by hydrothermal treatment of microemulsions and their activity for photocatalytic wet oxidation of phenol. *J. Phy. Chem. B* **2002**, *106*, 10674–10679.

19. Yin, H.; Wada, Y.; Kitamura, T.; Kambe, S.; Murasawa, S.; Mori, H.; Sakata, T.; Yanagida, S. Hydrothermal synthesis of nanosized anatase and rutile $TiO_2$ using amorphous phase $TiO_2$. *J. Mater. Chem.* **2001**, *11*, 1694–1703.

20. Wasa, K.; Kitabatake, M.; Adachi, H. *Thin Film Materials Technology: Sputtering of Control Compound Materials*; William Andrew Inc.: Norwich, CT, USA, 2004.

21. Laing, K.; Hampshire, J.; Teer, D.G.; Chester, G. The effect of ion current density on the adhesion and structure of coatings deposited by magnetron sputter ion plating. *Surf. Coat. Technol.* **1999**, *112*, 177–180.

22. Ohtani, T.; Ogawa, Y.; Nishimoto, S. Photocatalytic Activity of Amorphous-Anatase Mixture of Titanium(IV) Oxide Particles Suspended in Aqueous Solutions. *J. Phy.Chem. B* **1997**, *101*, 3746–3752.

23. Yu, J.; Zhao, X.; Du, J.; Chen, W. Preparation, microstructure and photocatalytic activity of the porous $TiO_2$ anatase coating by sol-gel processing. *J. Sol Gel Sci. Technol.* **2000**, *17*, 163–171.

24. Nam, H.; Amemyima, T.; Murabayashi, M.; Itoh, K. Photocatalytic Activity of Sol-Gel $TiO_2$ Thin Films on Various Kinds of Glass Substrates: The Effects of $Na^+$ and Primary Particle Size. *J. Phy. Chem. B* **2004**, *108*, 8254–8259.

25. Kubacka, A.; Colón G.; Fernández-García, M. Cationic (V, Mo, Nb, W) doping of $TiO_2$–anatase: A real alternative for visible light-driven photocatalysts. *Catal. Today* **2009**, *143*, 286–292.

26. Onifade, A.A.; Kelly, P.J. The influence of deposition parameters on the structure and properties of magnetron-sputtered titania coatings. *Thin Solid Films* **2006**, *494*, 8–12.

27. Stallard, J.; Poulat, S.; Teer, D.G. The study of the adhesion of a TiN coating on steel and titanium alloy substrates using a multi-mode scratch tester. *Tribol. Int.* **2006**, *39*, 159–166.

28. Ratova, M.; Kelly, P.J.; West, J.T.; Iordanova, I. Enhanced properties of magnetron sputtered photocatalytic coatings via transition metal doping. *Surf. Coat. Technol.* **2013**, *228*, S544–S549.

29. Christian, G.K.; Fryer, P.J.; Liu, W. How hygiene happens: Physics and chemistry of cleaning. *Int. J. Dairy Technol.* **2006**, *59*, 76–84.

30. Grönholm, L.; Wirtanen, G.; Ahlgren, K.; Nordström, K.; Sjöberg, A.-M. Screening of antimicrobial activities of disinfectants and cleaning agents against foodborne spoilage microbes. *Z. Lebensm. Unters. Forshung A* **1999**, *208*, 289–298.

31. Fujishima, A.; Rao, T.N.; Tryk, D.A. Titanium dioxide photocatalysis. *J. Photochem. Photobiol. C* **2000**, *1*, 1–21.

# 3

# Transparent, Adherent, and Photocatalytic SiO$_2$-TiO$_2$ Coatings on Polycarbonate for Self-Cleaning Applications

**Sanjay S. Latthe, Shanhu Liu, Chiaki Terashima, Kazuya Nakata and Akira Fujishima \***

Photocatalysis International Research Center, Research Institute for Science & Technology,
Tokyo University of Science, Noda, Chiba 278-8510, Japan;
E-Mails: sanjaylatthe@yahoo.com (S.S.L.); liushanhu@163.com (S.L.); terashima@rs.tus.ac.jp (C.T.);
nakata@rs.tus.ac.jp (K.N.)

\* Author to whom correspondence should be addressed; E-Mail: fujishima_akira@admin.tus.ac.jp

**Abstract:** Photocatalytic TiO$_2$ coatings are famously known for their excellent self-cleaning behavior, where very thin water layer formed on the superhydrophilic surface can easily wash-off the dirt particles while flowing. Here we report the preparation of the optically transparent, adherent, highly wettable towards water and photocatalytic SiO$_2$-TiO$_2$ coatings on polycarbonate (PC) substrate for self-cleaning applications. The silica barrier layer was applied on UV-treated PC substrate before spin coating the SiO$_2$-TiO$_2$ coatings. The effect of different vol% of SiO$_2$ in TiO$_2$ and its influence on the surface morphology, mechanical stability, wettability, and photocatalytic properties of the coatings were studied in detail. The coatings prepared from 7 vol% of SiO$_2$ in TiO$_2$ showed smooth, crack-free surface morphology and low surface roughness compared to the coatings prepared from the higher vol% of SiO$_2$ in TiO$_2$. The water drops on this coating acquires a contact angle less than 10° after UV irradiation for 30 min. All the coatings prepared from different vol% (7 to 20) of SiO$_2$ in TiO$_2$ showed high transparency in the visible range.

**Keywords:** photocatalytic; wettable; transparent; SiO$_2$-TiO$_2$ coatings; self-cleaning

## 1. Introduction

TiO$_2$ is one of the most widely researched photocatalytic semiconductor materials to date. When irradiated with UV light, TiO$_2$ can decompose the organic pollutants present on its surface, and in addition the surface turns to be highly hydrophilic [1]. Water spread out instantaneously by forming a thin layer on the superhydrophilic TiO$_2$ thin films and steadily carries away dust particles from the surface while flowing (Figure 1). Photocatalytic superhydrophilic thin films with excellent self-cleaning properties are receiving much research attention worldwide because such smart surfaces can save time and reduce maintenance costs [2]. Today, the optically transparent, superhydrophilic and photocatalytic TiO$_2$ thin films are finding increasing industrial applications. Many reports on application of self-cleaning TiO$_2$ thin films on glass substrates are available in the literature [3–6]. An attractive soft polymer material, polycarbonate (PC) has great demand in the optical industry. PC can replace heavy glass in the optical and electronic industries in the coming future due to its extremely low weight, durability, low-cost, and optical transparency [7]. Many attempts have been made to improve the scratch resistance and UV degradation property of the PC by coating the surface by numerous photocatalytic materials with suitable binders.

**Figure 1.** Schematic showing the self-cleaning phenomena on superhydrophilic surface.

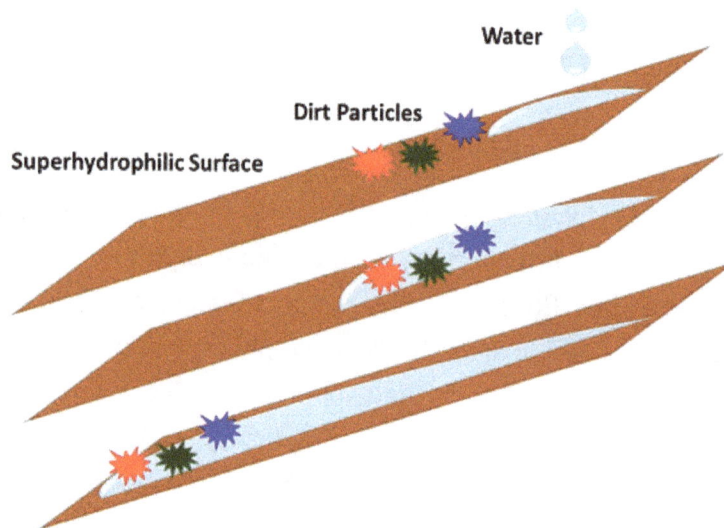

Hwang and coworkers [8] have prepared mechanically durable and optically transparent organic-inorganic SiO$_2$-TiO$_2$ nanocomposite coatings on PC by simple spray coating. The coating solution was prepared by incorporating nanosized TiO$_2$ sol into the silica polymeric sol. This nanocomposite coating also protected PC from photo-degradation under UV illumination. Lam *et al.* [9] reported the comparative study on the influence of NaOH etching and UVC irradiation on the mechanical stability of the TiO$_2$ thin films prepared on PC. They observed adherent TiO$_2$ film on UVC irradiated PC than NaOH treated PC due to increase in –OH and –COOH groups on the UVC-treated PC. However, the strong hydrophilic, antifogging, photocatalytic and self-cleaning properties were observed on the TiO$_2$ film coated on NaOH treated PC. Recently, Razan Fateh *et al.* have prepared optically transparent, highly hydrophilic, mechanically stable and photocatalytic TiO$_2$/SiO$_2$ [7,10], and TiO$_2$/ZnO coating [11] on PC for self-cleaning applications. In the present research work, we prepared transparent, adherent, highly wettable and self-cleaning $SiO_2$-$TiO_2$ coating on PC by simple spin coating. The coating showed faster wettability switching (hydrophilic to superhydrophilic) after 30 min. of UV irradiation. The effect of different vol% of $SiO_2$ sol in $TiO_2$ and its influence on the surface morphology, mechanical stability, wettability, and photocatalytic properties of the coatings were studied in detail.

## 2. Results and Discussion

### 2.1. Surface Microstructure, Roughness and Chemical Composition of the Coatings

The surface morphologies of the coatings prepared with different vol% of $SiO_2$ in $TiO_2$ are shown in Figure 2. The $SiO_2$-$TiO_2$ coating prepared with 7 vol% of $SiO_2$ in $TiO_2$ showed uniform, crack-free and smooth morphology (Figure 2a). However, significant cracks on the entire coating surface were observed with increase in $SiO_2$ concentration in $TiO_2$ (Figure 2b–d). Some small cracks were started to appear on the coating prepared from 10 vol% of $SiO_2$ in $TiO_2$ (Figure 2b), and these cracks goes bigger with increase in vol% of $SiO_2$. For the coating prepared with 20 vol% of $SiO_2$ in $TiO_2$, the cracks on the surface were abundant and coating material was popped out detaching from the substrate (Figure 2d). In the case of sol-gel coated films, during drying process, the capillary forces might have generated which provides cracks on the surface [12]. The surface roughness of the coatings were also studied (Figure 3). The $SiO_2$-$TiO_2$ coating prepared with 7 vol% of $SiO_2$ in $TiO_2$ showed the surface roughness of 67 nm (Figure 3a), whereas the surface roughness increases drastically to 98, 191 and 287 nm for the coating prepared from 10, 15 and 20 vol% of $SiO_2$ in $TiO_2$ (Figure 3b–d), respectively. This drastic increase in surface roughness may due to the increased density of cracks on the coating surface.

**Figure 2.** Field Emission Scanning Electron Microscope (FE-SEM) images of the coatings prepared from (**a**) 7; (**b**) 10; (**c**) 15; and (**d**) 20 vol% of $SiO_2$ in $TiO_2$.

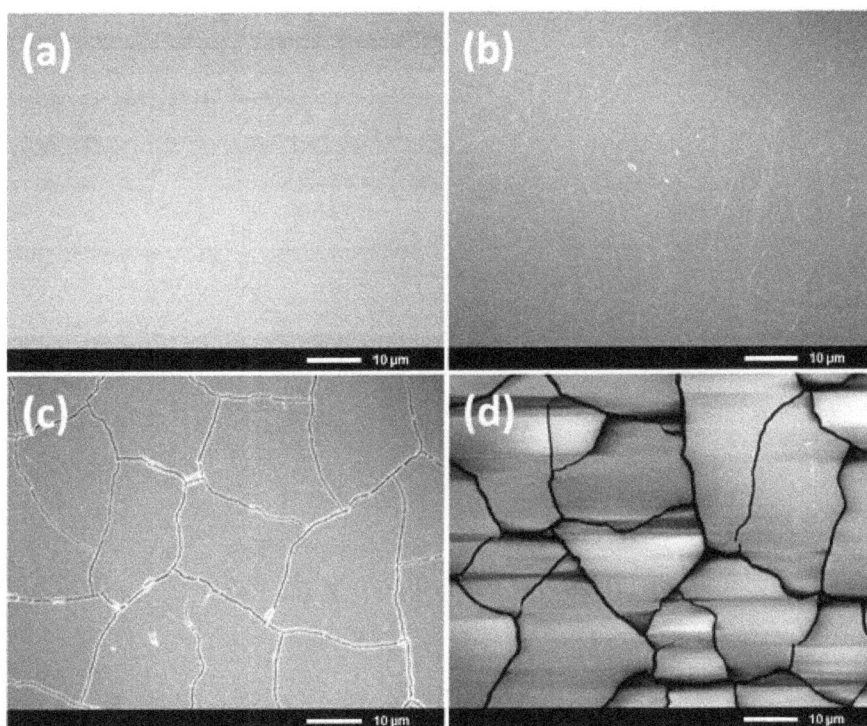

**Figure 3.** 3D Laser microscope images of the coatings prepared from (**a**) 7; (**b**) 10; (**c**) 15; and (**d**) 20 vol% of SiO$_2$ in TiO$_2$.

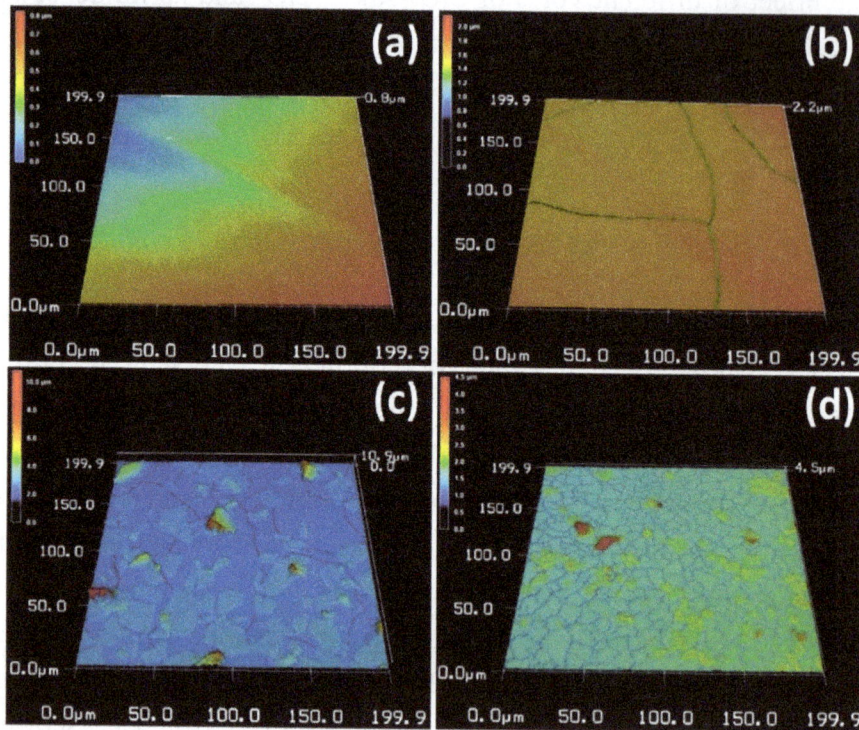

Figure 4 shows the FT-IR spectra of the coating prepared from 7 vol% of SiO$_2$ in TiO$_2$. The peaks observed in between 500 and 900 cm$^{-1}$ can be attributed to the characteristic vibrational modes of TiO$_2$ [13]. A peak observed near 954 cm$^{-1}$ is associated with Si-O-Ti vibration [14]. The absorption peak at 1118 cm$^{-1}$ confirms Si-O-Si linkage [14]. The absorption peaks near 3342 and 1630 cm$^{-1}$ can be attributed to the presence of stretching and bending vibrations of hydroxyl groups, respectively [15].

**Figure 4.** FT-IR spectra of the coatings prepared from 7 vol% of SiO$_2$ in TiO$_2$.

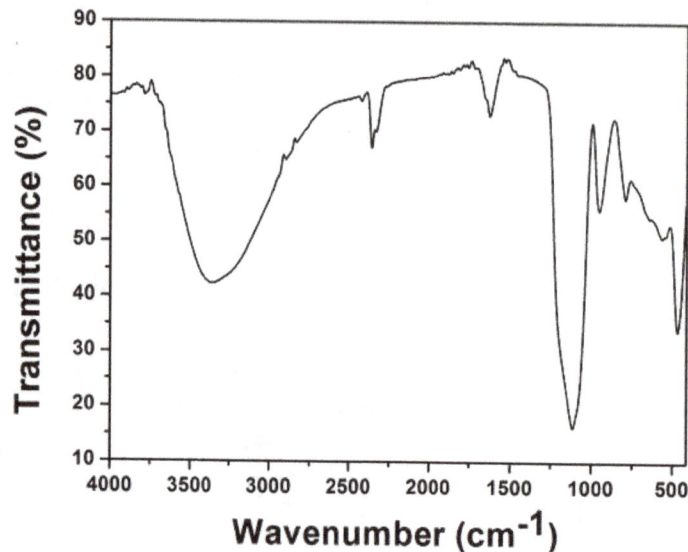

## 2.2. Superhydrophilic, Photocatalytic and Optical Properties of the Coatings

The wettability transition of the $SiO_2$-$TiO_2$ coating after UV illumination was studied. The prepared $SiO_2$-$TiO_2$ coatings were illuminated by UV light (365 nm, 2 mW/cm$^2$). The UV illumination creates structural changes in $TiO_2$ (transformation of $Ti^{4+}$ sites to $Ti^{3+}$ sites) [16] and oxidizes the organic contaminants present on the surface of $TiO_2$, which effectively transforms the wettability of the $TiO_2$ surface towards more hydrophilic [17]. This is called as photo-induced hydrophilicity on $TiO_2$ surface [18]. The water contact angles (WCA) on the coatings before and after exposure to the UV light were measured. The water drop volume of 5 μL was used to measure the water contact angles on the coating surface. The effect of UV exposure time on the wettability of the coatings was also studied. Figure 5 shows the wettability of the coatings prepared from 7 and 20 vol% of $SiO_2$ in $TiO_2$, before and after UV irradiation. Before UV illumination, the coating prepared from 7 and 20 vol% of $SiO_2$ in $TiO_2$ showed the WCA of 23° and 28°, respectively. After UV illumination of 30 min, the water drop immediately spread on the surface and the WCA drastically decreased to 8° on the coating prepared from 7 vol% of $SiO_2$ in $TiO_2$ and even for longer UV illumination time (5 h), the WCA remained in the range of 7°. The coating prepared from 20 vol% of $SiO_2$ in $TiO_2$ also showed decrease in WCA in the range of 10°, after longer UV illumination time. This slightly higher WCA in case of the coating prepared from 20 vol% of $SiO_2$ in $TiO_2$ is due to relatively high surface roughness provided by significant density of cracks present on the surface. Even after placing the UV irradiated coatings in dark for 3 months, the coatings showed stable wetting properties with WCA measured well below 10°. Figure 6 shows the optical photograph of water drops on bare PC substrate and the coating prepared from 7 vol% of $SiO_2$ in $TiO_2$ after 30 min. UV irradiation. Some of the water drops were colored blue using Methylene Blue for better visualization of water drops. The water drops on bare PC substrate maintain the contact angle of ~84°, whereas the water drops spreads on UV irradiated $SiO_2$-$TiO_2$ coating, confirming high affinity towards the water. In the case of $SiO_2$-$TiO_2$ coatings prepared by Fateh *et al.* [7], it needed more than 700 h of UV (A) irradiation to switch the wettability of the coatings in the superhydrophilic range.

**Figure 5.** Wetting properties of the coatings prepared from 7 and 20 vol% of $SiO_2$ in $TiO_2$ (Insets shows the shape of water drops on coatings).

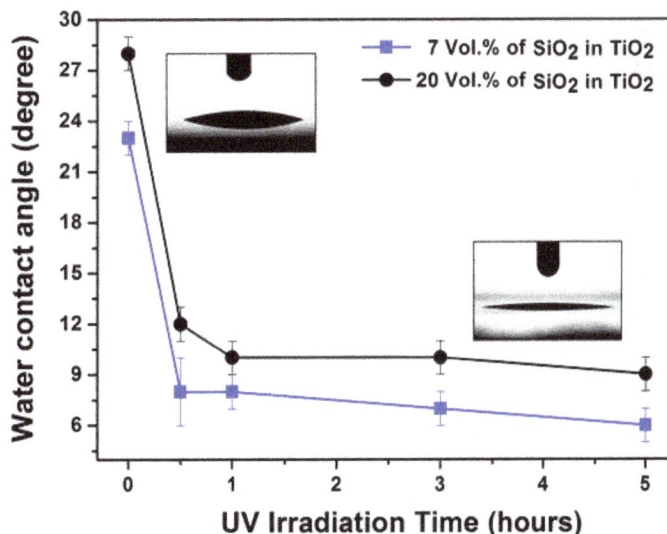

**Figure 6.** Optical photograph of water drops on bare PC substrate (**a**) and coating prepared from 7 vol% of $SiO_2$ in $TiO_2$ after 30 min. UV irradiation (**b**).

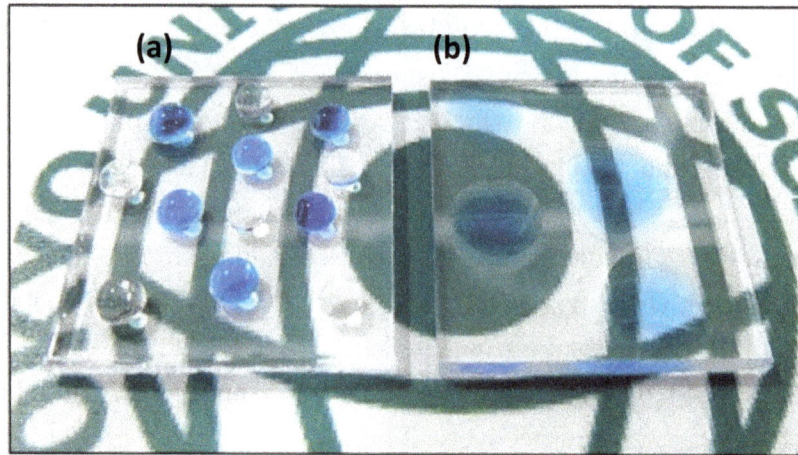

The $TiO_2$ is famously known for its excellent photocatalytic property, as it can degrade organic contamination under the illumination of UV light. We studied the photocatalytic degradation of the ODS monolayers after UV light illumination (365 nm, 2 mW/cm$^2$) by means of water contact angle measurements. At first, the bare PC substrate and coating prepared from 7 vol% of $SiO_2$ in $TiO_2$ were irradiated with UV light (365 nm, 2 mW/cm$^2$) for 1 h to make them superhydrophilic and hydrophobic ODS self-assembled monolayers (SAMs) were applied on them through vapor phase. To employ ODS SAMs on the surface, 2 mL of ODS in beaker was kept in the closed metal box containing the samples at 100 °C for 24 h. After application of ODS SAMs on the surface, the WCA on bare PC substrate and coating prepared from 7 vol% of $SiO_2$ in $TiO_2$ showed 95° and 50°, respectively (Figure 7). The ODS treated bare PC showed almost no change in WCA after 120 min of UV illumination, whereas the coating prepared from 7 vol% of $SiO_2$ in $TiO_2$ showed significant decrease in WCA as a function of UV illumination time and the surface becomes superhydrophilic after 120 min of UV illumination (Figure 7) confirming photocatalytic degradation of the ODS monolayers.

**Figure 7.** Wettability change on octadecyltrichlorosilane (ODS)-treated $SiO_2$-$TiO_2$ coatings under UV illumination.

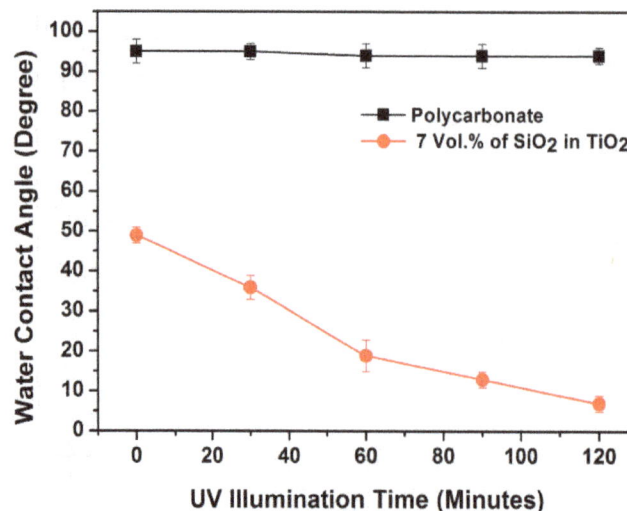

Optical transparency is prerequisite for the application of self-cleaning coating on transparent glass or plastic surface. Figure 8 shows the optical transmission of $SiO_2$-$TiO_2$ coatings prepared on the PC substrate from different vol% of $SiO_2$ in $TiO_2$. All the coatings are highly transparent and showed the optical transmittance values above 85% over the entire visible wavelength range. The optical transmission was gradually decreased from 89% to 85% with an increase in $SiO_2$ concentration in $TiO_2$. The increased surface roughness is responsible for the slight loss in optical transmission in the visible range. The $SiO_2$-$TiO_2$ coatings prepared by Hwang et al. [8] showed an optical transmission of ~87%, whereas all the coatings prepared by Fateh et al. [7,10] showed an optical transmission of >92%.

**Figure 8.** Optical transmission spectra of the $SiO_2$-$TiO_2$ coatings.

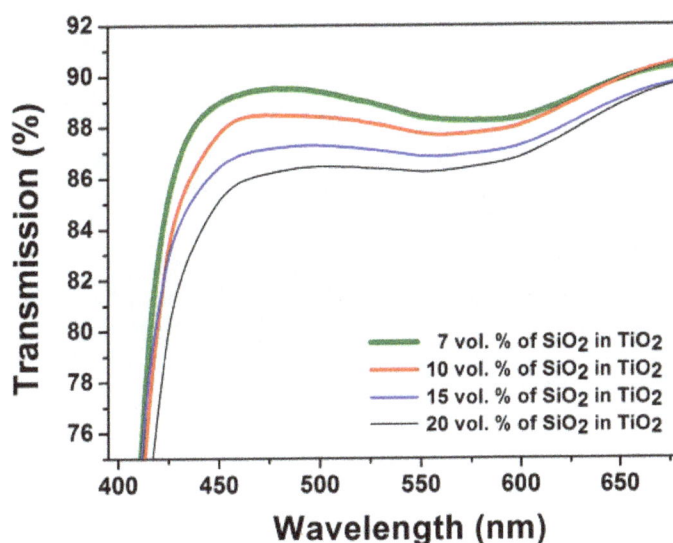

## 2.3. Mechanical Properties of the Coatings

The mechanical durability of the coatings is a very important criterion for industrial applications. The mechanical stability of the prepared $SiO_2$-$TiO_2$ coatings is depicted in Figure 9. The bare PC substrates get easily scratched at the applied force of less than 4.2 mN. For the coating prepared from 7 vol% of $SiO_2$ in $TiO_2$, the coating material was removed at the applied force of ~16.1 mN (Figure 9a), whereas this force was decreased to 12.2, 10.9 and 9.8 mN for the coating prepared from 10, 15 and 20 vol% of $SiO_2$ in $TiO_2$ (Figure 9b–d), respectively. The smooth, crack-free morphology and lower surface roughness on the coating resulted in the stable and adhesive coating on the PC substrate. Even rubbed by the fingers, the coating material was not easily removed. This is ascribed to the formation of an intermediate layer connecting the $TiO_2$ nanoparticles and the silica network. However, in the case of the coating prepared from higher concentration of $SiO_2$ in $TiO_2$, due to significant cracks and increased surface roughness, the coating material could be removed at relatively lower applied force.

**Figure 9.** Adhesion test performed on the coatings prepared from (**a**) 7; (**b**) 10; (**c**) 15; and (**d**) 20 vol% of SiO$_2$ in TiO$_2$.

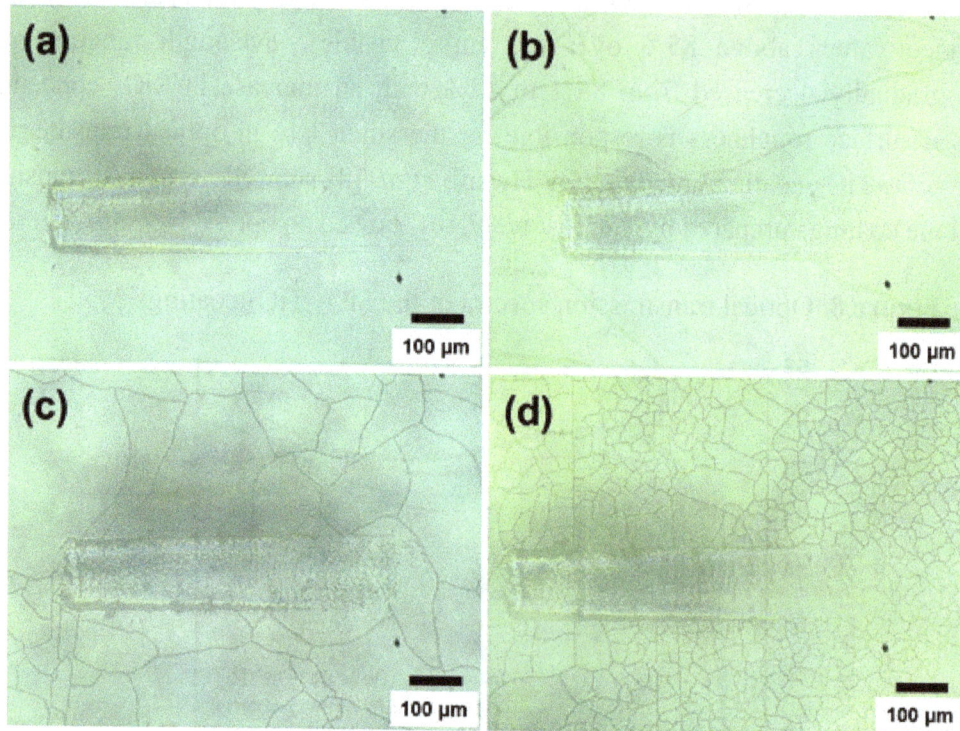

## 3. Experimental Section

### 3.1. Materials

Tetraethylorthosilicate (TEOS) and Octadecyltrichlorosilane (ODS) were purchased from Sigma Aldrich (St. Louis, MO, USA). Ethanol (99.5%) and nitric acid (69%) were purchased from Wako Pure Chemical Industries Ltd. (Kanto region, Japan). PC substrates (ECK, 100UU) were bought from Sumitomo Bakelite Co. Ltd. (Akita, Japan). The commercially available TiO$_2$ nanoparticle solution (NRC-300C) was purchased from Nippon Soda Co. Ltd. (Tokyo, Japan).

### 3.2. Self-Cleaning Coating on PC Substrates

PC substrates are known to be hydrophobic in nature and so the adhesion is not strong between the coating material and the PC substrate [19]. The hydrophobic nature of PC substrates can be transformed to strongly hydrophilic by UV irradiation due to occurrence of photo-Fries reaction on the surface [20]. At first, PC substrates were gently cleaned by using detergent and water and kept for ultrasonication in double distilled water for 30 min. After drying at room temperature, the PC substrates were illuminated by UV light (365 nm, 2 mW/cm$^2$) for 2 h. The PC substrates can photocatalytically degrade, if the TiO$_2$ coating is applied directly on the PC substrates. For this reason, the sol-gel processed silica layer was applied on the PC substrates prior to TiO$_2$ coating. A silica sol was prepared by adding 1.5 mL TEOS in 4 mL ethanol and this mixture was hydrolysed by adding 3 mL of double distilled water with 1 mL of nitric acid. This alcosol was stirred overnight and spin-coated on UV-treated PC substrates at 2000 rpm. This silica coating was annealed at 100 °C for 2 h. The above prepared silica sol was mixed at different vol% (7, 10, 15 and 20 vol%) in commercially available $TiO_2$ nanoparticle solution and spin-coated on $SiO_2$ pre-coated PC substrates with 2000 rpm and annealed in oven at 110 °C for 6 h. No change in PC substrate was observed after annealing at 110 °C, however for annealing above 120 °C, PC substrate start to bend in irregular shape due to softening. The schematic of coating preparation steps are shown in Figure 10.

**Figure 10.** Schematic showing coating preparation steps.

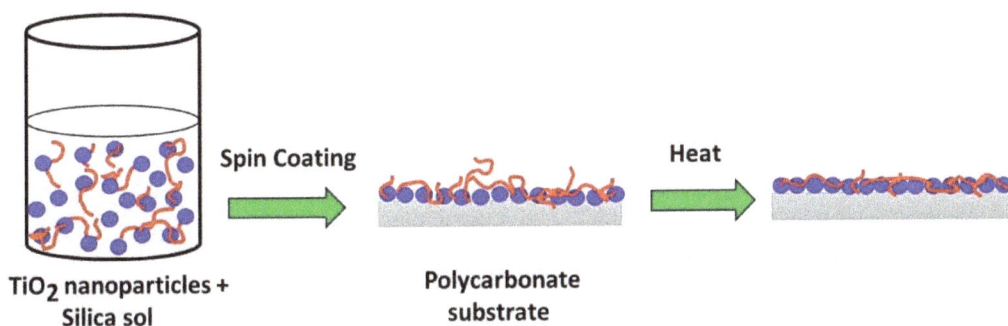

*3.3. Characterization Techniques*

The surface morphology was studied by Field Emission Scanning Electron Microscope (FE-SEM, JEOL, JSM-7600F; Tokyo, Japan). The surface roughness of the prepared coatings was evaluated by laser microscopy (KEYENCE, VK-X200 series; Itasca, IL, USA). The chemical composition was studied by Fourier transform infrared spectroscopy (FT-IR, JASCO, FT/IR-6100; Tokyo, Japan). The water contact angles (WCA) were measured at six different locations on same sample by using contact angle meter (KYOWA, Drop Master; Saitama, Japan) and average value is reported as a final contact angle value. The optical transmission of the coatings was measured by UV-VIS spectrophotometer (JASCO, V-670; Tokyo, Japan). The coatings were illuminated by UV light (365 nm, 2 mW/cm$^2$). The UV lamps were purchased from TOSHIBA (FL10BLB; Tokyo, Japan) and assembled with proper electric supply inside the wooden box covered by thick black cloth. The adhesion of the coating material on PC substrate was checked at five different spots by using Scratch tester (Nano-Layer Scratch Tester, CSR-2000; Rhesca, Tokyo, Japan). The adhesion of the coatings on PC substrate and the maximum force required to remove the coating material was calculated. The cantilever was moved from right to left side in contact with the coating surface. The force was gradually increased while moving towards left side. The maximum force at which the coating material was removed from the PC substrate was noted as critical force to damage the coating. The optical photographs were recorded using Canon Digital Camera (G 15 series).

## 4. Conclusions

The mechanically durable, optically transparent, photocatalytically active and superhydrophilic $SiO_2$-$TiO_2$ coatings are successfully prepared on PC substrates for self-cleaning applications. The uniform, crack-free coatings prepared from 7 vol% of $SiO_2$ in $TiO_2$ showed higher optical transmission in the visible range, strong superhydrophilicity and good scratch resistance properties. Such coatings can be applied on light-weight window and door polycarbonates for excellent self-cleaning applications.

## Acknowledgments

Authors Sanjay S. Latthe (P13067) and Shanhu Liu (P12345) are grateful for the financial support provided by the Japan Society for the Promotion of Science (JSPS), Japan, under Postdoctoral Fellowship for Foreign Researchers.

## Author Contributions

Sanjay S. Latthe conceived the idea and designed the structure of article. Sanjay S. Latthe and Shanhu Liu carried out the experiments and wrote the original manuscript and Chiaki Terashima, Kazuya Nakata and Akira Fujishima participated in the discussions and helped to revise it.

## Conflicts of Interest

The authors declare no conflict of interest.

## References

1.  Sakai, N.; Fukuda, K.; Shibata, T.; Ebina, Y.; Takada, K.; Sasaki, T. Photoinduced hydrophilic conversion properties of titania nanosheets. *J. Phys. Chem. B* **2006**, *110*, 6198–6203.

2.  Blossey, R. Self-cleaning surfaces—virtual realities. *Nat. Mater.* **2003**, *2*, 301–306.

3.  Weng, K.-W.; Huang, Y.-P. Preparation of $TiO_2$ thin films on glass surfaces with self-cleaning characteristics for solar concentrators. *Surf. Coat. Technol.* **2013**, *231*, 201–204.

4.  Xi, B.; Verma, L.K.; Li, J.; Bhatia, C.S.; Danner, A.J.; Yang, H.; Zeng, H.C. $TiO_2$ thin films prepared via adsorptive self-assembly for self-cleaning applications. *ACS Appl. Mater. Interfaces* **2012**, *4*, 1093–1102.

5.  Euvananont, C.; Junin, C.; Inpor, K.; Limthongkul, P.; Thanachayanont, C. $TiO_2$ optical coating layers for self-cleaning applications. *Ceram. Int.* **2008**, *34*, 1067–1071.

6.  Lai, Y.; Tang, Y.; Gong, J.; Gong, D.; Chi, L.; Lin, C.; Chen, Z. Transparent superhydrophobic/superhydrophilic $TiO_2$-based coatings for self-cleaning and anti-fogging. *J. Mater. Chem.* **2012**, *22*, 7420–7426.

7.  Fateh, R.; Dillert, R.; Bahnemann, D. Preparation and characterization of transparent hydrophilic photocatalytic $TiO_2/SiO_2$ thin films on polycarbonate. *Langmuir* **2013**, *29*, 3730–3739.

8.  Hwang, D.K.; Moon, J.H.; Shul, Y.G.; Jung, K.T.; Kim, D.H.; Lee, D.W. Scratch resistant and transparent UV-protective coating on polycarbonate. *J. Sol Gel Sci. Technol.* **2003**, *26*, 783–787.

9.  Lam, S.; Soetanto, A.; Amal, R. Self-cleaning performance of polycarbonate surfaces coated with titania nanoparticles. *J. Nanopart. Res.* **2009**, *11*, 1971–1979.

10. Fateh, R.; Ismail, A.A.; Dillert, R.; Bahnemann, D.W. Highly active crystalline mesoporous $TiO_2$ films coated onto polycarbonate substrates for self-cleaning applications. *J. Phys. Chem. C* **2011**, *115*, 10405–10411.

11. Fateh, R.; Dillert, R.; Bahnemann, D. Self-cleaning properties, mechanical stability, and adhesion strength of transparent photocatalytic $TiO_2$-ZnO coatings on polycarbonate. *ACS Appl. Mater. Interfaces* **2014**, *6*, 2270–2278.

12. Hatton, B.; Mishchenko, L.; Davis, S.; Sandhage, K.H.; Aizenberg, J. Assembly of large-area, highly ordered, crack-free inverse opal films. *Proc. Natl. Acad. Sci. USA* **2010**, *107*, 10354–10359.

13. Khanna, P.K.; Singh, N.; Charan, S. Synthesis of nano-particles of anatase-$TiO_2$ and preparation of its optically transparent film in PVA. *Mater. Lett.* **2007**, *61*, 4725–4730.

14. Kumar, D.A.; Shyla, J.M.; Xavier, F.P. Synthesis and characterization of $TiO_2/SiO_2$ nano composites for solar cell applications. *Appl. Nanosci.* **2012**, *2*, 429–436.

15. Rao, A.V.; Latthe, S.S.; Nadargi, D.Y.; Hirashima, H.; Ganesan, V. Preparation of MTMS based transparent superhydrophobic silica films by sol–gel method. *J. Colloid Interface Sci.* **2009**, *332*, 484–490.

16. Shultz, A.N.; Jang, W.; Hetherington, W.M., III; Baer, D.R.; Wang, L.-Q.; Engelhard, M.H. Comparative second harmonic generation and X-ray photoelectron spectroscopy studies of the UV creation and $O_2$ healing of $Ti^{3+}$ defects on (110) rutile $TiO_2$ surfaces. *Surf. Sci.* **1995**, *339*, 114–124.

17. Wang, R.; Hashimoto, K.; Fujishima, A.; Chikuni, M.; Kojima, E.; Kitamura, A.; Shimohigoshi, M.; Watanabe, T. Light-induced amphiphilic surfaces. *Nature* **1997**, *388*, 431–432.

18. Anandan, S.; Rao, T. N.; Sathish, M.; Rangappa, D.; Honma, I.; Miyauchi, M. Superhydrophilic graphene-loaded $TiO_2$ thin film for self-cleaning applications. *ACS Appl. Mater. Interfaces* **2012**, *5*, 207–212.

19. Aslan, K.; Holley, P.; Geddes, C.D. Metal-enhanced fluorescence from silver nanoparticle-deposited polycarbonate substrates. *J. Mater. Chem.* **2006**, *16*, 2846–2852.

20. Rivaton, A. Recent advances in bisphenol—A polycarbonate photodegradation. *Polym. Degrad. Stab.* **1995**, *49*, 163–179.

# 4

# A Study of the Abrasion of Squeegees Used in Screen Printing and Its Effect on Performance with Application in Printed Electronics

**Christopher O. Phillips \*, David G. Beynon, Simon M. Hamblyn, Glyn R. Davies, David T. Gethin and Timothy C. Claypole**

Welsh Centre for Printing and Coating, College of Engineering, Swansea University, Singleton Park, Swansea SA2 8PP, UK; E-Mails: d.g.beynon@swansea.ac.uk (D.G.B.); s.m.hamblyn@swansea.ac.uk (S.M.H.); g.r.davies@swansea.ac.uk (G.R.D.); d.t.gethin@swansea.ac.uk (D.T.G.); t.c.claypole@swansea.ac.uk (T.C.C.)

\* Author to whom correspondence should be addressed; E-Mail: c.o.phillips@swansea.ac.uk

---

**Abstract:** This article presents a novel method for accelerated wear of squeegees used in screen printing and describes the development of mechanical tests which allow more in-depth measurement of squeegee properties. In this study, squeegees were abraded on the screen press so that they could be used for subsequent print tests to evaluate the effect of wear on the printed product. Squeegee wear was found to vary between different squeegee types and caused increases in ink transfer and wider printed features. In production this will lead to greater ink consumption, cost per unit and a likelihood of product failure. This also has consequences for the production of functional layers, *etc.*, used in the construction of printed electronics. While more wear generally gave greater increases in ink deposition, the effect of wear differed, depending on the squeegee. There was a correlation between the angle of the squeegee wear and ink film thickness from a worn squeegee. An ability to resist flexing gave a high wear angle and presented a sharper edge at the squeegee/screen interface thus mitigating the effect of wear. There was also a good correlation between resistance to flexing and ink film thickness for unworn squeegees, which was more effective than a comparison based on Shore A hardness. Squeegee indentation at different force levels gave more information than a standard Shore A hardness test and the apparatus used was able to reliably measure reductions in surface hardness due to solvent absorption. Increases in ink deposition gave lower resistance in printed silver lines; however, the correlation between the amount of ink deposited and the resistance, remained the same for all levels of wear, suggesting that the wear regime designed for this study did not induce detrimental print defects such as line breakages.

**Keywords:** screen printing; squeegees; printed electronics; accelerated aging

## 1. Introduction

As well as conventional graphics printing, screen printing is increasingly being used for a large range of functional devices where thick deposits are required; including but not limited to solar cells [1], fuel cells [2], displays [3], Organic Light Emitting Diodes (OLEDs) [4], transistors [5] as well as sensors for gases [6], humidity [7] and biological materials [8]. There is an increasing requirement for more intricate features, with a high degree of control over their functional properties. The functionality of the various printed layers in terms of conductive, dielectric, insulating and light emitting properties, for example, will vary if the ink deposition is altered. Process consistency is therefore as vital as the understanding of the effect of process parameters when engaging in volume production. Process settings can be used to achieve product quality, but this needs to be maintained over the manufacturing production run. Where volumes are large or the printed inks contain abrasive elements, the squeegee condition will vary over its lifetime and a decision must be made when to replace it. Research to date has focused principally on the effect of process settings [9–12] and no work has been reported to explore squeegee deterioration during printing.

The squeegee (usually polyurethane) is used to transfer ink through the screen mesh onto a substrate (Figure 1) and there are a host of squeegee specific factors which influence the print quality and consistency. These can be categorized as either process setting effects, which are controlled by the selection of the squeegee, or duration effects which will have an impact over the course of the printing run and are less well understood and predictable. Process effects include surface hardness, bulk elastic modulus, mounting angle and edge profile of the squeegee. In the longer term, inks, and in particular functional materials such as conductive metallic particles, will abrade the squeegee over time thus changing its performance. A better understanding of squeegee abrasion and its consequent impact on the printed product are therefore required.

**Figure 1.** Schematic of screen printing process.

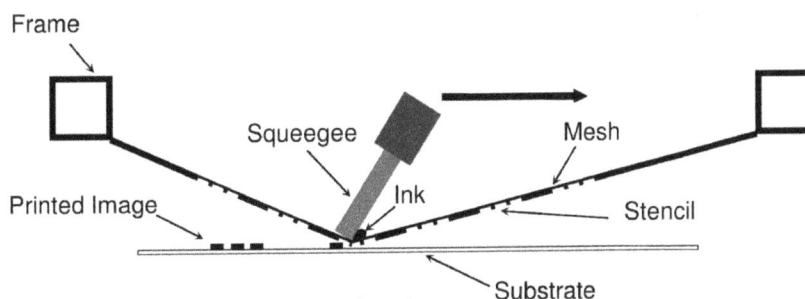

The effect of abrasion on print quality might vary depending on other squeegee factors. For this reason, other factors which influence the contact region between the squeegee and the screen are also

investigated. A large range of inks are used in screen printing, and there is a correspondingly large range of solvents used in these inks. There is gradual solvent absorption into the squeegee from the ink which causes swelling of the squeegee during the print run. Solvent absorption is typically evaluated by immersion of small pieces of squeegee [13] with changing mass being the most reliable indicator, rather than volume changes [14]. To assess the comparative uptake of solvent in the different squeegee types, absorption was tested via immersion in solvents used in carbon and silver conductive ink systems used in printed electronics. Solvent absorption has also been previously demonstrated to increase ink transfer by softening the squeegee [14]. However, the Shore A hardness test [15], which is the predominant method for assessing the surface hardness of the squeegee, cannot accurately measure squeegees that have been distorted through contact with ink or solvent as it has an 18 mm wide foot which must remain in contact with the squeegee. This required a method of measuring the resistance to indentation irrespective of the surface form. Furthermore, the Shore A method effectively measures the indentation made by a pin at only one single force or pressure (8.064 N for a 0.79 mm diameter pin). By applying a range of forces, a more comprehensive view of surface hardness should be revealed. Bulk mechanical properties influence how the squeegee bends in response to loading. Depending on the configuration of the screen printer, this will affect the pressure, contact and effective angle that the squeegee forms with the screen; thus affecting deposition. Mechanical properties are typically described according to tensile properties [16]. This is not necessarily representative of the multi-axial strain that occurs during printing. Therefore, the ability of a squeegee to resist flexing in a controlled manner was also investigated. Correlations between these measured parameters and print quality were explored.

Resistance to wear is typically established by controlled abrasion of a small piece of squeegee against a rotating drum mounted with an abrasive sheet [17]. The mass loss due to abrasion is then used to compare the abrasion resistance of the various squeegee materials. This has a number of disadvantages in that it cannot reproduce conditions during printing in terms of pressure, contact angle, resistance to flexing and sample geometry, as well as using a dry contact. However, it is not feasible, or cost effective, to wear squeegees by printing due to the time it would take, the wastage of both ink and substrate and the potentially uncontrolled wear that would result. This necessitated the development of an accelerated wear technique which was rapid and controlled but also allowed the effect on the print to be evaluated. Squeegee wear was then accurately measured using microscopy and image analysis techniques prior to printing in both worn and unworn states using a conductive silver ink. The resulting printed samples were then analyzed to compare the effect of wear on line geometry (ink film thickness, line width, overall deposition) and electrical resistance for printed silver lines. The relationship between the parameters of solvent absorption, surface hardness and resistance to flexing, wear characteristics and print quality were investigated.

## 2. Experimental Section

### 2.1. Materials

Six squeegees were obtained from commercial sources for wear testing and an additional squeegee was obtained for set-up tests and to act as a control in order to monitor process drift, during printing, without the influence of wear. All squeegees were obtained from different suppliers and spanned a range of costs. All squeegees had a square edge and most were nominally 70 to 80 Shore A hardness. Squeegee width was approximately 9 mm and height 50 mm. Squeegee 5 differed from the others in that it was composed of three layers, with a central core and two edge layers of nominally 75 Shore A. Shore A hardness [15] was measured for each squeegee, with 10 measurements taken at regular intervals over the length of the squeegee. The average and standard deviation for unused squeegees are presented in Table 1.

**Table 1.** Squeegees used in the experiments.

| Squeegee Number | Measured Shore A Hardness (Standard Deviation) |
|:---:|:---:|
| 1 | 74.2 (0.6) |
| 2 | 76.0 (0.0) |
| 3 | 75.8 (0.4) |
| 4 | 76.9 (1.2) |
| 5 | 78.7 (0.8) |
| 6 | 70.0 (0.8) |
| Control | 74.0 (0.0) |

Carbon and silver inks were used for wear and printing trials respectively and the solvents used in the manufacture of these inks were therefore used in the solvent absorption tests. Carbon paste screen ink (C2030519P4) was purchased from Gwent Electronic Materials Ltd., UK (GEM). The solvents used in the ink were 4-hydroxy-4-methyl-2-pentanone (diacetone alcohol, CAS: 123-42-2) and 3,3,5-trimethylcyclohex-2-enone (α-Isophorone, CAS: 78-59-1). The solvent blend used in the inks was also provided by GEM for testing the squeegees, in the same ratios used in the ink. A gel type flexible silver paste (C2080415D2) was purchased from GEM. The solvent used in the ink was ethylene glycol diacetate (CAS: 111-55-7). This was purchased in pure form from Sigma Aldrich (525200).

*2.2. Squeegee Characterization*

2.2.1. Solvent Absorption by Squeegees

The squeegees were subjected to immersion in the solvents used in the manufacture of the carbon and silver inks used in the wear and print tests. The squeegees were cut into 10 mm by 10 mm by 9 mm (squeegee thickness) pieces using a steel blade. For each squeegee type, five pieces were immersed and all squeegee types were immersed in the same dish at the same time. The squeegee pieces were weighed prior to immersion, using a mass balance accurate to 0.0001 g. The pieces were taken out of the solvent at regular intervals over a five hour period, patted dry, re-weighed and then placed back in the solvent. For each squeegee type, all five pieces were weighed at the same time and the mass was averaged.

2.2.2. Indentation and Deflection Tests

In order to test resistance to indentation, squeegees were cut into 15 mm by 15 mm sections and a Hounsfield HK10S tensile/compressive testing machine was used to indent the squeegee samples. A sample of squeegee was placed on a flat steel platform and a 1.1 mm diameter steel pin was gradually moved in to contact with the squeegee material and the force on the pin and its displacement were measured as indentation proceeded at a speed of 1 mm/min (Figure 2). As the pin engaged with the various squeegees at different points, the indentation was assumed to start once a force of 0.05 N was reached. Five repetitions were performed at different positions on each squeegee sample, with care taken not to indent at the edge of the squeegee. In order to measure the effect of solvent absorption on the hardness of the squeegees, only the squeegee surface subjected to indentation was exposed to the solvent. Squeegees samples were placed in 40 mm diameter watch glasses containing 1 mL of solvent, giving an immersion depth of approximately 2 mm. The reverse surface was undistorted and placed on the steel platform. Squeegee samples were placed in the solvent for 30 min, patted dry then tested immediately.

For deflection testing, squeegees were cut into 30 mm wide strips of 50 mm length (full squeegee strip width). The sample was supported on either end using two steel rods, which were 42 mm apart, and a rounded steel tool was pushed downwards in to the centre of the squeegee at a rate of 10 mm/min using the tensile/compressive testing apparatus. The force required to bring the tool downwards and deflect the squeegee was measured up to a maximum tool displacement of 10 mm (Figure 2). As the tool engaged with the various squeegees at different points, the flexing was assumed to start once a force of 0.1 N was reached. The test was performed six times for each squeegee using the same sample.

**Figure 2.** (a) Indentation and (b) deflection test methods for squeegees. Deflection test shows 10 mm tool displacement.

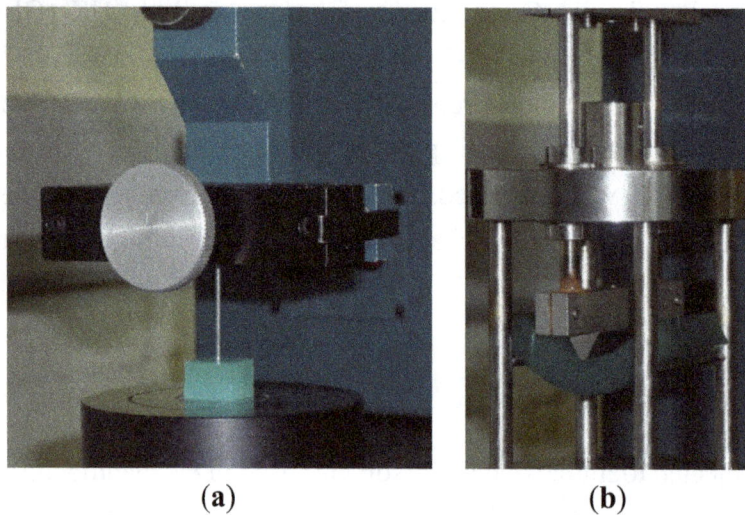

(a)                                    (b)

*2.3. Squeegee Wear Methodology*

In order for the wear to be representative of that achieved through printing, wear was performed using a screen printing press. Squeegees were subjected to controlled accelerated wear using various grades of silicon carbide ("wet and dry") abrasive papers lubricated with ink. This product was selected as it enabled a controlled and consistent means of wearing the squeegees and it was readily available in a range of standard grades.

The wear apparatus was designed specifically for this experiment and is shown in Figure 3. A stainless steel plate was attached to an aluminium screen printing frame (in place of the screen). Three different grades of silicon carbide abrasives were used; in order of declining roughness these were 1200, 2000 and 2500 grits (the lower the grit number, the higher the roughness—15.3, 10.3 and 8.4 µm average particle sizes respectively. For comparison, the approximate particle size in the silver ink, used in subsequent printing tests, was 2 to 3 µm). The abrasive sheets were cut into strips and placed side by side on the steel plate using a cushioned double sided tape. The full length of the sheets (280 mm) were used and they were cut into widths of 110 mm, for the 1200 and 2500 abrasives located at the sides, and 100 mm for the central strip of 2000 grit abrasive. The squeegees were cut to a length of 340 mm, allowing 10 mm overhang on either side of the abrasive. These dimensions were selected so that the worn squeegees could be used to print three identical test images from the same screen in the ensuing print tests. The parameters are summarized in Table 2.

**Figure 3.** Controlled wearing of a squeegee, using silicon carbide abrasive, on a screen printing press.

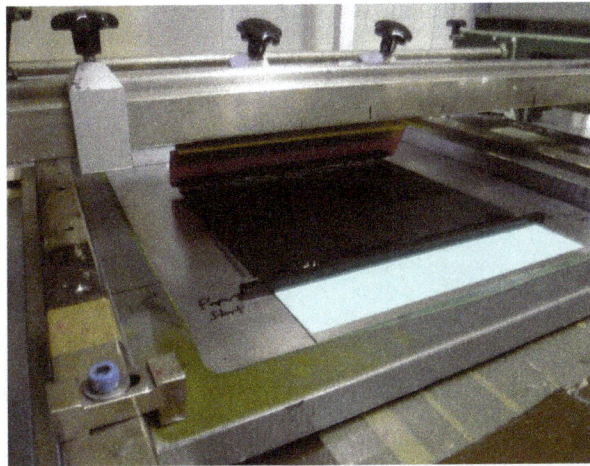

**Table 2.** Parameters used in wear experiment.

| Parameter | Setting |
|---|---|
| Printing machine | SveciaMatic SM |
| Abrasive types | 1200, 2000 and 2500 grit silicon carbide abrasive paper sheets |
| Abrasive sheet dimensions | 110 mm and 100 mm by 280 mm |
| Wear length | 260 mm per cycle |
| Number of wear cycles | 50 per squeegee |
| Screen frame | 580 mm × 580 mm Aluminium (510 mm × 510 mm internal) |
| Backing plate | 1 mm stainless steel |
| Backing tape | 3M E1715 (381 μm thickness) |
| Ink used | Carbon graphite paste (C2030519P4, Gwent Electronic Materials Ltd., UK) −40 grams per wear cycle |
| Squeegee dimensions | 9 mm (thickness approx) × 50 mm (height) × 340 mm (length) |
| Squeegee angle | 65° |
| Squeegee holder | Serilor®MACH straight upper with standard 54.1 mm jaws, Fimor, France. |
| Squeegee engagement | 21 mm (equivalent to kiss contact plus 8 mm) |
| Speed | 2.5 units (equivalent to approximately 0.35 m/s) |

In order to help lubricate the contact between the screen and squeegee and assist the transport of abraded particles (of squeegee and silicon carbide) away from contact area, carbon paste screen ink was spread over the abrasive sheet prior to wear. Dry abrasion, or use of a low volatility solvent alone, was found to be much more damaging to the squeegee in preliminary tests. A flow coat was not used as it would most likely damage the abrasive and would suffer abrasion itself. A 10 mm strip of squeegee material was attached to the adhesive tape at the end of the abrasive strips, where the squeegee lifted off after wear. The ink pooled at this point, as it was forced along the abrasive sheet by the squeegee, and the strip allowed a reservoir of ink to form that would recoat the squeegee at the end of each wear cycle. This ensured that a covering of ink remained on the squeegee; rather than having dry contact (Figure 3). This was confirmed upon cleaning the squeegee after wear; where a covering of ink was observed on the printing edge of the squeegee. The use of a lubricating ink also ensured that any solvent related swelling and softening; that would occur during printing, would also be factored in to the wear experiment. The settings used were the same as those used in the later printing trials, with the squeegee angle set to 65°, and are detailed in Table 2.

Each squeegee was reciprocated fifty times over the abrasives to cause it to wear, with bands of different levels of wear across the width from the different abrasive types. Both abrasives and ink were discarded after each cycle of fifty reciprocations to ensure consistency between squeegees. Following wearing, the squeegees were cleaned and left for a minimum of 48 h before wear measurement to allow absorbed solvent to escape and swelling to subside.

### 2.4. Measurement of Squeegee Wear

Images of squeegee wear were captured using a Leica stereo microscope with a CCD camera. The squeegees were measured from both the side and bottom of the squeegee and for both orientations three images were taken over each wear band. A sample image of a squeegee in both unworn and worn states is shown in Figure 4. Wear was evaluated using image analysis software (Image J 1.46r, U. S. National Institutes of Health). A rectangle was manually selected over the wear region, with the software outputting its dimensions. This was done five times in each image, giving a total of fifteen measurements per orientation per wear band. The microscope was calibrated using a tile with dots of known diameter. The amount of squeegee removed was then calculated as a triangle from the worn width of the squeegee from both orientations using Equation (1). Standard deviation (St. dev) was calculated using Equation (2) and the wear angle, that is the angle between the long side of the squeegee and the wear, calculated using Equation (3).

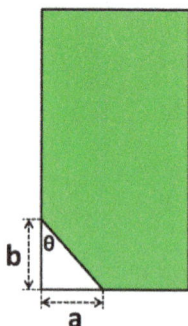

$$CSA \text{ removed} = \frac{a \times b}{2} \tag{1}$$

$$\text{St. dev in } CSA \text{ removed} = \frac{a}{2}\text{St. dev } b \times \frac{b}{2}\text{St. dev } a \tag{2}$$

$$\theta = \tan^{-1}\frac{a}{b} \tag{3}$$

where $CSA$ = cross-sectional area; $a$ = wear on bottom of squeegee; $b$ = wear on side of squeegee; $\theta$ = angle of wear.

The squeegees were used to print in both unworn and worn states using the settings listed in Table 3. Unworn and worn edges of the same squeegee were printed sequentially, before moving on to the next squeegee, with the squeegees printed in the order listed in Table 1 (*i.e.*, 1, 2, 3, 4, 5, 6). To alternate between unworn and worn edges, the squeegee holder was removed, rotated by 180° and replaced in the printing press. In addition to the test squeegees, a series of control prints were made with the unworn control squeegee both prior to and after printing with the other squeegees. This was performed in order to monitor any drifts in the printing process over time so that these could be distinguished from changes due to wear. All prints were performed on the same screen without changing over or cleaning between print cycles. None of the printing parameters were altered and the ink was kept in excess to deter drying in the mesh and replenished when required. A gel type ink (flexible silver paste C2080415D2, Gwent Electronic Materials, Pontypool, UK) was selected as this was stable over time and was not prone to drying in.

**Figure 4.** Microscope images of squeegee edge (**a**) before and (**b**) after wear. Images 3.84 mm × 2.46 mm.

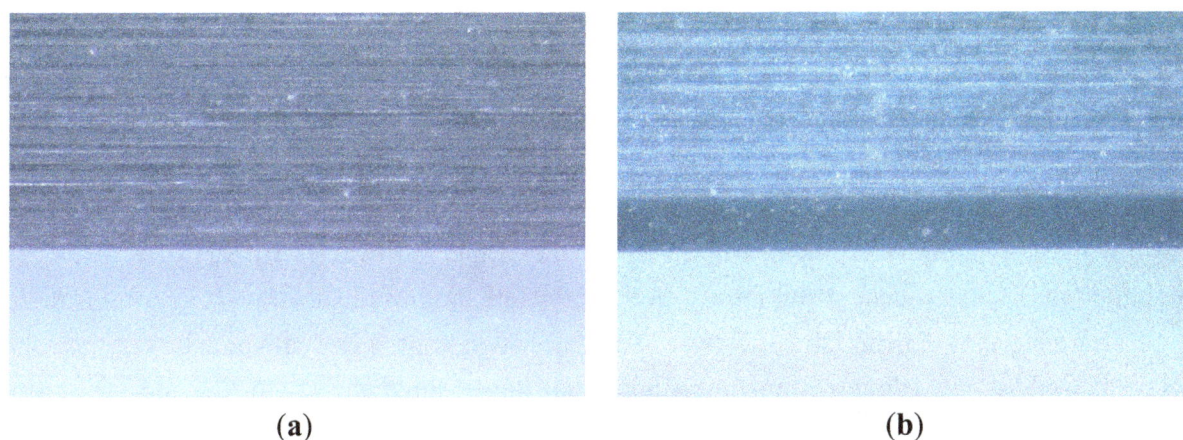

(**a**)                    (**b**)

**Table 3.** Parameters used in printing experiment.

| Parameter | Setting |
| --- | --- |
| Printing machine | SveciaMatic SM |
| Substrate | Melinex® 339, DuPont Teijin Films (330 μm thickness, 325 mm × 325 mm) opaque white |
| Screen frame | 800 mm × 800 mm Aluminium (700 mm × 700 mm internal) |
| Mesh | PET, 68 threads cm$^{-1}$, 55 μm thread diameter, 45° mesh angle |
| Ink used | Flexible silver paste C2080415D2 (Gwent Electronic Materials, Pontypool, UK) |
| Squeegee dimensions | 9 mm (thickness approx) × 50 mm (height) × 340 mm (length) (same as wear trial) |
| Squeegee angle | 65° (same as wear trial) |
| Squeegee engagement | 21 mm (equivalent to kiss contact plus 2 mm) |
| Squeegee holder | Serilor®MACH straight upper with standard 54.1 mm jaws, Fimor, France. (same as wear trial) |
| Snap off gap | 3 mm |
| Speed | 2.5 units (equivalent to approximately 0.35 m/s) |
| Flowcoat | 400 mm width |
| Flowcoat engagement | 6 units |
| Drying | Belt drying at 120 °C, with two passes at low speed giving approximately 3 min drying time |

## 2.5. Printing of Silver Ink Using Unworn and Worn Squeegees

The screen used for printing consisted of three bands of identical test images whose location coincided with the different wear bands. A range of different line widths in both print direction (perpendicular to the squeegee) and at 90° to the print direction (parallel to the squeegee) were included. A total of ten prints were made for each squeegee configuration, giving a total of 140 prints. Including changeover time, this took less than two hours and the condition of the ink did not change noticeably in that time. The dimensions of the printed features were measured using white light interferometry (Veeco NT9300, Veeco Instruments, Inc., Plainview, NY, USA). This allowed a full three-dimensional surface profile to be captured, so that line width, print thickness and local surface variations could be evaluated. Lines of 400 and 600 μm nominal width were measured both in the print direction and at 90° to the print direction. These lines were 30 mm in length. Measurements were taken on each of the bands for prints with worn and unworn squeegees. Three measurements were taken for each line and three print samples (Repetitions 8, 9 and 10) were measured (nine measurements per line per orientation). The number of samples was based on an analysis of the variation in control prints and there was not found to be any benefit in accuracy in using five prints instead of three. Measurement was performed using an automated method in which the measurement locations were the same for all sets of measurements. Five times magnification was used, giving a measurement area of 1.25 mm by 0.94 mm (a resolution of 640 × 480 pixels with sampling at 1.9 μm intervals).

Average line width and ink film thickness across the measured profiles (1.25 mm or 0.94 mm depending on the orientation) were evaluated using "WCPCLine" software written by WCPC. The software aligned the data, to account for any tilt in the substrate, and used substrate roughness data to precisely differentiate between ink and substrate. Standard deviations were calculated over the nine readings taken per line type (three readings per individual line × three sheets) to indicate variability between the measured lines (not variability within the lines).

The resistance of the lines was measured with a Keithley 2400 multimeter using the two point probe technique. Probes were applied to the contact pads at each end of the 30 mm long tracks and the resistance recorded. The reported resistance is the average of measurements over three samples with the probe contact resistance subtracted.

## 3. Results and Discussion

### 3.1. Solvent Absorption by Squeegees

The solvent absorption of the squeegee materials is shown in terms of the percentage change in mass [100 × (Mass−Original mass)/Original mass] for carbon ink and silver ink solvents in Figures 5 and 6 respectively.

For all squeegees, there was a substantial difference in solvent absorption between the different solvents; overall there was on average 3.4 times more mass of ethylene glycol diacetate absorbed than the solvent blend used in the carbon ink over a given time period. The rate of solvent uptake was fastest at the onset of immersion but remained reasonably stable between one and five hours after immersion. The squeegees kept absorbing solvent throughout the duration of the experiment, even when quite substantial amounts of solvent (up to 17% mass increase) were absorbed. Squeegee 2 gave the lowest amount of solvent absorption of all the squeegees and was particularly resistant to absorption of the solvent blend used in the carbon ink; increasing in mass by less than 1% after five hours immersion. The greatest solvent uptake was observed in Squeegees 1 and 6, followed by the control squeegee. The remaining squeegees (3, 4 and 5) were broadly similar and displayed intermediate absorption levels between those of the control squeegee and Squeegee 2. The levels of solvent uptake were higher than what would be observed during printing due to the immersive nature of the test, use of neat solvent, test duration and small specimen size. However, the test illustrates the wide range of responses of the different squeegees as well as the continual absorption of solvent over time.

For most of the squeegees there appeared to be a trend of increasing solvent uptake as the Shore A hardness decreased. However, Squeegee 1 showed higher levels of solvent absorption compared with squeegees of the same hardness while Squeegee 2 showed lower levels of absorption.

**Figure 5.** Percentage change in squeegee mass during immersion in solvent blend used in carbon (wear) ink.

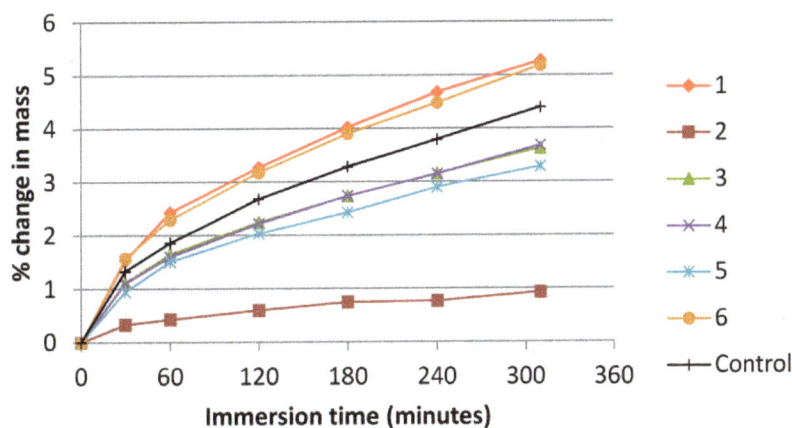

**Figure 6.** Percentage change in squeegee mass during immersion in ethylene glycol diacetate used in silver (print) ink.

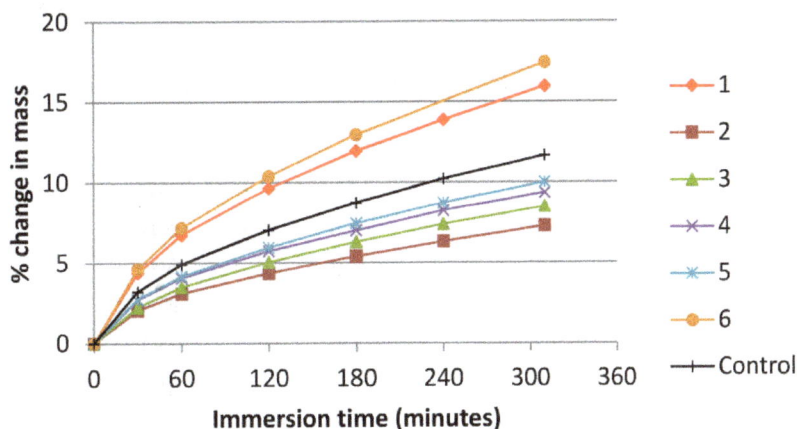

*3.2. Surface Hardness of Squeegees*

3.2.1. Untreated Squeegees

The force required for different levels of indentation is compared for all squeegees in Figure 7. The curves shown for each squeegee are the mean of five measurements. For the majority of the force-indentation curve, Squeegee 6 was the softest squeegee. The next softest was the control squeegee, followed by Squeegees 1 and 2. The remaining squeegees (3, 4 and 5) were the hardest and were fairly similar to one another. Plotting indentation *versus* Shore A hardness gave a linear relationship for a 10 N indentation force. However, in the low force part of the curve, Squeegee 1 was noticeably more resistant to indentation than the other squeegees and did not fit the trend. The Shore A hardness test will only cover a single force and does not give any information where low forces are concerned. As the indentation increases, the force-indentation curve for Squeegee 1 crosses over those of most of the other squeegees, so that at mid to high force levels, it appears to be one of the softer squeegees.

**Figure 7.** Force *vs.* indentation for untreated squeegees.

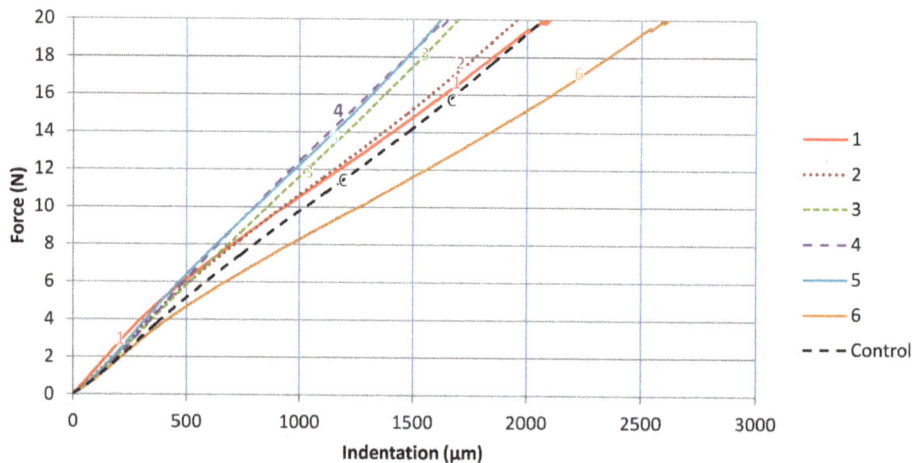

### 3.2.2. Solvent Treated Squeegees

The force required for different levels of indentation is compared for all squeegees when treated with the carbon ink and silver ink solvents in Table 4. In order to compare the squeegees more readily, indentation levels are shown for all squeegees before and after solvent treatment at 5 N indentation force.

**Table 4.** Indentation levels of squeegees at 5 N indentation force with and without solvent treatment.

| Squeegee | Untreated | Carbon Ink Solvent | | Ethylene Glycol Diacetate | |
|---|---|---|---|---|---|
| | Indentation (µm) | Indentation (µm) | Change (%) | Indentation (µm) | Change (%) |
| 1 | 386.8 | 444.2 | 14.8 | 501.0 | 29.5 |
| 2 | 421.5 | 416.0 | −1.3 | 500.6 | 18.8 |
| 3 | 430.6 | 456.2 | 5.9 | 467.4 | 8.5 |
| 4 | 432.6 | 432.8 | 0.0 | 481.0 | 11.2 |
| 5 | 393.8 | 422.2 | 7.2 | 490.0 | 24.4 |
| 6 | 538.6 | 579.6 | 7.6 | 631.4 | 17.2 |
| Control | 484.8 | 514.0 | 6.0 | 544.8 | 12.4 |

Solvent ingress softened the squeegees and the solvent used in the silver ink had a greater effect as it was absorbed by the squeegees in greater amounts. Squeegee 1, which absorbed solvent more readily than most of the other squeegees, showed the greatest percentage reduction in surface hardness as a result of solvent ingress. Squeegee 2 showed very little ingress of carbon ink solvent and no noticeable change in surface hardness resulted. When the absolute levels of indentation are compared, Squeegee 1 was found to be the hardest squeegee at a low indentation force of 3 N in both untreated and solvent treated states. This was not observed when force levels were increased as Squeegee 1 became softer in comparison with the other squeegees. This is similar to the observations made with the untreated squeegees.

In general, a greater reduction in hardness was associated with greater solvent ingress. However, there was not a straight-forward relationship between solvent uptake and loss in surface hardness. Squeegee 6, for example, absorbed more solvent than most of the other squeegees but this was not reflected in the changes in surface hardness, which were comparable to the changes seen in squeegees which absorbed less solvent. However, Squeegee 6 was already substantially softer than the others prior to solvent addition. Squeegee 2 showed very little mechanical response to carbon ink solvent but responded much more strongly to the silver ink solvent, despite absorbing less of this than any of the other squeegees.

### 3.3. Deflection of Squeegees

The relationship between squeegee deflection and force is shown in Figure 8. The data shown is the average of five measurements (reading one was discarded). As the squeegee is deflected, progressively more force is required to increase the bending angle. The force response was not smooth due to occasional slip at the contact between the squeegee and the supporting rods. The data suggests that at large amounts of deflection, Squeegee 5 was the most resilient to bending, followed by 4, 2, 1 and 3, the control squeegee and finally Squeegee 6. However, when looking at smaller levels of deflection, which are more representative of screen printing, Squeegee 1 appeared to be the most resistant to bending. This has parallels with the observations made in the indentation testing, and there is a good correlation between the indentation produced at 5 N with the force required to deflect the squeegee to 3 mm (approximately 8°). Likewise there is a correlation at higher forces; and the correlation between Shore A hardness and deflection improves at higher amounts of deflection. This suggests that Shore A hardness is a more suitable indicator for squeegee behavior when under high deflection. However, behavior under low deflection, or indentation force, cannot always be inferred from Shore A hardness data. This is particularly true of Squeegee 1, which shows higher comparative hardness at low indentation than at higher indentations.

The data from squeegee testing was compared with ink film thickness data from the print tests for unworn squeegees (print methods detailed in Table 3, ink film thickness data in Table 6). The optimum predictor of ink thickness was found to be the force measured at 3 mm squeegee deflection. With the exception of Squeegee 6, there was a general pattern of increasing ink deposition with decreasing resistance to deflection (Figure 9). This is due to the squeegee exerting a greater pressure on the screen and being forced in to the open areas of the mesh. Excluding Squeegee 6 there was an $R^2$ value of 0.92, indicating a good correlation. When considering indentation force, again with the exception of Squeegee 6, there was a general pattern of increasing ink deposition with decreasing resistance to indentation. The correlation was less effective with an $R^2$ value of 0.77 at an indentation force of 5 N. Finally, Shore A hardness gave the worst correlation with ink film thickness with an $R^2$ value of 0.47. Squeegee 1 did not fit the pattern when Shore A was used, for reasons outlined previously, which further demonstrates that Shore A hardness cannot necessarily reflect the behavior of a squeegee during printing. Analogous to selecting a softer squeegee, the softening of squeegees by solvent absorption has been shown to cause increased ink deposition [14]. This should increase over time as the squeegee progressively absorbs solvent.

**Figure 8.** Force *vs.* deflection tool displacement for squeegees.

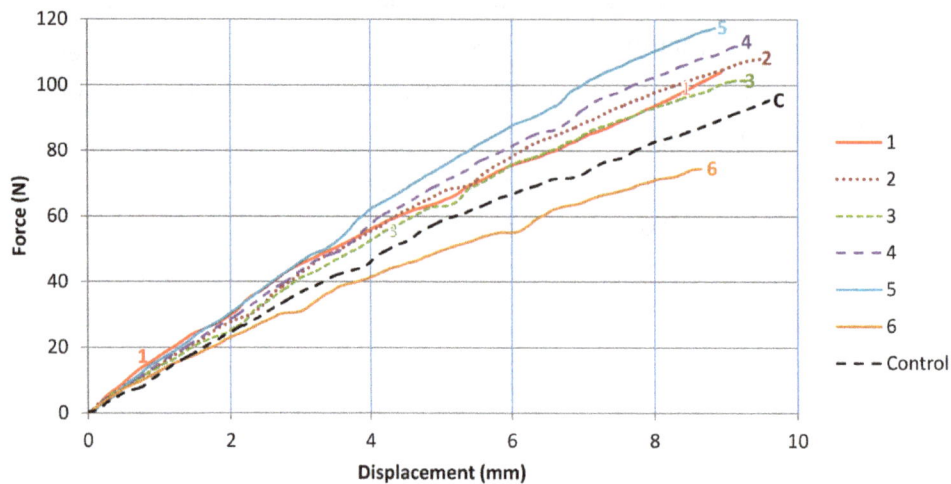

**Figure 9.** Ink film thickness *vs.* force required to deflect squeegee ($R^2$ 0.92 when Squeegee 6 not included).

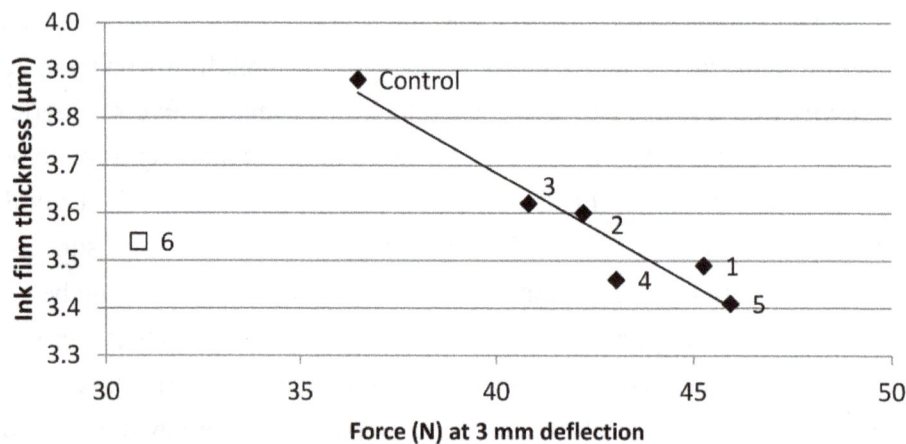

## 3.4. Squeegee Wear

The amount of wear, in terms of cross-sectional area removed from the squeegee and wear angle, is shown for the three wear bands (1200, 2000 and 2500) and for each squeegee in Table 5. The squeegee removal is also shown graphically in Figure 10. The roughest abrasive (1200 grit) gave the highest amount of wear, while the less rough papers (2000 and 2500 grit) gave less wear but were fairly similar to each other. For the roughest abrasive (1200 grit), the lowest amount of wear was observed in Squeegee 3, followed by 1 and 4, though all three were broadly similar with between 0.057 and 0.070 mm$^2$ removed. Squeegees 6 and 5 gave more wear than 1, 3 and 4, and performed similarly with 0.111 and 0.112 mm$^2$ removed. The most wear was observed in Squeegee 2 with 0.193 mm$^2$ removed; significantly more than any of the other squeegees. For the 2000 grit abrasive, Squeegees 3 and 4 gave the least wear, followed by 1, 5, 6 and finally 2. For the 2500 grit abrasive, Squeegee 3 gave the least wear, followed by 4, 5, 1, 6 and finally 2. For both 2000 and 2500 abrasives, Squeegees 2 and 6 gave substantially more wear than Squeegees 1, 3, 4 and 5. Overall, across all the abrasive types, the least wear was observed in Squeegee 3. Squeegees 2 and 6 were inferior to the other squeegees in terms of their resistance to wear.

**Table 5.** Squeegee cross-sectional area removed and wear angle after 50 wear cycles with different silicon carbide abrasives. Standard deviation shown in parentheses.

| Squeegee Number | Removal (mm²) | Wear Angle (°) | Removal (mm²) | Wear Angle (°) | Removal (mm²) | Wear Angle (°) |
|---|---|---|---|---|---|---|
| | 1200 grit | | 2000 grit | | 2500 grit | |
| 1 | 0.066 (0.005) | 49.3 | 0.019 (0.001) | 51.9 | 0.025 (0.002) | 53.7 |
| 2 | 0.193 (0.011) | 44.4 | 0.053 (0.002) | 43.2 | 0.049 (0.003) | 43.8 |
| 3 | 0.057 (0.004) | 46.2 | 0.010 (0.001) | 46.4 | 0.010 (0.001) | 46.2 |
| 4 | 0.070 (0.004) | 48.3 | 0.010 (0.001) | 49.0 | 0.011 (0.001) | 48.1 |
| 5 | 0.112 (0.005) | 46.3 | 0.023 (0.002) | 49.7 | 0.024 (0.002) | 47.9 |
| 6 | 0.111 (0.004) | 45.5 | 0.048 (0.003) | 47.8 | 0.043 (0.002) | 47.0 |

**Figure 10.** Squeegee cross-sectional area removed after 50 wear cycles with different silicon carbide abrasives. Error bars show standard deviations.

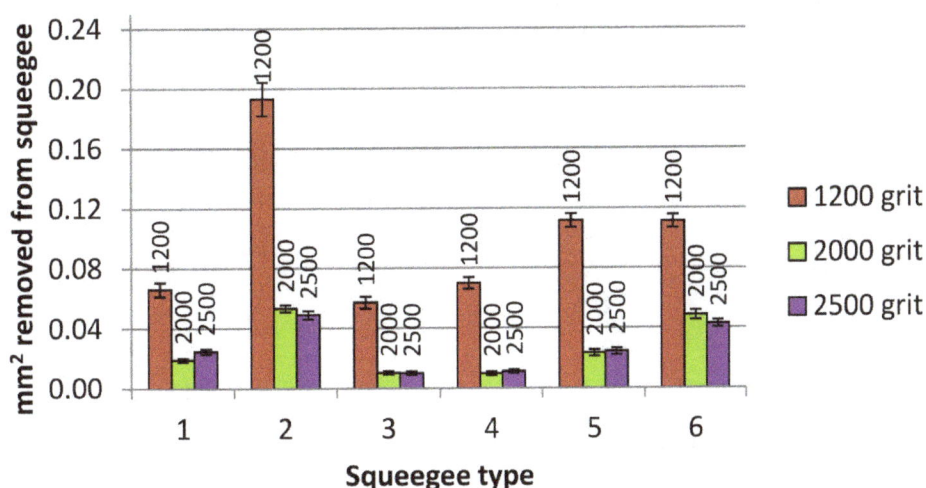

The angle of wear differed depending on the squeegee type. Squeegee 1 had the highest wear angle, with more wear apparent from the bottom of the squeegee than from the face. The lowest wear angles were observed on Squeegee 2, with the other squeegees showing intermediate wear angles. There appeared to be a rough correlation between the ability of a squeegee to flex and the wear angle; the squeegees most resistant to bending tended to give the highest wear angles. The ratios of wear for the roughest abrasive to wear with the other abrasive materials was not consistent. So for example, Squeegee 4 showed seven times more wear with 1200 grit abrasive than for 2000 grit abrasive but Squeegee 6 showed only 2.3 times more wear with 1200 grit abrasive than for 2000 grit abrasive. The other squeegees showed intermediate wear ratios for the different abrasives.

## 3.5. Geometry and Electrical Resistance of Printed Silver Lines Using Unworn Squeegees

The geometry of the printed lines is described in terms of the average ink film thickness over the width of the line and the average line width over the measured length of the line. The dry ink contained in that line is thus ink film thickness multiplied by line width. This can be used as an indicator of ink consumption.

For prints made on the unworn squeegees, the ink film thickness varied between the different line orientations and widths, the different positions along the squeegee and between the different squeegees. The average data for all measurements on each unworn squeegee is shown in Table 6. The average film thicknesses ranged from 3.41 to 3.62 μm (*i.e.*, a 6% increase from thinnest to thickest ink film) for the test squeegees, and were higher for the prints made with the control squeegee. The correlations with squeegee flexure, hardness, *etc.*, are detailed previously. There was a general decline in line width over the course of the experiment as demonstrated by the lower line widths recorded in the Control end when compared with Control start. Ink film thickness was also lower in the final control prints than in the starting control prints (by 3%). This suggested that there was some drying in the mesh during the printing that reduced ink transfer. However, the effect should be minimal between prints made with unworn and worn edges of the same squeegee.

**Table 6.** Ink film thickness, line width and line resistance for unworn squeegees. Averaged over all measurements.

| Squeegee Number | Ink Film Thickness (μm) | 400 μm Line Width (μm) | 600 μm Line Width (μm) | 400 μm Line Resistance (Ω) | 600 μm Line Resistance (Ω) |
|---|---|---|---|---|---|
| 1 | 3.49 | 346.9 | 541.8 | 7.93 | 4.72 |
| 2 | 3.60 | 334.5 | 531.5 | 8.24 | 4.65 |
| 3 | 3.62 | 338.6 | 534.1 | 8.17 | 4.79 |
| 4 | 3.46 | 323.5 | 525.0 | 8.80 | 5.00 |
| 5 | 3.41 | 316.9 | 513.3 | 9.38 | 5.41 |
| 6 | 3.54 | 313.2 | 510.2 | 9.33 | 5.40 |
| Control start | 3.88 | 380.5 | 577.1 | 5.87 | 3.74 |
| Control end | 3.76 | 326.8 | 524.4 | 8.17 | 4.66 |
| Percentage drift | −3.1% | −14.1% | −9.1% | +39.2% | +24.6% |

The mean standard deviation in ink film thickness over all sets of nine measured lines (for the various squeegee type, line width and orientation combinations) was 0.12 μm which is 3.3% of the mean ink film thickness. For line width this was 9.5 and 7.6 μm for 400 and 600 μm lines respectively (2.9% and 1.4% of the mean line widths). For line resistance this was 0.22 and 0.17 Ω for 400 and 600 μm lines respectively (2.6% and 3.4% of the mean resistances). The orientation of the printed lines affected the ink film thickness. Lines produced in the print direction (perpendicular to the squeegee) tended to have a greater amount of ink deposition than those printed at 90° to the print direction. This is described in more detail in the following section.

### 3.6. Geometry and Electrical Resistance of Printed Silver Lines Using Worn Squeegees

The effect of wear on print geometry is shown in terms of the percentage change in printed line thickness and line width when moving from unworn to worn squeegees [*i.e.*, 100 × (thickness worn−thickness unworn)/thickness unworn]. The average effect of wear on the ink film thickness, width, overall ink deposition (cross-sectional area) and resistance of printed lines are shown for each squeegee and abrasive type in Table 7. The change in ink film thickness is illustrated graphically in Figure 11.

**Table 7.** Percentage change in printed ink film thickness, line width, ink deposition and line resistance as a result of squeegee wear. Data as average for all measured lines.

| Squeegee | Change in Ink Film Thickness (%) | | | Change in Line Width (%) | | |
|---|---|---|---|---|---|---|
| | 1200 grit | 2000 grit | 2500 grit | 1200 grit | 2000 grit | 2500 grit |
| 1 | 21.9 | −3.0 | 0.2 | 3.0 | −0.2 | 0.5 |
| 2 | 38.5 | 29.6 | 30.1 | 7.9 | 5.4 | 6.7 |
| 3 | 36.0 | 12.4 | 13.1 | 5.9 | 0.6 | 1.1 |
| 4 | 44.0 | 13.6 | 19.6 | 6.9 | 0.7 | 1.9 |
| 5 | 29.2 | 14.0 | 17.5 | 5.9 | −0.3 | 1.9 |
| 6 | 33.3 | 37.1 | 31.9 | 4.7 | 5.4 | 7.6 |

| Squeegee | Change in Deposition (%) | | | Change in Line Resistance (%) | | |
|---|---|---|---|---|---|---|
| | 1200 grit | 2000 grit | 2500 grit | 1200 grit | 2000 grit | 2500 grit |
| 1 | 25.5 | −3.1 | 0.8 | −21.1 | 3.1 | 1.5 |
| 2 | 49.5 | 36.6 | 38.8 | −33.2 | −29.8 | −27.8 |
| 3 | 44.0 | 13.0 | 14.5 | −34.1 | −33.2 | −17.8 |
| 4 | 53.9 | 14.4 | 21.9 | −37.0 | −12.7 | −19.1 |
| 5 | 36.8 | 13.7 | 19.7 | −29.3 | −17.4 | −22.4 |
| 6 | 39.5 | 44.6 | 41.9 | −30.3 | −37.8 | −32.3 |

**Figure 11.** Percentage change in printed ink film thickness as a result of squeegee wear (average for all measured lines).

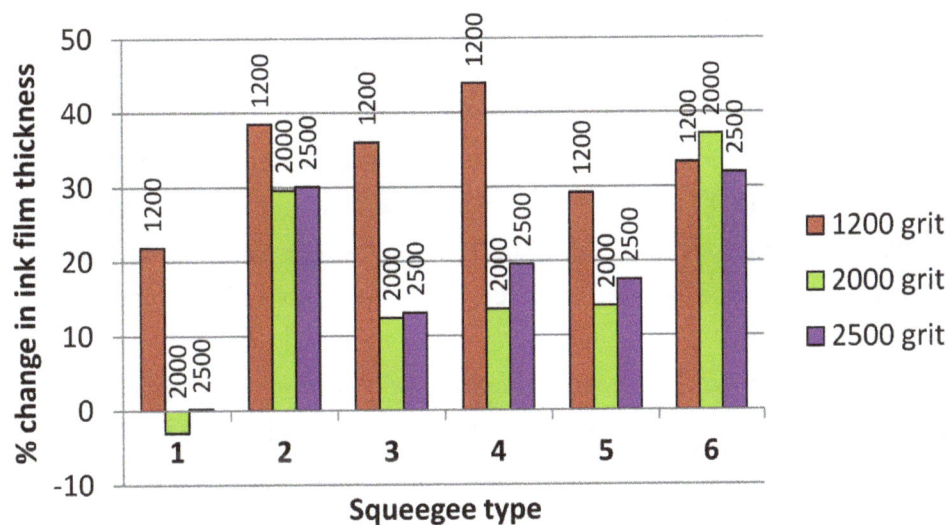

Worn squeegees, for the most part, gave greater ink film thickness than unworn squeegees. This was most severe in the higher levels of wear given by the roughest, 1200 grit abrasive, with Squeegee 1 showing the lowest increase in ink film thickness due to wear (average 21.9% increase overall). This was followed by Squeegees 5, 6, 3, 2 and 4 with overall increases up to 44%. For the mid roughness abrasive, Squeegee 1 showed a small decrease of 3% in ink film thickness due to wear while the other squeegees all increased their ink film thickness, in varying amounts, between 12.4% and 37.1%. For the smoothest abrasive, Squeegee 1 showed only a negligible increase in average ink film thickness of 0.2% while the other squeegees all increased their ink film thickness, in varying amounts, between 13.1% and 31.9%. There was a print defect in Squeegee 3 for the mid wear range lines at 90° to the print direction. This caused break-up in the lines and the formation of satellite drops of ink around the line but was not observed with any of the other squeegees, in either worn or unworn states, or for lines printed in the print direction. Squeegee 1 was the best performing in terms of maintaining consistency in the print as a result of wear and there was a marked contrast between its performance and that of the other squeegees. Of the remaining squeegees, 3, 4 and 5 were substantially better than 2 and 6 when using the 2000 and 2500 abrasives but this was not the case for the roughest 1200 abrasive. The wear levels for the roughest abrasive would be unlikely to be tolerated in practice. Variability within the individual sets of measurements was similar to that measured for the unworn squeegees (standard deviation of 0.13 μm, 3.0% of the mean).

Squeegee wear tended to influence line width in a similar way to ink film thickness but the effect on overall deposition was generally lower. Squeegee wear also tended, for the most part, to increase the width of the printed lines up to a maximum of around 8%, depending on the squeegee and abrasive type, but showed decreased line width in some instances. Squeegee 1 showed the least variation in line width between worn and unworn squeegees. Overall, Squeegees 3, 4 and 5 gave intermediate behavior, while 2 and 6 generally gave the greatest increase in printed line width. The 2000 and 2500 abrasives gave only marginal changes in line width for Squeegees 1, 3, 4 and 5, while Squeegees 2 and 6 showed more substantial changes for these abrasives. Variability within the individual sets of measurements was similar to that measured for the unworn squeegees (standard deviation of 9.7 μm and 8.9 μm for 400 and 600 μm lines respectively: 2.8% and 1.7% of the mean line widths respectively).

Ink film thickness, line width, and hence ink deposition generally increased with the amount of wear on the squeegee. However, the dominant factor in the deposition was the change in ink film thickness rather than the width of the line. In line with the trends for ink film thickness and line width, Squeegee 1 gave the smallest changes in ink deposition between worn and unworn states. For, the roughest abrasive, an increase in ink deposition of 25.5% was recorded for Squeegee 1, while the other squeegees showed increases between 36.8% and 53.9%. For the mid roughness abrasive, Squeegee 1 showed a small decrease of 3% in ink deposition due to wear while the other squeegees all increased deposition, in varying amounts, between 13% and 44.6%. For the smoothest abrasive, Squeegee 1 showed only a negligible change in deposition (+0.8%) while the other squeegees all increased deposition, in varying amounts, between 14.5% and 41.9%. The small reduction in deposition observed in Squeegee 1 for the 2000 abrasive was most likely within the inherent variability in the process and the gradual drying in the mesh (demonstrated by the change in the control prints). There was not an intermediate increase in ink deposition for the 2000 grit abrasive. However, this abrasive did not produce intermediate wear levels (Figure 10).

For squeegees worn with the roughest, 1200, abrasive, there was a reduction in electrical resistance for all printed lines. The average reduction was between 21% (Squeegee 1) and 37% of the initial values, depending on the squeegee type. This was due to the increase in ink deposition from the worn squeegees, primarily due to the increased ink film thickness but also increased line width, as described previously. Reductions in ink film deposition gave the higher resistances noted for Squeegee 1 when using the 2000 and 2500 abrasives. For the mid roughness abrasive, Squeegee 1 showed a small increase of 3% in line resistance due to wear while the other squeegees all showed a reduction in resistance, in varying amounts between 12.7% and 37.8%. For the smoothest abrasive, Squeegee 1 showed only a very small increase in resistance of 1.5% while the other squeegees all gave reduced resistances, in varying amounts between 17.8% and 32.3%. For worn squeegees mean standard deviations in resistance were 0.13 $\Omega$ (2% of mean) and 0.20 $\Omega$ (5.2% of mean) for 400 and 600 µm lines respectively.

The effect of wear on the print varied depending on the orientation of the printed lines. Ink film thickness in unworn and worn states, as well as the percentage change in printed ink film thickness resulting from squeegee wear, is shown both in the print direction and at 90° to the print (parallel to the squeegee) in Table 8.

**Table 8.** Printed ink film thickness (µm) in unworn and worn states and percentage change in ink film thickness as a result of squeegee wear for lines printed in the print direction and at 90° (parallel to squeegee).

| | | Printed Ink Film Thickness (µm) | | | | | | | | | | |
| | | 1200 grit | | | | 2000 grit | | | | 2500 grit | | | |
| | Squeegee | Line Width 400 µm | | Line Width 600 µm | | Line Width 400 µm | | Line Width 600 µm | | Line Width 400 µm | | Line Width 600 µm | |
| | | Print | 90° | Print | 90° | Print | 90° | Print | 90° | Print | 90° | Print | 90° |
|---|---|---|---|---|---|---|---|---|---|---|---|---|---|
| | Unworn | 3.69 | 3.11 | 3.91 | 3.47 | 3.58 | 3.15 | 3.62 | 3.45 | 3.55 | 3.14 | 3.65 | 3.57 |
| 1 | Worn | 4.26 | 3.98 | 4.72 | 4.27 | 3.36 | 3.13 | 3.47 | 3.42 | 3.46 | 3.24 | 3.54 | 3.69 |
| | Change (%) | 15.4 | 28.2 | 20.8 | 23.1 | −6.1 | −0.5 | −4.3 | −1.0 | −2.6 | 3.4 | −3.1 | 3.2 |
| | Unworn | 4.02 | 3.12 | 4.21 | 3.60 | 3.51 | 3.26 | 3.69 | 3.62 | 3.58 | 3.08 | 3.93 | 3.62 |
| 2 | Worn | 5.00 | 4.69 | 5.68 | 5.21 | 4.49 | 4.24 | 4.90 | 4.62 | 4.45 | 4.29 | 4.94 | 4.76 |
| | Change (%) | 24.2 | 50.1 | 34.9 | 44.6 | 27.9 | 30.2 | 32.8 | 27.4 | 24.3 | 39.0 | 25.5 | 31.5 |
| | Unworn | 3.61 | 3.39 | 4.13 | 3.56 | 3.74 | 3.26 | 3.83 | 3.59 | 3.64 | 3.17 | 4.07 | 3.45 |
| 3 | Worn | 5.05 | 4.46 | 5.48 | 4.98 | 4.16 | N/A | 4.35 | N/A | 3.97 | 3.76 | 4.21 | 4.18 |
| | Change (%) | 39.9 | 31.4 | 32.7 | 40.0 | 11.1 | N/A | 13.7 | N/A | 9.1 | 18.6 | 3.5 | 21.1 |
| | Unworn | 3.75 | 2.98 | 3.80 | 3.38 | 3.56 | 3.44 | 3.57 | 3.61 | 3.41 | 2.98 | 3.71 | 3.35 |
| 4 | Worn | 5.01 | 4.46 | 5.59 | 4.94 | 4.05 | 3.64 | 4.25 | 4.17 | 3.86 | 3.70 | 4.21 | 4.27 |
| | Change (%) | 33.5 | 49.4 | 47.1 | 46.1 | 14.0 | 5.9 | 19.1 | 15.5 | 13.2 | 24.2 | 13.5 | 27.4 |
| | Unworn | 3.69 | 3.35 | 3.81 | 3.73 | 3.45 | 3.10 | 3.39 | 3.39 | 3.33 | 2.91 | 3.37 | 3.40 |
| 5 | Worn | 4.76 | 4.23 | 5.23 | 4.63 | 3.96 | 3.54 | 4.07 | 3.63 | 3.62 | 3.53 | 4.04 | 4.09 |
| | Change (%) | 29.0 | 26.4 | 37.3 | 24.1 | 15.0 | 14.1 | 20.0 | 6.9 | 8.7 | 21.2 | 19.9 | 20.2 |
| | Unworn | 3.97 | 3.37 | 4.30 | 3.57 | 3.71 | 3.07 | 3.50 | 3.17 | 3.37 | 3.21 | 3.71 | 3.60 |
| 6 | Worn | 4.97 | 4.61 | 5.40 | 5.18 | 4.49 | 4.43 | 4.83 | 4.60 | 4.35 | 4.30 | 4.87 | 4.78 |
| | Change (%) | 25.4 | 36.8 | 25.7 | 45.2 | 21.0 | 44.4 | 38.1 | 44.9 | 29.3 | 34.0 | 31.3 | 32.8 |

For lines printed at 90°, there tended to be a greater increase in ink deposition (as a proportion of unworn ink film thickness) as a result of wear than for lines produced in the print direction. However, it should be noted that ink film thicknesses in both unworn and worn states were lower for lines printed at 90°. In this orientation, it is postulated that there is greater "scooping-out" as the squeegee is less restricted by the stencil after travelling over the edge of the stencil. For lines printed at 90°, there was also a greater increase in ink film thickness on the leading edge of the line when compared with that on the trailing edge. However, for lines deposited in the print direction, wear related increases in ink film thickness were observed more equally on both edges of the lines. Regardless of line orientation, the middle of the line was lower than the edges. This is illustrated in Figure 12, sample graphs of the cross-sectional profiles of different orientation lines before and after wear. It is postulated that this greater ink deposition after wear is due to the worn edge of the squeegee no longer being able to deform into the mesh to the same extent as the unworn squeegee. This effect varied depending on the squeegee type, the amount of wear and the orientation of the lines.

**Figure 12.** Sample images of variation in ink deposition (cross-section) as a result of squeegee wear at different orientations: (**a**) Parallel to print direction (90° to squeegee); (**b**) 90° to print direction (parallel to squeegee); using 1200 grit abrasive, Squeegee 4. Note this effect varies depending on squeegee/abrasive types.

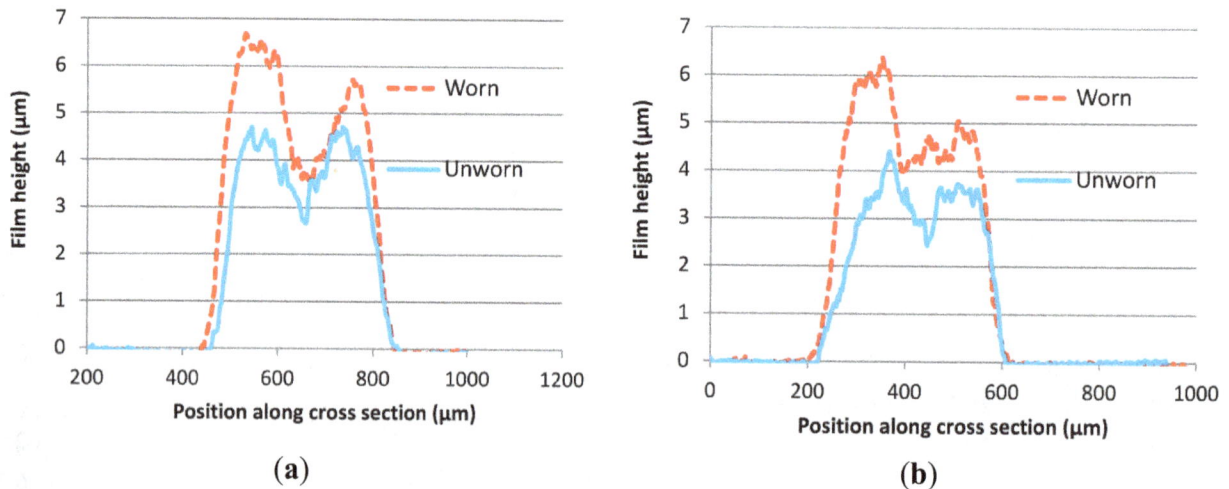

(a)                                                                           (b)

The relationship between the ink deposition (cross-sectional area of the printed lines—ink film thickness × line width) and the reciprocal of the measured line resistance is shown in Figure 13. The relationship was very similar regardless of whether the lines were printed with the worn or unworn squeegees. There was a linear relationship with high $R^2$ values. While the worn squeegees gave a general increase in ink deposition, which gave a reduction in line resistance, there was no deviation from the relationship which would suggest print defects, such as broken lines, which would lead to higher than expected resistances. The data confirms that resistance measurements are sufficient to accurately characterize the amount of ink deposition for silver prints made with unworn and worn squeegees.

**Figure 13.** Correlations between ink deposition (line cross-sectional area) and line resistance.

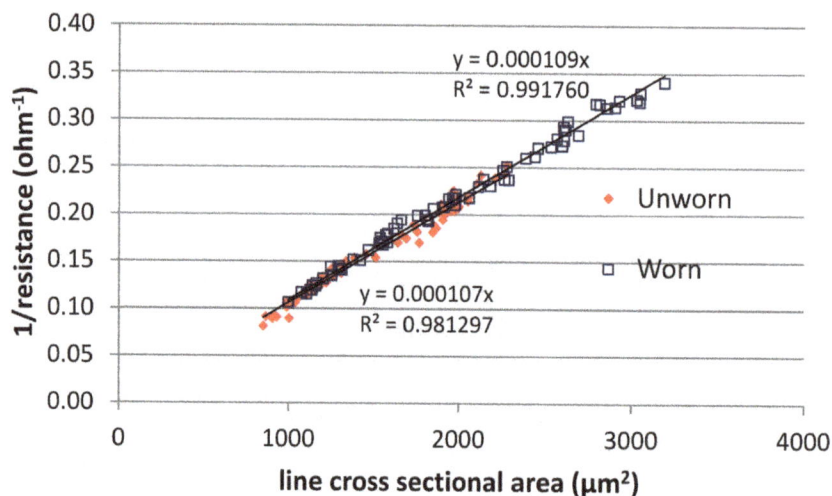

## 3.7. Discussion

The indentation and bending tests described in this report have been able to characterize the squeegees in a more effective manner than traditional Shore A hardness measurement. Measuring indentation at a lower pressure than a Shore A hardness meter gives data which is more representative of what occurs during printing, while squeegee deflection tests offered an even better correlation with ink film thickness. However different values will be obtained if any of the settings such as pin or deflection tool geometry are altered. The method of immersing only one surface of the squeegee in solvent, coupled with the indentation method, allowed solvent related softening of the squeegees to be reliably measured. Solvent absorption softens the squeegee which should lead to an increase in ink transfer as indicated both by the correlation between hardness and ink transfer and by previous work [14]. Squeegee hardness after exposure to solvent could also be linked with wear characteristics and there appeared to be a correlation between squeegee hardness and the amount of wear, with harder squeegees wearing more. However, this was only observed at a 5 N indentation force and did not apply at higher forces, or indeed when using Shore A hardness. It was also only apparent when solvent treated indentation data was used and this observation should therefore be treated with caution. Both squeegee hardness and pressure should influence the amount of wear but these factors will interact, with a harder squeegee having a higher pressure. The squeegee contact needs to be better understood so that wear can be anticipated from measurable squeegee parameters. Furthermore, each squeegee responded differently to the various abrasives. The amount of wear from one abrasive could not be used to anticipate the wear from another and there was not a consistent ratio of wear for a certain abrasive against wear for another.

There were distinct differences in the amount of wear observed in the different squeegees and a large range in the effects of this wear on the printed lines. There was a general trend of increasing wear levels giving greater levels of ink film thickness and line width and hence reduced line resistance for silver lines. The effect of wear on ink film thickness was more significant than the effect on line width. During production, such an increase in ink transfer would lead to an increase in ink consumption as well as a variation in the quality of the printed features. In the case of functional screen printing for electronics or sensors, this would have an effect on the functional of the end product, while for graphics the appearance of the product would be affected. This would have cost implications in terms of ink consumption but would also lead to greater product failure and rejection.

White *et al.* [9] state that ink flux through the screen is proportional to the square root of the squeegee tip curvature; provided other factors remain unchanged. Although this is based on modeling using a Newtonian fluid, when screen printing inks are usually shear thinning, worn squeegees should give greater ink transfer as their sharp edges are gradually rounded. It is proposed that the change in line geometry is due to a reduced ability for the squeegee to deform into the mesh and displace the ink from the mesh. However, the relationship between the amount of wear and ink deposition was not straightforward and depended on the squeegee. Squeegee 1 suffered similar levels of wear to other squeegees yet it was much more effective at maintaining consistency in the print. Even with the wear suffered in 2000 and 2500 abrasives, the squeegee remained usable. The others squeegees all showed substantial increases in ink deposition (up to 44%) indicating that they would not be useable at this point and would consume much higher amounts of ink in the printing process.

The higher comparative resistance to both indentation and flexing at lower forces, previously noted, suggested that Squeegee 1 would print with a high pressure which would give a reduced ink film thickness. This also influences the angle of wear, which changes the effective squeegee angle during printing and in turn also affects the ink film thickness. The influence of wear angles on ink film thickness is shown in Figure 14. The ink film thickness appears to be affected by a combination of squeegee wear angle and the amount of squeegee removed during wear. Squeegee 1 had both the highest wear angles and the lowest ink film thickness after wear, despite other squeegees having lower levels of wear. The combination of resistance to bending and a high wear angle presents a sharper edge and higher pressure at the squeegee/screen interface. This reduces the ink film thickness.

**Figure 14.** Ink film thickness *vs.* wear angle in worn squeegees.

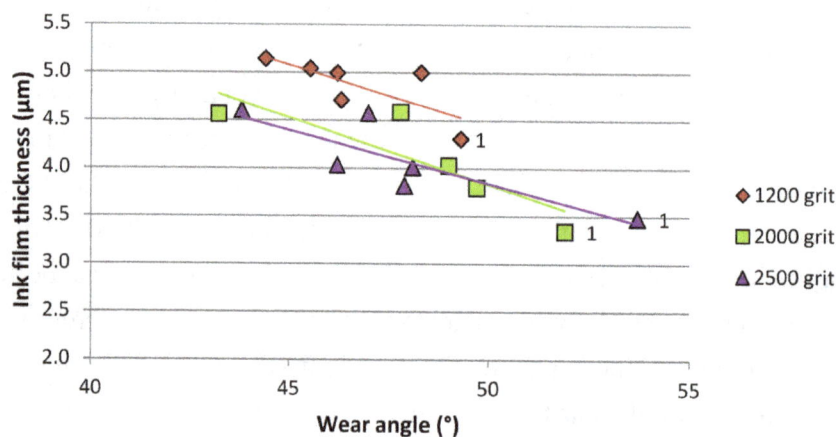

The controlled wear did not, apart from one line orientation for Squeegee 3, cause any print defects, such as breakages, pinholes in the lines, or satellite drops of ink around lines that would be detrimental to their electrical performance. This highlights the benefit of a controlled wear methodology rather than testing squeegees worn through printing which might suffer nicks or other uneven damage and cause broken lines. Wear trials performed using both silver and carbon inks on a screen with a blocked mesh, did not show such levels of wear, even after thousands of cycles.

The roughest abrasive material gave very high levels of wear which would not be tolerated in practice. This was reduced when using smoother abrasives but still gave substantial increases in ink consumption for most squeegees. There was a drift in the print characteristics over the duration of the printing experiment with a small reduction in ink film thickness was observed but a more noticeable reduction in line width. This also gave an increase in line resistance. Assuming this is a gradual effect related to ink build-up in the mesh, the drift anticipated between the sequential prints of an unworn and worn state of a given squeegee is only a small pro-rata proportion of this. Averaged over all features this would be of the order of 0.25% for ink film thickness and less than 1% for line width. This would not be significant for most of the observations, where large changes in deposition were observed.

The 2000 and 2500 grit abrasives did not appear to differ substantially from one another, either in their ability to abrade the squeegee or change the ink deposition after wear. The particle size of the intermediate 2000 grit abrasive was closer to that of the 2500 grit than the 1200 grit abrasive. There might also be some variations in pressure due to the positions of the different bands which will affect wear. Preliminary analysis of the surface topography of the abrasives using white light interferometry

[Veeco NT2000 (Veeco Instruments, Inc., Plainview, NY, USA) with array size 305 μm × 232 μm] suggested that roughness characteristics, in terms of average and root mean squared roughness surface roughness ($R_a$ and $R_q$ respectively) were very similar for 2000 and 2500 grit abrasives (Figure 15). An alternative intermediate 1500 grit abrasive has a particle size profile much closer to the 1200 grit abrasive and hence would be expected to give wear results more similar to that.

**Figure 15.** Surface roughness data for abrasives obtained using white light interferometry. Five measurements per abrasive type with error bars showing standard deviations.

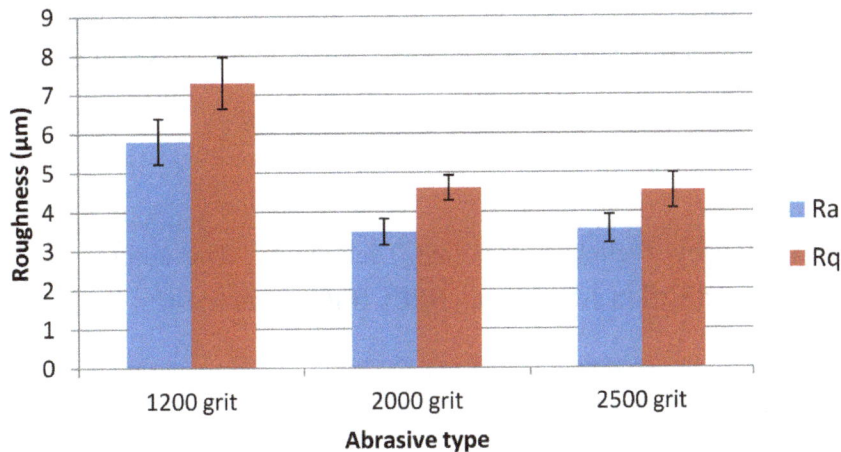

The ink used in the wear testing can be selected to match a particular application. The various inks use different solvents which will affect how the squeegee abrades. This is particularly relevant for novel formulations whose effect on squeegee material is unknown.

During printing, squeegee wear would be expected to be inconsistent and localized, due to varying topography from the patterning in the screen and possible build-up of material in certain areas over time. This would then give rise to localized variations in ink film thickness within the printed sheet. Although these experiments do not simulate the localized defects that would occur during printing, the findings are applicable in terms of the consequences of wear on the print.

## 4. Conclusions

A reliable accelerated wear test has been developed for squeegees used in screen printing. Mechanical tests have also been developed which allow more in-depth measurement of squeegee properties than currently used tests. These measurements have subsequently been used to establish correlations with print quality both before and after wear. Squeegee wear differed between different squeegee types and caused increases in ink transfer and wider printed lines. This will lead to greater ink consumption and therefore cost per unit and an increasing likelihood of product failure or rejection, particularly for functional layers used in printed electronics. While more wear generally gave greater increases in ink deposition, the effect of wear differed, depending on the squeegee and the orientation of the line. There was a correlation between the angle of the squeegee wear and ink film thickness from a worn squeegee. A higher ability to resist flexing gave a higher wear angle. This in turn presented a sharper edge at the squeegee/screen interface thus mitigating the effect of wear.

Increases in ink deposition gave lower electrical resistance in printed silver lines; however, the correlation between the amount of ink deposit and the resistance remained the same, for all levels of wear. This suggested that the wear regime designed for this study did not induce detrimental print defects such as line breakages. Therefore resistance can be used as a rapid indicator of changes in conductive ink transfer.

Squeegee indentation at different force levels gave more information than a standard Shore A hardness test and the apparatus used was able to reliably measure reductions in surface hardness due to solvent ingress. Indentation data obtained at low forces was a better indicator of the likely effect of squeegee hardness on ink deposition than Shore A. However, the mechanical resistance of the squeegee to deflection was found to be the most effective predictor of ink film thickness as it indicates the pressure at the squeegee-screen interface.

## Acknowledgments

This work was part funded by Trelleborg Applied Technology. The authors wish to thank Paul Habberfield from Trelleborg and Eifion Jewell from SPECIFIC, Swansea University for their help and advice with this work.

## Author Contributions

Christopher Phillips prepared the manuscript, performed wear and print tests, measured and assessed squeegee characteristics. David Beynon worked on print tests, surface profiling and electrical measurement of print samples. Simon Hamblyn worked on wear tests and wear method development. Glyn Davies, David Gethin and Timothy Claypole contributed towards development of the squeegee wear and characterization methods.

## Conflicts of Interest

The authors declare no conflict of interest.

## References

1.  Krebs, F.C.; Jørgensen, M.; Norrman, K.; Hagemann, O.; Alstrup, J.; Nielsen, T.D.; Fyenbo, J.; Larsen, K.; Kristensen, J. A complete process for production of flexible large area polymer solar cells entirely using screen printing—First public demonstration. *Sol. Energy Mater. Sol. Cells* **2009**, *93*, 422–441.
2.  Rotureau, D.; Viricelle, J.-P.; Pijolat, C.; Caillol, N; Pijolat, M. Development of a planar SOFC device using screen-printing technology. *J. Eur. Ceram. Soc.* **2005**, *25*, 2633–2636.
3.  Shi, Y.S.; Zhu, C.-C.; Wang, Q.; Li, X. Large area screen-printing cathode of CNT for FED. *Diam. Relat. Mater.* **2003**, *12*, 1449–1452.
4.  Pardo, D.A.; Jabbour, G.E.; Peyghambarian, N. Application of screen printing in the fabrication of organic light-emitting devices. *Adv. Mater.* **2000**, *12*, 1249–1252.

5.  Lee, M.-Y.; Lee, M.-W.; Park, J.-E.; Park, J.-S.; Song, C.-K. A printing technology combining screen-printing with a wet-etching process for the gate electrodes of organic thin film transistors on a plastic substrate. *Microelectron. Eng.* **2010**, *87*, 1922–1926.

6.  Viricelle, J.-P.; Riviere, B.; Pijolat, C. Optimization of $SnO_2$ screen-printing inks for gas sensor applications. *J. Eur. Ceram. Soc.* **2005**, *25*, 2137–2140.

7.  Qi, Q.; Zhang, T.; Yu, Q.J.; Wang, R.; Zeng, Y.; Liu, L.; Yang, H.B. Properties of humidity sensing ZnO nanorods-base sensor fabricated by screen-printing. *Sens. Actuators B Chem.* **2008**, *133*, 638–643.

8.  Crouch, E.; Cowell, D.C.; Hoskins, S.; Pittson, R.W.; Hart, J.P. Amperometric, screen-printed, glucose biosensor for analysis of human plasma samples using a biocomposite water-based carbon ink incorporating glucose oxidase. *Anal. Biochem.* **2005**, *347*, 17–23.

9.  White, G.S.; Breward, C.J.W.; Howell, P.D.; Young, R.J.S. A model for the screen-printing of Newtonian fluids. *J. Eng. Math.* **2006**, *54*, 49–70.

10. Fox, I.J.; Bohan, M.F.J.; Claypole, T.C.; Gethin, D.T. Film thickness prediction in halftone screen-printing. *Proc. Inst. Mech. Eng. E* **2003**, *217*, 345–359.

11. Owczarek, J.A.; Howland, F.L. A study of the off-contact screen printing process. II. Analysis of the model of the printing process. *IEEE Trans. Compon. Hybrids Manuf. Technol.* **1990**, *13*, 368–375.

12. Pan, J.; Tonkay, G.L.; Quintero, A. Screen printing process design of experiments for fine line printing of thick film ceramic substrates. *J. Electron. Manuf.* **1999**, *9*, 203–213.

13. *BS ISO 1817:2011 Rubber, vulcanized or thermoplastic—Determination of the effect of liquids*; British Standards Institution: London, UK, 2011.

14. Jewell, E.H.; Claypole, T.C.; Gethin, D.T. The effect of exposure to inks and solvents on squeegee performance. *Surf. Coat. Int. B* **2004**, *87*, 253–260.

15. *BS ISO 7619–1:2010 Rubber, vulcanized or thermoplastic—Determination of indentation hardness. Part 1: Durometer method (Shore hardness)*; British Standards Institution: London, UK, 2010.

16. *BS ISO 37:2011 Rubber, vulcanized or thermoplastic—Determination of tensile stress-strain properties*; British Standards Institution: London, UK, 2011.

17. *BS ISO 4649:2010 Rubber, vulcanized or thermoplastic—Determination of abrasion resistance using a rotating cylindrical drum device*; British Standards Institution: London, UK, 2010.

# Thermal Performance of Hollow Clay Brick with Low Emissivity Treatment in Surface Enclosures

**Roberto Fioretti \* and Paolo Principi**

Department of Industrial Engineering and Mathematical Sciences, Università Politecnica delle Marche, Via Brecce Bianche 12, 60131 Ancona, Italy; E-Mail: p.principi@univpm.it

\* Author to whom correspondence should be addressed; E-Mail: r.fioretti@univpm.it

External Editor: Alessandro Lavacchi

**Abstract:** External walls made with hollow clay brick or block are widely used for their thermal, acoustic and structural properties. However, the performance of the bricks frequently does not conform with the minimum legal requirements or the values required for high efficiency buildings, and for this reason, they need to be integrated with layers of thermal insulation. In this paper, the thermal behavior of hollow clay block with low emissivity treatment on the internal cavity surfaces has been investigated. The purpose of this application is to obtain a reduction in the thermal conductivity of the block by lowering the radiative heat exchange in the enclosures. The aims of this paper are to indicate a methodology for evaluating the thermal performance of the brick and to provide information about the benefits that should be obtained. Theoretical evaluations are carried out on several bricks (12 geometries simulated with two different thermal conductivities of the clay), using a finite elements model. The heat exchange procedure is implemented in accordance with the standard, so as to obtain standardized values of the thermal characteristics of the block. Several values of emissivity are hypothesized, related to different kinds of coating. Finally, the values of the thermal transmittance of walls built with the evaluated blocks have been calculated and compared. The results show how coating the internal surface of the cavity provides a reduction in the thermal conductivity of the block, of between 26% and 45%, for a surface emissivity of 0.1.

**Keywords:** low emissivity coating; hollow brick; heat transfer; energy saving; improved thermal performance

## 1. Introduction

Energy consumption in the building sector has increased in recent years, representing approximately 40% of the overall energy required and generating 36% of greenhouse gases in Europe [1]. In European countries, the reduction in energy consumption for buildings is addressed by legislation, with a view to increasing energy efficiency, reducing greenhouse gas emissions and promoting the use of renewable energy. The target for 2020 is to have new buildings that do not require energy in the use stage (heating, cooling, illumination, hot water and other electric uses) [2].

One of the most widely used technologies for the building envelope involves walls made with hollow clay bricks or blocks, used for their good characteristics, such as high durability and excellent fire, thermal and sound insulation.

On the market there are several kinds of bricks, with different dimensions, with or without cavities, made with clay, concrete or wood and characterized by different thermal properties. The thermal properties that influence energy consumption for maintaining indoor microclimate comfort are the thermal transmittance and the thermal inertia of the envelopes. The parameter that mainly influences energy loss during the cold seasons is the thermal transmittance of the envelopes, and some countries, including Italy, have imposed a limit value [3]. Thermal inertia mainly influences energy consumption during the warm seasons and thermal comfort, and walls built with block normally have acceptable values. Nowadays, clay blocks are a widespread solution for building external walls. The thermal insulation provided by these products is a result of the geometry of the brick and of the small pores present in the clay. Some studies evaluate the possibility of increasing the thermal resistance of the block through a change in the configuration of the enclosures [4–9] or by filling the enclosures with insulation material, such as perlite, mineral wood, polystyrene and other substances with low thermal conductivity [10].

Low emissivity coating, treatment or film is currently used to reduce radiative energy transmission in several fields comprising the building sector. For transparent surfaces, low emissivity coating is often used to reduce the $U$-value of the glass, with an infrared coating on the internal surface of the double or triple glazing [11–13]. Another application is on external walls or the roof, using a low emissivity and absorption treatment in order to reduce both the radiative heat exchange with the external environment and the solar absorption. One innovative application that uses low emissivity surfaces, in this case aluminum film, is thin multi-foil insulation that consists of a series of low emissivity films separated by wadding foams [14–17].

In a numerical analysis on the thermal behavior of fired clay hollow bricks [18], the reduction in the emissivities of cavities was investigated as a solution for improving the thermal resistance of the brick. The analysis was carried out on small and large-sized bricks, for different values of surface emissivity (between 0.3 and 0.7). The study demonstrates that the improvement depends on the configuration and the dimensions of the brick. For cavity surfaces with an emissivity of 0.3 applied on hollow bricks, this variation ranges between approximately 10% and 50%, depending on the kind of brick and the dimensions of the holes.

One theoretical and experimental analysis [19] investigated the increase in the thermal performance of a thermal block. In that study, an improvement in performance through the use of low emissivity coating and phase change material (PCM) was tested, by means of a theoretical analysis (conducted using FEM software) and an experimental analysis carried out with a thermo flux meter device. The results demonstrate the validity of these technical solutions for increasing thermal resistance (low emissivity coating) and thermal inertia (PCM) and the accuracy of the theoretical method used for the evaluation.

The present paper proposes the improvement in the energy performance of hollow clay block through the use of a low emissivity coating technique applied on the block by covering the enclosures with low emission and absorption paint, thereby reducing the overall heat transfer coefficient value of the block. For the study, several blocks and emissivities of the coating are hypothesized in order to obtain a large number of values of improvement. This technology is compared with that of filling the enclosures with an insulation material. A comparison between the results obtained and the previous evaluations is also discussed.

## 2. Theoretical Evaluation

The evaluation of the thermal performance of the blocks with low emissivity treatment on the surface of the cavities was carried out using the COMSOL software [20]. To verify the thermal performance and the benefits provided by the low emissivity coating, a two-dimensional steady-state simulation was performed. The calculation method was previously validated [19] in accordance with ISO 10211-1 (Annex A) [21] and EN 1745 (Annex D) [22], and the results are reported in Table 1.

**Table 1.** Results of the numerical method and software validation, reprinted with permission from [19], copyright Elsevier 2012.

| Standard | Number of elements of the mesh | Degrees of freedom | Value comparison | Value standard | Value program | Accuracy/ Error | Max accuracy/Error |
|---|---|---|---|---|---|---|---|
| ANNEX D.4 case 1—EN 1745 | 21,024 | 42,313 | conductance | 0.6258 W/m²·K | 0.6274 W/m²·K | 0.25% | 2% |
| ANNEX A case 1—UNI 10211-1 | 8,192 | 16,577 | temperatures | all 28 values in the grid | | <0.1 K | 0.1 K |
| ANNEX A case 2—UNI 10211-1 | 16,572 | 33,815 | temperatures | all 9 values in the grid | | <0.1 K | 0.1 K |
| | | | thermal flux | 9.5 W/m² | 9.49 W/m² | 0.01 W/m² | 0.1 W/m² |

To simulate the heat exchange inside the cavities, a simplified and standardized method using equivalent conductivity was used. The algorithm used was developed according to the calculation procedure suggested in EN 1745 (Annex D) and EN ISO 6946 (Annex B-C) [23], which combines the effects of heat conduction, convection and radiation through the use of equivalent conductivity, thus approaching the problem exclusively from the conduction heat transfer perspective. The equivalent conductivity of the core is obtained by the following equation:

$$\lambda_{eq} = \frac{d}{R_g} \qquad (1)$$

where the thermal resistance of the cavity may be calculated using the following equation:

$$R_g = \frac{1}{h_a + 1/2 \cdot E \cdot h_{ro}(1 + \sqrt{1 + d^2/b^2} - d/b)} \qquad (2)$$

where E is the emittance of the surface:

$$E = \frac{1}{1/\varepsilon_1 + 1/\varepsilon_2 - 1} \qquad (3)$$

which depends on the emissivity of the surfaces of the cavity and where $\varepsilon_1$ and $\varepsilon_2$ are, respectively, the emissivities of the emitting and receiving surfaces.

The convective heat exchange coefficient $h_a$ is the maximum value between 1.25 W/m²·K and 0.025/$d$, where $d$ is the width of the void in the direction of the thermal flux. The lowest heat exchange coefficient $h_a$ is for a value of $d$ equal to or less than 0.02 m.

The coefficient $h_{ro}$ is the radiative heat exchange coefficient of the black body, and the value depends on the mean temperature of the cavity. For this evaluation the temperature hypothesized was 10 °C [the mean value between the standard internal temperature (20 °C) and the standard external temperature during the winter (0 °C)], and the corresponding value is equal to 5.1 W/m²·K (Table A.1 UNI EN ISO 6946) [23].

The main assumptions considered for the simulation process are:

- two-dimensional model;
- steady-state condition;
- all thermophysical properties kept constant;
- isotropic conductivity;
- equivalent conductivity (convective and radiative part);
- thermal resistance of the voids calculated according to B.3 of EN ISO 6946:1996;
- no mass transfer;
- emissivity values of the emitting and receiving surfaces are equal ($\varepsilon_1 = \varepsilon_2$);
- thermal resistance of the holes with non-rectangular shape is assumed equal to a rectangular hole with the same area and the same dimensional ratio (d/b) [23].

Steady-state analysis to determine the benefits of applying low emissivity coating was performed on several kinds of blocks characterized by different dimensions, geometry and void fractions. To have a sufficient sample in order to be able to discuss the benefits provided by the application, 12 different block geometries (Figure 1) were evaluated, with several values of total thickness and the number and dimensions of holes. The simulation was also performed with two different values of thermal conductivity of the material, 0.90 W/m·K and 0.45 W/m·K, the first related to a high density clay of 2400 kg/m³ and the second to a clay with pore density of 1500 kg/m³. In order to verify the effect of surface emissivity on the overall thermal resistance of the block, the emissivity was varied from 0.9 to 0.1. An emissivity of 0.9 corresponds to the thermal performance of the standard block without coating.

**Figure 1.** Images of the evaluated blocks (plan view).

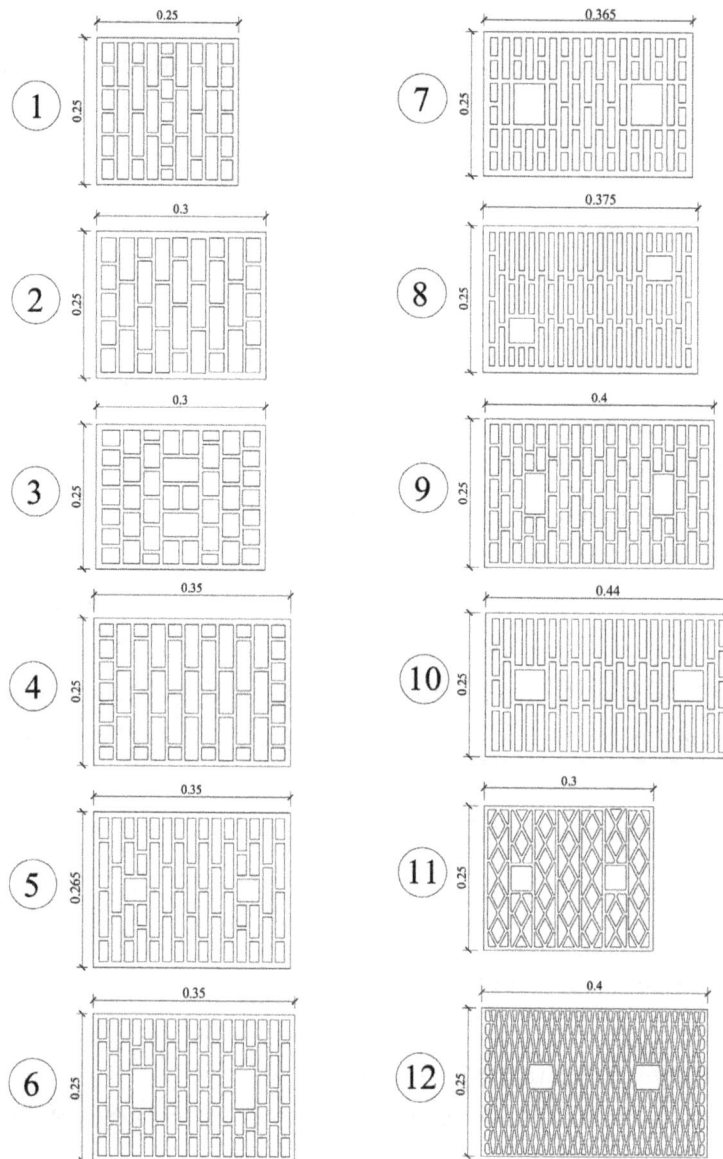

In order to set a reachable limit value, an evaluation without radiative heat exchange was carried out. This case is not real, because it corresponds to an emissivity equal to zero, but it is the lowest value of heat exchange in the cavities, in the case of only a convective heat exchange. The geometric and thermal characteristics of the blocks are reported in Table 2, which also lists the number of the elements of the meshes and the number of degrees of freedom of the model built in the software. Figure 1 shows the geometry and the dimensions of the investigated blocks.

Alternative simulations were carried out on blocks with EPS filling (extruded polystyrene; thermal conductivity of 0.04 W/m·K) in the voids in order to obtain a comparison with the B.A.T. (best available technology) of a high thermal performance block.

Finally, the thermal transmittances (*U*-value) of a wall built with the evaluated block units have been calculated using the procedure set down in UNI EN 1745 [22] and UNI EN ISO 6946 [23]. The *U*-value is the most important parameter used to verify the conformity of the wall to national legislation [24–26] and to evaluate the thermal heat loss and energy consumption. To evaluate the *U*-value, a sample wall has been hypothesized, built with an internal layer of plaster (thickness = 0.02 m; thermal conductivity = 0.4 W/m·K), masonry made with block and a horizontal mortar joint (height = 0.015 m; thermal conductivity of 1.0 W/m·K) and a layer of external plaster (thickness = 0.02 m; thermal conductivity = 0.5 W/m·K). The boundary conditions to evaluate the $U$-value were $R_{se}$ = 0.04 m²·K/W and $R_{si}$ = 0.13 m²·K/W in accordance with the Standard UNI EN ISO 6946 [23].

**Table 2.** Characteristics and thermal proprieties of the investigated blocks.

| Sample | Length (m) | Width (m) | Height (m) | Number of voids | Rows of voids | Void fraction | Mesh | Degrees of freedom | Conductivity of the material (W/m·K) |
|---|---|---|---|---|---|---|---|---|---|
| 1A | 0.25 | 0.25 | 0.25 | 39 | 9 | 61.97% | 14,144 | 28,469 | 0.9 |
| 1B | | | | | | | | | 0.45 |
| 2A | 0.25 | 0.3 | 0.25 | 34 | 9 | 65.19% | 13,532 | 27,251 | 0.9 |
| 2B | | | | | | | | | 0.45 |
| 3A | 0.25 | 0.3 | 0.25 | 44 | 8 | 62.91% | 13,640 | 27,461 | 0.9 |
| 3B | | | | | | | | | 0.45 |
| 4A | 0.25 | 0.35 | 0.25 | 45 | 11 | 65.58% | 17,152 | 34,517 | 0.9 |
| 4B | | | | | | | | | 0.45 |
| 5A | 0.265 | 0.35 | 0.25 | 57 | 15 | 57.25% | 17,888 | 35,977 | 0.9 |
| 5B | | | | | | | | | 0.45 |
| 6A | 0.25 | 0.35 | 0.25 | 77 | 17 | 59.27% | 20,872 | 41,977 | 0.9 |
| 6B | | | | | | | | | 0.45 |
| 7A | 0.25 | 0.365 | 0.25 | 64 | 17 | 50.55% | 27,800 | 55,949 | 0.9 |
| 7B | | | | | | | | | 0.45 |
| 8A | 0.25 | 0.375 | 0.25 | 67 | 21 | 46.03% | 28,152 | 56,641 | 0.9 |
| 8B | | | | | | | | | 0.45 |
| 9A | 0.25 | 0.4 | 0.25 | 86 | 19 | 60.48% | 26,100 | 52,529 | 0.9 |
| 9B | | | | | | | | | 0.45 |
| 10A | 0.25 | 0.45 | 0.25 | 61 | 21 | 55.95% | 21,856 | 44,027 | 0.9 |
| 10B | | | | | | | | | 0.45 |
| 11A | 0.25 | 0.30 | 0.25 | 96 | 21 | 60.63% | 32,516 | 65,235 | 0.9 |
| 11B | | | | | | | | | 0.45 |
| 12A | 0.25 | 0.40 | 0.25 | 252 | 41 | 57.38% | 44,876 | 90,037 | 0.9 |
| 12B | | | | | | | | | 0.45 |

## 3. Results and Discussion

In order to verify the impact of the low emissivity coating on the brick thermal conductivity, the numerical simulations were conducted for different values of emissivity ranging from 0.1 to 0.9. The evaluation shows that the equivalent thermal resistances of the simulated blocks, in comparison with their initial resistance, increase with the decrease in cavity surface emissivity. This decrease in heat exchange is consequent to a reduction in the radiative heat exchange in the voids, due to the low emissivity coating. The reduction in the radiative part causes a decrease in the equivalent thermal conductivity of the cavity, while the convective heat coefficient remains the same in the different evaluations.

As shown in Table 3, the change in the emissivity of the cavity surfaces leads to a reduction in the equivalent conductivity of the cavities, which is dependent on the emissivities and the dimensions (*b*) and (*d*) of the cavities. The equivalent conductivity of all three types of cavities drops by between 55% and 60%, when the emissivity is reduced to 0.1, assuming a value between 0.028 and 0.077 W/m·K (while the value in the uncoated block was between 0.062 and 0.232 W/m·K). The best value attainable without radiative heat exchange was 0.025–0.066 W/m·K, corresponding to a reduction of 59%–75%.

**Table 3.** Dimensions and equivalent thermal conductivity of the cavities.

| Cavity | $d$ (m) | $b$ (m) | $h_a$ (W/m²·K) | $\lambda_{eq}$ (W/m·K) | | | | | | | | | |
|---|---|---|---|---|---|---|---|---|---|---|---|---|---|
| | | | | $\varepsilon=0.9$ | $\varepsilon=0.8$ | $\varepsilon=0.7$ | $\varepsilon=0.6$ | $\varepsilon=0.5$ | $\varepsilon=0.4$ | $\varepsilon=0.3$ | $\varepsilon=0.2$ | $\varepsilon=0.1$ | No radiation |
| 1-1 | 0.02 | 0.033 | 1.250 | 0.090 | 0.078 | 0.068 | 0.059 | 0.052 | 0.045 | 0.039 | 0.034 | 0.029 | 0.025 |
| 1-2 | 0.02 | 0.073 | 1.250 | 0.099 | 0.085 | 0.073 | 0.064 | 0.055 | 0.047 | 0.041 | 0.035 | 0.030 | 0.025 |
| 1-3 | 0.02 | 0.017 | 1.250 | 0.082 | 0.071 | 0.063 | 0.055 | 0.048 | 0.042 | 0.037 | 0.033 | 0.029 | 0.025 |
| 1-4 | 0.02 | 0.031 | 1.250 | 0.089 | 0.078 | 0.067 | 0.059 | 0.051 | 0.045 | 0.039 | 0.034 | 0.029 | 0.025 |
| 2-1 | 0.026 | 0.041 | 1.250 | 0.117 | 0.101 | 0.088 | 0.077 | 0.067 | 0.058 | 0.051 | 0.044 | 0.038 | 0.033 |
| 2-2 | 0.026 | 0.072 | 1.250 | 0.125 | 0.108 | 0.093 | 0.081 | 0.070 | 0.061 | 0.052 | 0.045 | 0.038 | 0.033 |
| 2-3 | 0.026 | 0.033 | 1.250 | 0.113 | 0.098 | 0.086 | 0.075 | 0.065 | 0.057 | 0.050 | 0.043 | 0.038 | 0.033 |
| 3-1 | 0.031 | 0.027 | 1.250 | 0.128 | 0.111 | 0.097 | 0.085 | 0.075 | 0.066 | 0.058 | 0.051 | 0.0e44 | 0.039 |
| 3-2 | 0.029 | 0.017 | 1.250 | 0.113 | 0.099 | 0.087 | 0.077 | 0.068 | 0.060 | 0.053 | 0.047 | 0.041 | 0.036 |
| 3-3 | 0.029 | 0.04 | 1.250 | 0.128 | 0.111 | 0.096 | 0.084 | 0.073 | 0.064 | 0.056 | 0.049 | 0.042 | 0.036 |
| 3-4 | 0.04 | 0.064 | 1.250 | 0.180 | 0.156 | 0.135 | 0.118 | 0.103 | 0.090 | 0.078 | 0.068 | 0.058 | 0.050 |
| 4-1 | 0.024 | 0.021 | 1.250 | 0.099 | 0.086 | 0.075 | 0.066 | 0.058 | 0.051 | 0.045 | 0.039 | 0.034 | 0.030 |
| 4-2 | 0.024 | 0.031 | 1.250 | 0.105 | 0.091 | 0.079 | 0.069 | 0.060 | 0.053 | 0.046 | 0.040 | 0.035 | 0.030 |
| 4-3 | 0.024 | 0.072 | 1.250 | 0.116 | 0.100 | 0.087 | 0.075 | 0.065 | 0.056 | 0.049 | 0.042 | 0.036 | 0.030 |
| 5-1 | 0.016 | 0.036 | 1.563 | 0.080 | 0.070 | 0.061 | 0.054 | 0.047 | 0.042 | 0.037 | 0.032 | 0.029 | 0.025 |
| 5-2 | 0.016 | 0.077 | 1.563 | 0.086 | 0.074 | 0.065 | 0.057 | 0.050 | 0.043 | 0.038 | 0.033 | 0.029 | 0.025 |
| 5-3 | 0.016 | 0.055 | 1.563 | 0.083 | 0.073 | 0.063 | 0.056 | 0.049 | 0.043 | 0.038 | 0.033 | 0.029 | 0.025 |
| 5-4 | 0.038 | 0.038 | 1.250 | 0.160 | 0.139 | 0.121 | 0.106 | 0.093 | 0.082 | 0.072 | 0.063 | 0.055 | 0.048 |
| 6-1 | 0.015 | 0.033 | 1.667 | 0.076 | 0.067 | 0.059 | 0.052 | 0.046 | 0.041 | 0.036 | 0.032 | 0.028 | 0.025 |
| 6-2 | 0.015 | 0.055 | 1.667 | 0.080 | 0.070 | 0.061 | 0.054 | 0.047 | 0.042 | 0.037 | 0.032 | 0.029 | 0.025 |
| 6-3 | 0.015 | 0.045 | 1.667 | 0.079 | 0.069 | 0.060 | 0.053 | 0.047 | 0.041 | 0.037 | 0.032 | 0.028 | 0.025 |
| 6-4 | 0.035 | 0.068 | 1.250 | 0.161 | 0.140 | 0.121 | 0.105 | 0.092 | 0.080 | 0.069 | 0.060 | 0.051 | 0.044 |
| 7-1 | 0.012 | 0.035 | 2.083 | 0.068 | 0.060 | 0.053 | 0.047 | 0.042 | 0.038 | 0.034 | 0.031 | 0.028 | 0.025 |
| 7-2 | 0.012 | 0.07 | 2.083 | 0.071 | 0.063 | 0.055 | 0.049 | 0.044 | 0.039 | 0.035 | 0.031 | 0.028 | 0.025 |
| 7-3 | 0.053 | 0.07 | 1.250 | 0.232 | 0.201 | 0.175 | 0.153 | 0.134 | 0.117 | 0.102 | 0.089 | 0.077 | 0.066 |
| 8-1 | 0.01 | 0.031 | 2.500 | 0.061 | 0.054 | 0.049 | 0.044 | 0.040 | 0.036 | 0.033 | 0.030 | 0.027 | 0.025 |
| 8-2 | 0.01 | 0.065 | 2.500 | 0.064 | 0.057 | 0.051 | 0.045 | 0.041 | 0.037 | 0.033 | 0.030 | 0.027 | 0.025 |
| 8-3 | 0.01 | 0.081 | 2.500 | 0.064 | 0.057 | 0.051 | 0.046 | 0.041 | 0.037 | 0.033 | 0.030 | 0.028 | 0.025 |
| 8-4 | 0.042 | 0.042 | 1.250 | 0.176 | 0.153 | 0.134 | 0.117 | 0.103 | 0.090 | 0.079 | 0.069 | 0.060 | 0.053 |
| 9-1 | 0.015 | 0.033 | 1.667 | 0.076 | 0.067 | 0.059 | 0.052 | 0.046 | 0.041 | 0.036 | 0.032 | 0.028 | 0.025 |
| 9-2 | 0.015 | 0.054 | 1.667 | 0.080 | 0.070 | 0.061 | 0.054 | 0.047 | 0.042 | 0.037 | 0.032 | 0.029 | 0.025 |
| 9-3 | 0.015 | 0.045 | 1.667 | 0.079 | 0.069 | 0.060 | 0.053 | 0.047 | 0.041 | 0.037 | 0.032 | 0.028 | 0.025 |
| 9-4 | 0.034 | 0.069 | 1.250 | 0.158 | 0.136 | 0.118 | 0.103 | 0.089 | 0.078 | 0.067 | 0.058 | 0.050 | 0.043 |
| 10-1 | 0.013 | 0.045 | 1.923 | 0.073 | 0.064 | 0.056 | 0.050 | 0.044 | 0.040 | 0.035 | 0.031 | 0.028 | 0.025 |
| 10-2 | 0.013 | 0.065 | 1.923 | 0.074 | 0.065 | 0.057 | 0.051 | 0.045 | 0.040 | 0.036 | 0.032 | 0.028 | 0.025 |
| 10-3 | 0.013 | 0.079 | 1.923 | 0.075 | 0.066 | 0.058 | 0.051 | 0.045 | 0.040 | 0.036 | 0.032 | 0.028 | 0.025 |
| 10-4 | 0.052 | 0.052 | 1.250 | 0.218 | 0.190 | 0.166 | 0.145 | 0.128 | 0.112 | 0.098 | 0.086 | 0.075 | 0.065 |

**Table 3.** *Cont.*

| Cavity | d (m) | b (m) | $h_a$ (W/m²·K) | $\lambda_{eq}$ (W/m·K) | | | | | | | | | |
|--------|-------|-------|----------------|ε = 0.9|ε = 0.8|ε = 0.7|ε = 0.6|ε = 0.5|ε = 0.4|ε = 0.3|ε = 0.2|ε = 0.1|No radiation|
| 11-1 | 0.017 | 0.031 | 1.47 | 0.081 | 0.071 | 0.062 | 0.055 | 0.048 | 0.042 | 0.037 | 0.033 | 0.029 | 0.025 |
| 11-2 | 0,009 | 0.025 | 2.78 | 0.057 | 0.051 | 0.046 | 0.042 | 0.038 | 0.035 | 0.032 | 0.029 | 0.027 | 0.025 |
| 11-3 | 0.010 | 0.054 | 2.50 | 0.063 | 0.056 | 0.050 | 0.045 | 0.041 | 0.037 | 0.033 | 0.030 | 0.027 | 0.025 |
| 11-4 | 0.036 | 0.041 | 1.25 | 0.154 | 0.134 | 0.117 | 0.102 | 0.089 | 0.078 | 0.069 | 0.060 | 0.052 | 0.045 |
| 11-5 | 0.010 | 0.022 | 2.50 | 0.059 | 0.053 | 0.048 | 0.043 | 0.039 | 0.035 | 0.032 | 0.030 | 0.027 | 0.025 |
| 12-1 | 0.008 | 0.018 | 3.13 | 0.053 | 0.047 | 0.043 | 0.039 | 0.036 | 0.033 | 0.031 | 0.029 | 0.027 | 0.025 |
| 12-2 | 0.006 | 0.019 | 4.17 | 0.047 | 0.043 | 0.039 | 0.036 | 0.034 | 0.032 | 0.030 | 0.028 | 0.026 | 0.025 |
| 12-3 | 0.006 | 0.039 | 4.17 | 0.048 | 0.044 | 0.040 | 0.037 | 0.034 | 0.032 | 0.030 | 0.028 | 0.026 | 0.025 |
| 12-4 | 0.044 | 0.042 | 1.25 | 0.184 | 0.160 | 0.140 | 0.122 | 0.107 | 0.094 | 0.083 | 0.072 | 0.063 | 0.055 |

Figure 2 shows the value corresponding to the thermal conductivity of EPS (red line). For most of the enclosures, the value of equivalent thermal conductivity obtained with an emissivity equal to 0.01 is less than the thermal conductivity of the EPS. In particular, if the thickness (*d*) of the cavity is less than 0.029 m, the equivalent thermal conductivity is lower than the value of EPS.

**Figure 2.** Impact of surface emissivity on the variation in equivalent thermal conductivity of the cavities [red line = thermal conductivity of EPS (extruded polystyrene)].

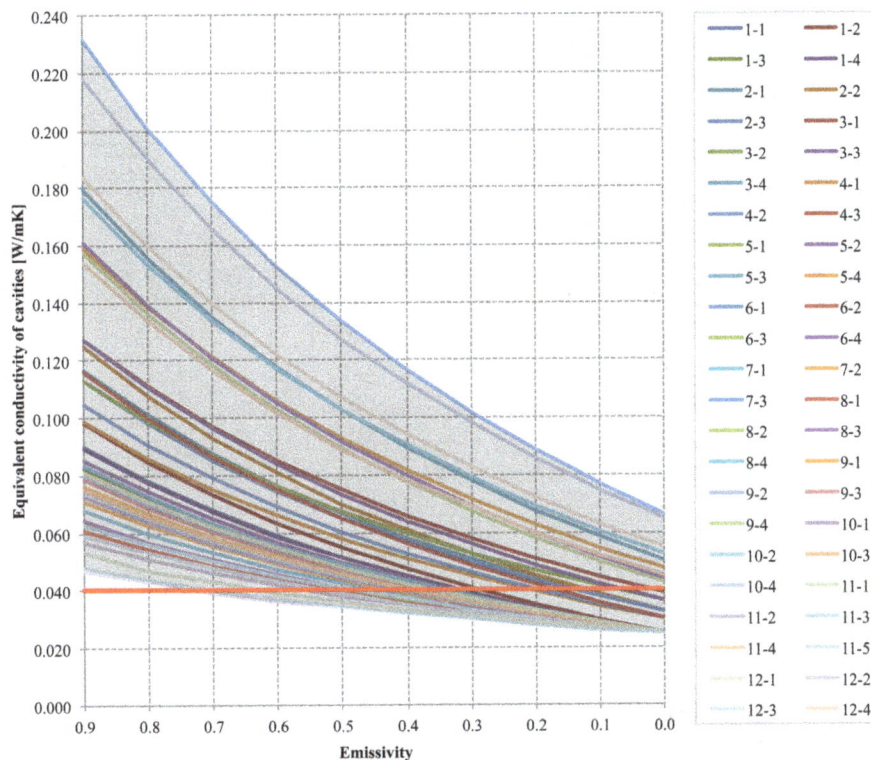

Values of thermal resistance of the evaluated block are shown in Table 4. A reduction in the equivalent thermal conductivity of the enclosures produces an increase in the thermal resistance value of the block of +32%–82% (Table 5), from values of 2.65–0.96 m²·K/W (ε = 0.9) to values of 4.20–1.34 m²·K/W (ε = 0.1) (Table 4).

**Table 4.** Thermal resistance of the evaluated block (m$^2$·K/W).

| Sample | $\varepsilon = 0.90$ | $\varepsilon = 0.80$ | $\varepsilon = 0.70$ | $\varepsilon = 0.60$ | $\varepsilon = 0.50$ | $\varepsilon = 0.40$ | $\varepsilon = 0.30$ | $\varepsilon = 0.20$ | $\varepsilon = 0.10$ | No radiation | EPS |
|---|---|---|---|---|---|---|---|---|---|---|---|
| 1A | 1.04 | 1.11 | 1.18 | 1.25 | 1.33 | 1.38 | 1.45 | 1.51 | 1.57 | 1.64 | 1.45 |
| 1B | 1.46 | 1.58 | 1.71 | 1.84 | 1.97 | 2.12 | 2.25 | 2.39 | 2.54 | 2.70 | 2.26 |
| 2A | 1.14 | 1.22 | 1.31 | 1.39 | 1.47 | 1.55 | 1.63 | 1.70 | 1.79 | 1.82 | 1.76 |
| 2B | 1.55 | 1.70 | 1.84 | 1.99 | 2.15 | 2.30 | 2.47 | 2.64 | 2.83 | 2.92 | 2.77 |
| 3A | 0.96 | 1.02 | 1.07 | 1.12 | 1.17 | 1.21 | 1.25 | 1.29 | 1.34 | 1.35 | 1.34 |
| 3B | 1.39 | 1.49 | 1.59 | 1.70 | 1.80 | 1.90 | 2.00 | 2.09 | 2.20 | 2.30 | 2.27 |
| 4A | 1.39 | 1.50 | 1.59 | 1.70 | 1.79 | 1.89 | 1.98 | 2.08 | 2.17 | 2.25 | 2.11 |
| 4B | 1.90 | 2.08 | 2.25 | 2.44 | 2.62 | 2.82 | 3.01 | 3.22 | 3.43 | 3.68 | 3.28 |
| 5A | 1.55 | 1.65 | 1.75 | 1.84 | 1.93 | 2.03 | 2.12 | 2.21 | 2.29 | 2.37 | 2.09 |
| 5B | 2.15 | 2.33 | 2.51 | 2.68 | 2.87 | 3.07 | 3.25 | 3.45 | 3.63 | 3.84 | 3.20 |
| 6A | 1.57 | 1.66 | 1.76 | 1.84 | 1.93 | 2.00 | 2.08 | 2.17 | 2.24 | 2.30 | 2.06 |
| 6B | 2.22 | 2.39 | 2.57 | 2.73 | 2.92 | 3.08 | 3.25 | 3.45 | 3.58 | 3.78 | 3.21 |
| 7A | 1.39 | 1.45 | 1.53 | 1.59 | 1.65 | 1.71 | 1.76 | 1.82 | 1.86 | 1.91 | 1.75 |
| 7B | 2.03 | 2.16 | 2.30 | 2.43 | 2.55 | 2.68 | 2.80 | 2.93 | 3.04 | 3.17 | 2.79 |
| 8A | 1.57 | 1.64 | 1.71 | 1.77 | 1.84 | 1.91 | 1.97 | 2.02 | 2.07 | 1.88 | 1.88 |
| 8B | 2.28 | 2.41 | 2.53 | 2.69 | 2.82 | 2.95 | 3.09 | 3.21 | 3.33 | 3.45 | 2.90 |
| 9A | 1.76 | 1.87 | 1.97 | 2.07 | 2.17 | 2.25 | 2.34 | 2.43 | 2.49 | 2.59 | 2.34 |
| 9B | 2.48 | 2.67 | 2.87 | 3.06 | 3.27 | 3.45 | 3.65 | 3.86 | 4.03 | 4.24 | 3.58 |
| 10A | 1.87 | 1.97 | 2.08 | 2.17 | 2.27 | 2.36 | 2.44 | 2.52 | 2.61 | 2.68 | 2.40 |
| 10B | 2.65 | 2.84 | 3.04 | 3.23 | 3.43 | 3.62 | 3.80 | 3.99 | 4.20 | 4.38 | 3.73 |
| 11A | 1.70 | 1.78 | 1.85 | 1.93 | 1.99 | 2.05 | 2.12 | 2.17 | 2.23 | 2.28 | 2.03 |
| 11B | 2.33 | 2.49 | 2.63 | 2.79 | 2.91 | 3.05 | 3.20 | 3.33 | 3.47 | 3.58 | 3.02 |
| 12A | 2.77 | 2.89 | 3.01 | 3.11 | 3.21 | 3.29 | 3.37 | 3.46 | 3.55 | 3.60 | 3.07 |
| 12B | 3.73 | 3.93 | 4.16 | 4.34 | 4.55 | 4.70 | 4.86 | 5.05 | 5.24 | 5.35 | 4.28 |

**Table 5.** Thermal resistance of the evaluated block: percentage of improvement.

| Sample | $\varepsilon = 0.90$ | $\varepsilon = 0.80$ | $\varepsilon = 0.70$ | $\varepsilon = 0.60$ | $\varepsilon = 0.50$ | $\varepsilon = 0.40$ | $\varepsilon = 0.30$ | $\varepsilon = 0.20$ | $\varepsilon = 0.10$ | No radiation | EPS |
|---|---|---|---|---|---|---|---|---|---|---|---|
| 1A | 100% | 107% | 114% | 120% | 128% | 133% | 139% | 145% | 151% | 157% | 140% |
| 1B | 100% | 109% | 118% | 126% | 135% | 145% | 154% | 164% | 174% | 185% | 155% |
| 2A | 100% | 107% | 115% | 122% | 129% | 135% | 143% | 149% | 156% | 160% | 154% |
| 2B | 100% | 109% | 119% | 128% | 138% | 148% | 159% | 170% | 182% | 188% | 178% |
| 3A | 100% | 106% | 111% | 116% | 121% | 126% | 130% | 134% | 139% | 140% | 139% |
| 3B | 100% | 107% | 115% | 122% | 130% | 137% | 144% | 151% | 159% | 166% | 164% |
| 4A | 100% | 107% | 114% | 122% | 129% | 136% | 142% | 149% | 156% | 162% | 151% |
| 4B | 100% | 109% | 118% | 128% | 138% | 148% | 158% | 169% | 180% | 193% | 172% |
| 5A | 100% | 107% | 113% | 119% | 125% | 132% | 137% | 143% | 148% | 154% | 135% |
| 5B | 100% | 109% | 117% | 125% | 134% | 143% | 151% | 161% | 169% | 179% | 149% |
| 6A | 100% | 106% | 112% | 117% | 123% | 128% | 133% | 138% | 143% | 147% | 131% |
| 6B | 100% | 108% | 116% | 123% | 131% | 139% | 147% | 155% | 162% | 171% | 145% |
| 7A | 100% | 105% | 110% | 115% | 119% | 123% | 127% | 131% | 134% | 138% | 126% |
| 7B | 100% | 106% | 113% | 119% | 125% | 132% | 138% | 144% | 150% | 156% | 137% |
| 8A | 100% | 105% | 109% | 113% | 118% | 122% | 126% | 129% | 132% | 120% | 120% |
| 8B | 100% | 106% | 111% | 118% | 124% | 130% | 136% | 141% | 147% | 151% | 127% |
| 9A | 100% | 106% | 112% | 117% | 123% | 128% | 133% | 138% | 141% | 147% | 132% |
| 9B | 100% | 108% | 116% | 123% | 132% | 139% | 147% | 156% | 162% | 171% | 144% |
| 10A | 100% | 106% | 111% | 116% | 122% | 126% | 130% | 135% | 140% | 143% | 129% |
| 10B | 100% | 107% | 115% | 122% | 130% | 137% | 143% | 151% | 158% | 165% | 141% |
| 11A | 100% | 105% | 109% | 114% | 117% | 121% | 125% | 128% | 131% | 134% | 120% |
| 11B | 100% | 107% | 113% | 119% | 125% | 131% | 137% | 143% | 149% | 154% | 129% |
| 12A | 100% | 104% | 108% | 112% | 116% | 119% | 122% | 125% | 128% | 130% | 111% |
| 12B | 100% | 105% | 111% | 116% | 122% | 126% | 130% | 135% | 140% | 143% | 115% |

Table 6 shows the values of the equivalent thermal conductivity of the analyzed block, which is derived by dividing the thickness of a block by its thermal resistance. While for the standard block ($\varepsilon = 0.9$), the equivalent thermal conductivity obtained is between 0.311 and 0.157 W/m·K, for the improved block ($\varepsilon = 0.1$), the values are 0.224–0.096 W/m·K, corresponding to a reduction of 24%–45% (Table 7). The equivalent conductivity of the block without radiative heat exchange was 0.222–0.091 W/m·K, corresponding to a reduction of 17%–48%. For the block with polystyrene, the value of equivalent conductivity calculated is 0.224–0.150 W/m·K. All of the blocks evaluated with a low emissivity coating of 0.1 have a thermal behavior equal to or higher than the block simulated with the polystyrene inside the void. The block with EPS has the same behavior as the block with low emissivity coating for a surface emissivity between 0.4 and 0.1. The mean value of the reduction in the equivalent thermal conductivity obtained for an emissivity equal to 0.1 is 35%, compared with 30% for the block with EPS inside the cavity.

**Table 6.** Equivalent thermal conductivity of the evaluated block (W/m·K).

| Sample | $\varepsilon = 0.90$ | $\varepsilon = 0.80$ | $\varepsilon = 0.70$ | $\varepsilon = 0.60$ | $\varepsilon = 0.50$ | $\varepsilon = 0.40$ | $\varepsilon = 0.30$ | $\varepsilon = 0.20$ | $\varepsilon = 0.10$ | No radiation | EPS |
|---|---|---|---|---|---|---|---|---|---|---|---|
| 1A | 0.240 | 0.225 | 0.211 | 0.201 | 0.188 | 0.181 | 0.173 | 0.166 | 0.159 | 0.153 | 0.172 |
| 1B | 0.171 | 0.158 | 0.146 | 0.136 | 0.127 | 0.118 | 0.111 | 0.104 | 0.098 | 0.093 | 0.111 |
| 2A | 0.263 | 0.245 | 0.229 | 0.216 | 0.204 | 0.194 | 0.184 | 0.176 | 0.168 | 0.164 | 0.170 |
| 2B | 0.193 | 0.177 | 0.163 | 0.151 | 0.140 | 0.130 | 0.121 | 0.114 | 0.106 | 0.103 | 0.108 |
| 3A | 0.311 | 0.295 | 0.280 | 0.268 | 0.257 | 0.248 | 0.239 | 0.232 | 0.224 | 0.222 | 0.224 |
| 3B | 0.217 | 0.201 | 0.188 | 0.177 | 0.167 | 0.158 | 0.150 | 0.143 | 0.136 | 0.130 | 0.132 |
| 4A | 0.251 | 0.234 | 0.220 | 0.206 | 0.195 | 0.185 | 0.177 | 0.168 | 0.161 | 0.155 | 0.166 |
| 4B | 0.184 | 0.169 | 0.156 | 0.144 | 0.133 | 0.124 | 0.116 | 0.109 | 0.102 | 0.095 | 0.107 |
| 5A | 0.227 | 0.212 | 0.200 | 0.190 | 0.181 | 0.172 | 0.165 | 0.158 | 0.153 | 0.147 | 0.167 |
| 5B | 0.163 | 0.150 | 0.140 | 0.131 | 0.122 | 0.114 | 0.108 | 0.101 | 0.096 | 0.091 | 0.109 |
| 6A | 0.223 | 0.210 | 0.199 | 0.190 | 0.181 | 0.175 | 0.168 | 0.162 | 0.156 | 0.152 | 0.170 |
| 6B | 0.158 | 0.147 | 0.136 | 0.128 | 0.120 | 0.114 | 0.108 | 0.102 | 0.098 | 0.092 | 0.109 |
| 7A | 0.263 | 0.251 | 0.239 | 0.230 | 0.222 | 0.214 | 0.207 | 0.201 | 0.196 | 0.191 | 0.209 |
| 7B | 0.180 | 0.169 | 0.159 | 0.150 | 0.143 | 0.136 | 0.130 | 0.125 | 0.120 | 0.115 | 0.131 |
| 8A | 0.239 | 0.229 | 0.219 | 0.212 | 0.203 | 0.197 | 0.190 | 0.185 | 0.181 | 0.200 | 0.200 |
| 8B | 0.165 | 0.155 | 0.148 | 0.139 | 0.133 | 0.127 | 0.121 | 0.117 | 0.112 | 0.109 | 0.129 |
| 9A | 0.227 | 0.214 | 0.203 | 0.194 | 0.185 | 0.178 | 0.171 | 0.164 | 0.161 | 0.155 | 0.171 |
| 9B | 0.161 | 0.150 | 0.139 | 0.131 | 0.122 | 0.116 | 0.110 | 0.104 | 0.099 | 0.094 | 0.112 |
| 10A | 0.241 | 0.228 | 0.217 | 0.207 | 0.198 | 0.191 | 0.185 | 0.179 | 0.173 | 0.168 | 0.187 |
| 10B | 0.170 | 0.158 | 0.148 | 0.139 | 0.131 | 0.124 | 0.119 | 0.113 | 0.107 | 0.103 | 0.121 |
| 11A | 0.177 | 0.169 | 0.162 | 0.155 | 0.151 | 0.146 | 0.142 | 0.138 | 0.134 | 0.132 | 0.148 |
| 11B | 0.129 | 0.121 | 0.114 | 0.108 | 0.103 | 0.098 | 0.094 | 0.090 | 0.087 | 0.084 | 0.099 |
| 12A | 0.144 | 0.139 | 0.133 | 0.129 | 0.124 | 0.122 | 0.119 | 0.116 | 0.113 | 0.111 | 0.130 |
| 12B | 0.107 | 0.102 | 0.096 | 0.092 | 0.088 | 0.085 | 0.082 | 0.079 | 0.076 | 0.075 | 0.093 |

**Table 7.** Equivalent thermal conductivity of the evaluated block: percentage of reduction.

| Sample | ε = 0.90 | ε = 0.80 | ε = 0.70 | ε = 0.60 | ε = 0.50 | ε = 0.40 | ε = 0.30 | ε = 0.20 | ε = 0.10 | No radiation | EPS |
|--------|----------|----------|----------|----------|----------|----------|----------|----------|----------|--------------|------|
| 1A  | 0% | −6% | −12% | −17% | −22% | −25% | −28% | −31% | −34% | −36% | −28% |
| 1B  | 0% | −8% | −15% | −21% | −26% | −31% | −35% | −39% | −43% | −46% | −35% |
| 2A  | 0% | −7% | −13% | −18% | −22% | −26% | −30% | −33% | −36% | −37% | −35% |
| 2B  | 0% | −8% | −16% | −22% | −28% | −32% | −37% | −41% | −45% | −47% | −44% |
| 3A  | 0% | −5% | −10% | −14% | −17% | −20% | −23% | −26% | −28% | −29% | −28% |
| 3B  | 0% | −7% | −13% | −18% | −23% | −27% | −31% | −34% | −37% | −40% | −39% |
| 4A  | 0% | −7% | −13% | −18% | −22% | −26% | −30% | −33% | −36% | −38% | −34% |
| 4B  | 0% | −8% | −15% | −22% | −27% | −33% | −37% | −41% | −45% | −48% | −42% |
| 5A  | 0% | −6% | −12% | −16% | −20% | −24% | −27% | −30% | −32% | −35% | −26% |
| 5B  | 0% | −8% | −14% | −20% | −25% | −30% | −34% | −38% | −41% | −44% | −33% |
| 6A  | 0% | −6% | −11% | −15% | −19% | −22% | −25% | −27% | −30% | −32% | −24% |
| 6B  | 0% | −7% | −14% | −19% | −24% | −28% | −32% | −36% | −38% | −41% | −31% |
| 7A  | 0% | −5% | −9% | −13% | −16% | −19% | −21% | −24% | −26% | −27% | −21% |
| 7B  | 0% | −6% | −12% | −16% | −20% | −24% | −27% | −31% | −33% | −36% | −27% |
| 8A  | 0% | −4% | −8% | −12% | −15% | −18% | −20% | −23% | −24% | −17% | −17% |
| 8B  | 0% | −6% | −10% | −15% | −19% | −23% | −26% | −29% | −32% | −34% | −22% |
| 9A  | 0% | −6% | −11% | −15% | −19% | −22% | −25% | −28% | −29% | −32% | −24% |
| 9B  | 0% | −7% | −14% | −19% | −24% | −28% | −32% | −36% | −38% | −41% | −31% |
| 10A | 0% | −5% | −10% | −14% | −18% | −21% | −23% | −26% | −28% | −30% | −22% |
| 10B | 0% | −7% | −13% | −18% | −23% | −27% | −30% | −34% | −37% | −39% | −29% |
| 11A | 0% | −4% | −8% | −12% | −14% | −17% | −20% | −22% | −24% | −25% | −16% |
| 11B | 0% | −6% | −11% | −16% | −20% | −23% | −27% | −30% | −33% | −35% | −23% |
| 12A | 0% | −4% | −8% | −11% | −14% | −16% | −18% | −20% | −22% | −23% | −10% |
| 12B | 0% | −5% | −10% | −14% | −18% | −21% | −23% | −26% | −29% | −30% | −13% |

Table 8 shows the $U$-value of the walls built with the evaluated blocks, calculated considering a hypothetical configuration of the layers (external plaster, masonry with horizontal mortar and internal plaster). While for the standard block (ε = 0.9), the value of thermal transmittance is between 0.32–0.87 W/m²·K, for the improved block (ε = 0.1), the values are 0.26–0.69 W/m²·K, corresponding to a reduction of 18%–37% (Table 9). For the block with polystyrene, the $U$-value calculated is 0.29–0.69 W/m²·K.

The comparison with the two previous evaluations [18,19] offers insight into whether the percentages of improvement found in this paper could be extended to other kinds of brick or block, with different geometric and material properties. Figure 3 shows the values of reduction obtained in this work compared with the values obtained in a previous analysis, using the same methodology, but carried out on a different block, and an analysis carried out on walls made with traditional bricks. The results obtained in the previous study [19] do not differ greatly from the values obtained in the present work, with a variation of less than 3% with respect to the mean values.

On the contrary, the comparison made with a previous analysis [18] carried out on walls with traditional brick indicates a substantial difference. In fact, only four of the eight walls analyzed have a behavior that is comparable with the mean values obtained in the present work (difference < 6%), while for the remaining walls, the values obtained are clearly different.

The possible causes of these differences are:

- evaluation made considering not only the brick but also the mortar;
- dimension of the cavity and the block;
- percentage of voids;
- number of voids;
- thermal conductivity of the clay.

Further analysis has been suggested in order to evaluate the performance of a large number of samples that could represent most of the blocks or bricks available on the market. The evaluation of the impact of the surface emissivity on the *U*-value of the wall could also be useful in order to investigate energy consumption and economic benefits, under real conditions.

**Table 8.** Overall thermal transmittance of the wall built with the evaluated block ($W/m^2 \cdot K$).

| Sample | $\varepsilon = 0.90$ | $\varepsilon = 0.80$ | $\varepsilon = 0.70$ | $\varepsilon = 0.60$ | $\varepsilon = 0.50$ | $\varepsilon = 0.40$ | $\varepsilon = 0.30$ | $\varepsilon = 0.20$ | $\varepsilon = 0.10$ | No radiation | EPS |
|---|---|---|---|---|---|---|---|---|---|---|---|
| 1A | 0.84 | 0.80 | 0.76 | 0.74 | 0.70 | 0.68 | 0.66 | 0.64 | 0.63 | 0.61 | 0.66 |
| 1B | 0.66 | 0.62 | 0.59 | 0.56 | 0.53 | 0.51 | 0.49 | 0.47 | 0.45 | 0.43 | 0.49 |
| 2A | 0.77 | 0.74 | 0.70 | 0.67 | 0.65 | 0.62 | 0.60 | 0.58 | 0.56 | 0.55 | 0.57 |
| 2B | 0.62 | 0.58 | 0.55 | 0.52 | 0.49 | 0.47 | 0.45 | 0.43 | 0.41 | 0.40 | 0.41 |
| 3A | 0.87 | 0.84 | 0.81 | 0.78 | 0.76 | 0.74 | 0.72 | 0.71 | 0.69 | 0.69 | 0.69 |
| 3B | 0.67 | 0.64 | 0.61 | 0.58 | 0.56 | 0.54 | 0.52 | 0.50 | 0.48 | 0.47 | 0.47 |
| 4A | 0.66 | 0.63 | 0.60 | 0.58 | 0.55 | 0.53 | 0.51 | 0.50 | 0.48 | 0.47 | 0.49 |
| 4B | 0.53 | 0.50 | 0.47 | 0.44 | 0.42 | 0.40 | 0.38 | 0.36 | 0.35 | 0.33 | 0.36 |
| 5A | 0.62 | 0.59 | 0.56 | 0.54 | 0.52 | 0.50 | 0.49 | 0.47 | 0.46 | 0.45 | 0.49 |
| 5B | 0.48 | 0.46 | 0.43 | 0.41 | 0.39 | 0.38 | 0.36 | 0.35 | 0.34 | 0.32 | 0.37 |
| 6A | 0.61 | 0.58 | 0.56 | 0.54 | 0.52 | 0.51 | 0.50 | 0.48 | 0.47 | 0.46 | 0.50 |
| 6B | 0.47 | 0.45 | 0.43 | 0.41 | 0.39 | 0.38 | 0.36 | 0.35 | 0.34 | 0.33 | 0.36 |
| 7A | 0.66 | 0.64 | 0.62 | 0.60 | 0.59 | 0.57 | 0.56 | 0.55 | 0.54 | 0.52 | 0.56 |
| 7B | 0.50 | 0.48 | 0.46 | 0.44 | 0.43 | 0.41 | 0.40 | 0.39 | 0.38 | 0.37 | 0.40 |
| 8A | 0.61 | 0.59 | 0.57 | 0.55 | 0.54 | 0.52 | 0.51 | 0.50 | 0.49 | 0.48 | 0.53 |
| 8B | 0.46 | 0.44 | 0.43 | 0.41 | 0.40 | 0.38 | 0.37 | 0.36 | 0.35 | 0.34 | 0.39 |
| 9A | 0.55 | 0.53 | 0.51 | 0.49 | 0.47 | 0.46 | 0.45 | 0.44 | 0.43 | 0.42 | 0.45 |
| 9B | 0.43 | 0.41 | 0.39 | 0.37 | 0.35 | 0.34 | 0.33 | 0.31 | 0.31 | 0.30 | 0.33 |
| 10A | 0.52 | 0.50 | 0.48 | 0.47 | 0.45 | 0.44 | 0.43 | 0.42 | 0.41 | 0.40 | 0.43 |
| 10B | 0.40 | 0.38 | 0.37 | 0.35 | 0.34 | 0.32 | 0.31 | 0.30 | 0.29 | 0.28 | 0.32 |
| 11A | 0.58 | 0.56 | 0.55 | 0.53 | 0.52 | 0.51 | 0.50 | 0.49 | 0.48 | 0.47 | 0.51 |
| 11B | 0.46 | 0.44 | 0.43 | 0.41 | 0.40 | 0.39 | 0.37 | 0.36 | 0.35 | 0.35 | 0.39 |
| 12A | 0.40 | 0.39 | 0.37 | 0.37 | 0.36 | 0.35 | 0.35 | 0.34 | 0.33 | 0.33 | 0.37 |
| 12B | 0.32 | 0.31 | 0.30 | 0.29 | 0.28 | 0.28 | 0.27 | 0.26 | 0.26 | 0.25 | 0.29 |

**Figure 3.** Comparison between the impacts of surface emissivity on the variation in equivalent thermal conductivity of the blocks obtained in the present work (grey lines) and in a previous investigation (red point [18], blue point [19]).

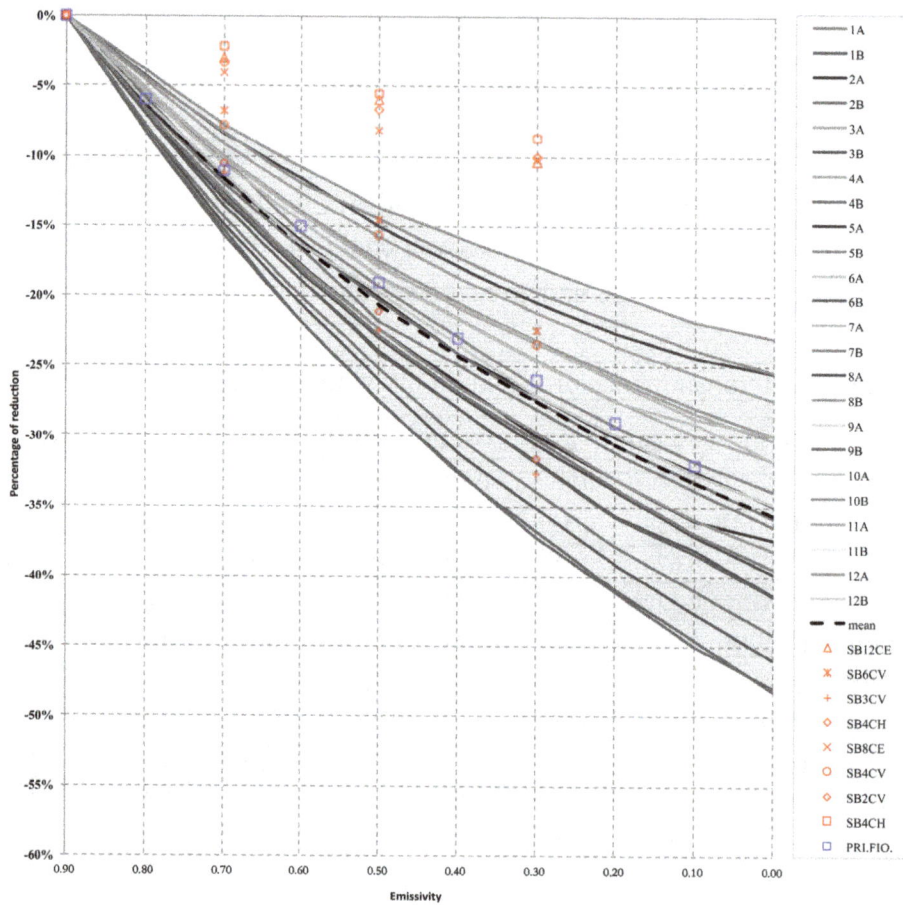

**Table 9.** Overall thermal transmittance of the wall built with the evaluated block: percentage of reduction.

| Sample | ε = 0.90 | ε = 0.80 | ε = 0.70 | ε = 0.60 | ε = 0.50 | ε = 0.40 | ε = 0.30 | ε = 0.20 | ε = 0.10 | No radiation | EPS |
|---|---|---|---|---|---|---|---|---|---|---|---|
| 1A | 0% | −5% | −9% | −12% | −16% | −18% | −21% | −23% | −25% | −27% | −21% |
| 1B | 0% | −6% | −11% | −15% | −19% | −23% | −26% | −29% | −32% | −35% | −26% |
| 2A | 0% | −5% | −9% | −13% | −16% | −19% | −22% | −25% | −27% | −28% | −27% |
| 2B | 0% | −6% | −11% | −16% | −21% | −24% | −28% | −31% | −34% | −36% | −34% |
| 3A | 0% | −4% | −7% | −10% | −13% | −15% | −17% | −19% | −21% | −21% | −21% |
| 3B | 0% | −5% | −10% | −13% | −17% | −20% | −23% | −26% | −28% | −30% | −30% |
| 4A | 0% | −5% | −9% | −13% | −17% | −20% | −23% | −25% | −28% | −29% | −26% |
| 4B | 0% | −6% | −11% | −16% | −21% | −25% | −28% | −31% | −34% | −37% | −32% |
| 5A | 0% | −5% | −9% | −12% | −15% | −18% | −20% | −23% | −25% | −27% | −20% |
| 5B | 0% | −6% | −11% | −15% | −19% | −22% | −25% | −28% | −31% | −33% | −25% |
| 6A | 0% | −4% | −8% | −11% | −14% | −16% | −18% | −21% | −23% | −24% | −18% |
| 6B | 0% | −5% | −10% | −14% | −18% | −21% | −24% | −27% | −29% | −31% | −23% |
| 7A | 0% | −3% | −7% | −9% | −12% | −14% | −16% | −18% | −19% | −21% | −16% |
| 7B | 0% | −4% | −9% | −12% | −15% | −18% | −21% | −23% | −25% | −27% | −20% |

**Table 9.** *Cont.*

| Sample | ε = 0.90 | ε = 0.80 | ε = 0.70 | ε = 0.60 | ε = 0.50 | ε = 0.40 | ε = 0.30 | ε = 0.20 | ε = 0.10 | No radiation | EPS |
|--------|----------|----------|----------|----------|----------|----------|----------|----------|----------|--------------|-----|
| 8A  | 0% | −3% | −6%  | −9%  | −11% | −13% | −15% | −17% | −18% | −19% | −12% |
| 8B  | 0% | −4% | −7%  | −11% | −14% | −17% | −20% | −22% | −24% | −25% | −16% |
| 9A  | 0% | −4% | −8%  | −11% | −14% | −16% | −19% | −21% | −22% | −24% | −19% |
| 9B  | 0% | −5% | −10% | −14% | −18% | −21% | −24% | −27% | −29% | −31% | −23% |
| 10A | 0% | −4% | −8%  | −11% | −13% | −16% | −18% | −20% | −22% | −23% | −17% |
| 10B | 0% | −5% | −10% | −13% | −17% | −20% | −23% | −25% | −28% | −30% | −22% |
| 11A | 0% | −3% | −6%  | −9%  | −11% | −13% | −15% | −16% | −18% | −19% | −12% |
| 11B | 0% | −4% | −8%  | −12% | −14% | −17% | −20% | −22% | −24% | −25% | −16% |
| 12A | 0% | −3% | −6%  | −8%  | −10% | −11% | −13% | −14% | −16% | −17% | −7%  |
| 12B | 0% | −4% | −7%  | −10% | −12% | −14% | −16% | −18% | −20% | −21% | −9%  |

## 4. Conclusions

This paper presents the evaluation of the thermal performance of hollow clay bricks and blocks with low emissivity treatment of the internal cavity surfaces.

The numerical evaluation was carried out in order to determine the increase in the thermal resistance provided by low emissivity coating applied on the void surfaces of the brick. For the evaluation, a set of 20 samples was investigated, characterized by different shapes and the thermal conductivity of the materials. Each sample was simulated for surface emissivities of 0.9–0.1. The values obtained with the numerical simulation were compared with the standard block and with the block insulated with EPS in the voids.

The results showed that a low emissivity coating leads to an increase in the thermal resistance provided by the block, through the reduction in the radiative heat exchange in the voids. Using the low emissivity coating, it is possible to reduce the equivalent thermal conductivity of the cavities by between 55%–70% (ε = 0.1) to a value of 0.027 W/m·K, which is lower than the thermal conductivity of common thermal insulation.

The effect of the reduction in the thermal heat exchange in the void has a significant impact on the thermal resistance of the block. In fact, the results show that the application of low emissivity coating could reduce the equivalent thermal conductivity of the block by at least 24% (for an emissivity of 0.1). The consequent reduction in the *U*-value of a wall built with the improved blocks is 18%–37%.

In all cases, the thermal behavior of the treated block is improved with respect to the standard block, and with an emissivity of 0.1, it is better than the block insulated with polystyrene.

The theoretical procedure to calculate the thermal performance of the block is easily repeatable and is employed to verify conformity with the current standard.

Further studies are suggested to include a wide range of blocks, to test the real emissivity of the coated surface with different kinds of coating and to evaluate performance decay in real conditions.

## Author Contributions

Numerical evaluations have been conducted by Roberto Fioretti. Analysis and interpretation of the results as well as conclusions have been conducted by Roberto Fioretti and Paolo Principi. The manuscript has been written by Roberto Fioretti and Paolo Principi.

## Abbreviations

| | | |
|---|---|---|
| $\lambda_{eq}$ | equivalent conductivity of the cavity | W/m·K |
| $\lambda_{equ}$ | equivalent conductivity of the block | W/m·K |
| $d$ | size of cavity in the direction of thermal flux | m |
| $b$ | size of cavity across the thermal flux | m |
| $R_g$ | overall thermal resistance of the cavity | m²·K/W |
| $R_{se}$ | external surface resistance | m²·K/W |
| $R_{si}$ | internal surface resistance | m²·K/W |
| $h_a$ | convective coefficient | W/m²·K |
| $h_{ro}$ | radiative coefficient | W/m²·K |
| $E$ | emittance | |
| $\varepsilon$ | emissivities of the surfaces | |
| $\varepsilon_1$ | emissivities of the emitting surfaces | |
| $\varepsilon_2$ | emissivities of the receiving surface | |
| FEM | finite elements method | |
| EPS | extruded polystyrene | |

## Conflicts of Interest

The authors declare no conflict of interest.

## References

1.  *Directive 2010/31/EU of the European Parliament and of the Council of 19 May 2010 on the Energy Performance of Buildings*; European Commission: Brussels, Belgium, 2010.

2.  *Directive 2012/27/EU of the European Parliament and of the Council of 25 October 2012 on Energy Efficiency, Amending Directives 2009/125/EC and 2010/30/EU and Repealing Directives 2004/8/EC and 2006/32/EC Text with EEA Relevance*; European Commission: Brussels, Belgium, 2012.

3.  *Decreto Legislativo 29 dicembre 2006, n. 311. Disposizioni correttive ed integrative al Decreto Legislativo 19 agosto 2005, n. 192, recante attuazione della direttiva 2002/91/CE, relativa al rendimento energetico nell'edilizia*; Gazzetta Ufficiale della Repubblica Italiana: Roma, Italy, 2006. (In Italian)

4.  Del Coz Díaz, J.J.; García Nieto, P.J.; Betegón Biempica, C.; Prendes Gero, M.B. Analysis and optimization of the heat-insulating light concrete hollow brick walls design by the finite element method. *Appl. Therm. Eng.* **2007**, *27*, 1445–1456.

5.  Morales, M.P.; Juárez, M.C.; López-Ochoa, L.M.; Doménech, J. Study of the geometry of a voided clay brick using non-rectangular perforations to optimize its thermal properties. *Energy Build.* **2011**, *43*, 2494–2498.

6.  Morales, M.P.; Juárez, M.C.; Muñoz, P.; Gómez, J.A. Study of the geometry of a voided clay brick using rectangular perforations to optimize its thermal properties. *Appl. Therm. Eng.* **2011**, *31*, 2063–2065.

7. Sun, J.; Fang, L.; Han, J. Optimization of concrete hollow brick using hybrid genetic algorithm combining with artificial neural networks. *Int. J. Heat Mass Transf.* **2012**, *53*, 5509–5518.

8. Li, L.P.; Wu, Z.G.; He, Y.L.; Lauriat, G.; Tao, W.Q. Optimization of the configuration of 290 × 140 × 90 hollow clay bricks with 3-D numerical simulation by finite volume method. *Energy Build.* **2008**, *40*, 1790–1798.

9. Del Coz Dìaz, J.J.; Garcìa Nieto, P.J.; Suàrez Sierra, J.L.; Betegòn Biempica, C. Nonlinear thermal optimization of external light concrete multi-holed brick walls by the finite element method. *Int. J. Heat Mass Transf.* **2008**, *51*, 1530–1541.

10. Zukowski, M.; Haese, G. Experimental and numerical investigation of a hollow brick filled with perlite insulation. *Energy Build.* **2010**, *42*, 1402–1408.

11. Schaefer, C.; Bräuer, G.; Szczyrbowski, J. Low emissivity coatings on architectural glass. *Surf. Coat. Technol.* **1997**, *1*, 37–45.

12. Martin-Palma, R.J.; Vazquez, L.; Martinez-Duart, J.M.; Malats-Riera. Silver-based low-emissivity coatings for architectural windows: Optical and structural properties. *Sol. Energy Mater. Sol. Cells* **1998**, *1–2*, 55–66.

13. Miyazaki, M.; Ando, E. Durability improvement of Ag-based low-emissivity coatings. *J. Non-Cryst. Solids* **1994**, *3*, 245–249.

14. Pasztory, Z.; Peralta, P.N.; Peszlen, I. Multi-layer heat insulation system for frame construction buildings. *Energy Build.* **2011**, *43*, 713–717.

15. Saber, H.H.; Maref, W.; Swinton, M.C.; St-Onge, C. Thermal analysis of above-grade wall assembly with low emissivity materials and furred airspace. *Build. Environ.* **2011**, *7*, 1403–1414.

16. Saber, H.H.; Maref, W.; Swinton, M.C. Thermal response of basement wall systems with low-emissivity material and furred airspace. *J. Build. Phys.* **2012**, *35*, 353–371.

17. Saber, H.H.; Maref, W.; Sherrer, G.; Swinton, M.C. Numerical modeling and experimental investigations of thermal performance of reflective insulations. *J. Build. Phys.* **2012**, *36*, 163–177.

18. Bouchair, A. Steady state theoretical model of fired clay hollow bricks for enhanced external wall thermal insulation. *Build. Environ.* **2008**, *43*, 1603–1618.

19. Principi, P.; Fioretti, R. Thermal analysis of the application of pcm and low emissivity coating in hollow bricks. *Energy Build.* **2012**, *51*, 131–142.

20. *COMSOL Multiphysics*; Modeling and simulation software. Avaiable online: http://www.comsol.it/ (accessed on 28 July 2014).

21. *EN ISO 10211-1:1995 Thermal Bridges in Building Construction—Heat Flows and Surface Temperatures—Part 1: General Calculation Methods*; European Committee for Standardization: Brussels, Belgium, 1995.

22. *BS EN 1745:2002 Masonry and Masonry Products. Methods for Determining Design Thermal Values*; European Committee for Standardization: Brussels, Belgium, 2002.

23. *EN ISO 6946:1996 Building Components and Building Elements—Thermal Resistance and Thermal Transmittance—Calculation Method;* European Committee for Standardization: Brussels, Belgium, 1996.

24. *Attuazione Della Direttiva 2002/91/CE Relativa al Rendimento Energetico Nell'edilizia*; Italian Legislative Decree No. 192/2005; Gazzetta Ufficiale della Repubblica italiana: Roma, Italy, 2005. (In Italian)

25. *Disposizioni Correttive ed Integrative al Decreto Legislativo 19 Agosto 2005, n. 192, Rec Ante Attuazione Della Direttiva 2002/91/CE Relativa al Rendimento Energetico Nell'edilizia;* Italian Legislative Decree No. 311/2006; Gazzetta Ufficiale della Repubblica italiana: Roma, Italy, 2006. (In Italian)

26. *Regolamento di Attuazione Dell'articolo 4, Comma1, Lettere a) e b), del decreto19 agosto 2005, n.192, Concernente Attuazione Della Direttiva 2002/91/CE sul Rendimento Energetico in Edilizia;* Italian Decree No. 59/2009; Gazzetta Ufficiale della Repubblica italiana: Roma, Italy, 2009. (In Italian)

# 6

# Laboratory and Field Studies of Poly(2,5-bis(*N*-methyl-*N*-hexylamino)phenylene vinylene) (BAM-PPV): A Potential Wash Primer Replacement for Army Military Vehicles

Peter Zarras [1,*], Christopher E. Miller [2], Cindy Webber [1], Nicole Anderson [1] and John D. Stenger-Smith [1]

[1] Naval Air Warfare Center Weapons Division (NAWCWD), Polymer Science & Engineering Branch (Code 4L4200D), 1900 N. Knox Road (Stop 6303), China Lake, CA 93555-6106, USA; E-Mails: cynthia.webber@navy.mil (C.W.); nicole.anderson@navy.mil (N.A.); john.stenger-smith@navy.mil (J.D.S.-S.)

[2] Army Research Laboratory, Coatings and Corrosion, Building 4600, ARSRD-ARL-WM-SG Aberdeen Proving Ground, MD 21005-5069, USA; E-Mail: christopher.e.miller44.civ@mail.mil

* Author to whom correspondence should be addressed; E-Mail: peter.zarras@navy.mil

**Abstract:** In this study, an electroactive polymer (EAP), poly(2,5-bis(*N*-methyl-*N*-hexylamino)phenylene vinylene) (BAM-PPV), was tested as an alternative to current hexavalent chromium (Cr(VI))-based Army wash primers. BAM-PPV was tested in both laboratory and field studies to determine its adhesive and corrosion-inhibiting properties when applied to steel and aluminum alloys. The Army Research Laboratory (ARL) tests showed that BAM-PPV combined with an epoxy primer and the Army chemical agent-resistant coating (CARC) topcoat met Army performance requirements for military coatings. After successful laboratory testing, the BAM-PPV was then field tested for one year at the Aberdeen Test Center (ATC). This field testing showed that BAM-PPV incorporated into the Army military coating survived with no delamination of the coating and only minor corrosion on the chip sites.

**Keywords:** poly(2,5-bis(*N*-methyl-*N*-hexylamino)phenylene vinylene (BAM-PPV); wash primer; adhesion; corrosion-inhibiting; hexavalent chromium (Cr(VI)); grit blasting; field studies

---

## 1. Introduction

The Army wash primer (DOD-P-15328D) that is currently used on Army military vehicles was developed during World War II. It is a two-component system consisting of a zinc chromate rust inhibiting pigment in a flexible adhering polymer solution activated by phosphoric acid just before use [1,2]. The Army wash primer is used as the pretreatment coating for surfaces, such as moist steel, aluminum, galvanized steel, brass and stainless steel [3]. The Army wash primer chemically alters the surface of these metals and consequently provides an excellent surface for receiving primers or topcoats. The phosphate treatment provides a non-metallic barrier against corrosion and ensures a roughened surface. This roughness provides more surface area for the adhesion of the polyvinyl butyral (pvb) resin when the Army wash primer cures. The reaction of the Army wash primer with the substrate surface produces a phosphate layer and a crosslinked resin containing organocomplexes that adheres strongly to the surface of the phosphate layer. The Army wash primer acts as a single-coat, self-stratifying, multi-layer coating, which consists of: (1) an unpigmented resin layer at the free surface; (2) a pvb-phosphoric acid-chromium layer; and (3) a phosphate layer at the metal surface (Figure 1) [4,5]. A solution of hexavalent chromium (Cr(VI)) ions is then added to passivate the metal surface. The zinc chromate is basic and reacts with the phosphoric acid to form a mixture of chromium in both cationic and anionic forms. The Cr(VI) is available for passivating the phosphate surface, insuring a three-dimensional organic-metallic polymer coating. This coating is capable of inhibiting corrosion and promoting adhesion between the metal substrate and epoxy primer layer.

**Figure 1.** Schematic diagram of multi-layer Cr(VI)-based Army wash primer (DOD-P-15328D) coating on metal substrate.

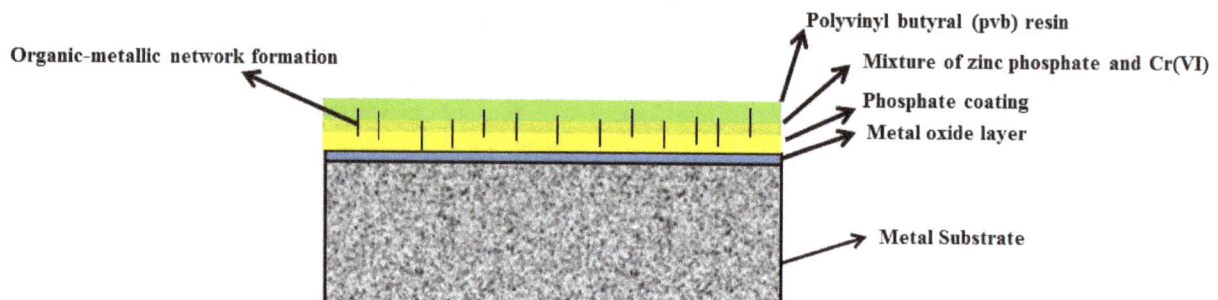

Next, the Army wash primer is coated with an epoxy primer, followed by the Army's chemical agent resistant coating (CARC) polyurethane topcoat (Figure 2).

Cr(VI) has been shown to inhibit corrosion through a release of unreacted and available Cr(VI) when exposed to a corrosive environment via "a self-healing mechanism" [6–9]. The storage and release of soluble Cr(VI) enables chromate conversion coatings (CCC's) and Cr(VI)-epoxy primers to passivate defects and scratches [10,11]. Unfortunately, Cr(VI), which is used in Army wash primer coatings, has been identified as a toxic and carcinogenic material [12–15]. In order to meet new federal and state environmental regulations and to protect worker safety, alternative coating systems that reduce or eliminate Cr(VI) are urgently needed.

**Figure 2.** Schematic diagram of DOD Cr(VI)-based Army wash primer with military coating [epoxy primer and chemical agent resistant coating (CARC) topcoat] on a metal substrate.

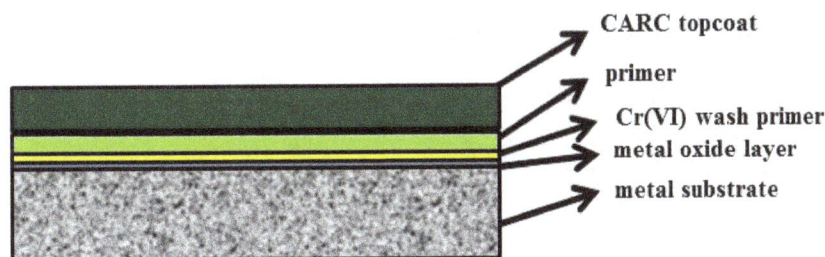

Electroactive polymers (EAPs) have been investigated at the laboratory scale and commercialized as viable corrosion-inhibiting coatings [16–20]. Polyaniline (PANI) is one of the most studied EAPs for corrosion inhibition. PANI-dispersions in an epoxy formulation under the trade name CORRPASSIV™ have been commercialized. This product has found widespread use in numerous countries throughout Europe and Asia. PANI-dispersions are used in diverse applications and in a variety of corrosive environments, such as waste water treatment plants, construction materials (bridges) and commercial steel structures. PANI has shown corrosion-inhibition via a passivation mechanism on steel substrates. PANI dispersions and PANI-containing lacquers were used to coat steel samples to provide primer layers of between 0.3 and 20 μm. These PANI dispersive layers were observed to shift the corrosion potential in the direction of the "noble region" [21].

There are several reports of EAPs used as replacements for the Cr(VI)-based wash primer [22,23]. These studies focused on laboratory testing of the EAP via electrochemical and ASTM methods, which provided evidence for corrosion protection of the underlying metal. Poly(2,5-bis(*N*-methyl-*N*-hexylamino)phenylene vinylene) (BAM-PPV) has been studied and tested at the Naval Air Warfare Center Weapons Division (NAWCWD) for its corrosion-inhibiting properties. The mechanism of BAM-PPV's corrosion-inhibition was investigated via electrochemical methods. BAM-PPV coated onto AA 2024-T3 alloy provided corrosion inhibition via a surface passivation mechanism [24]. BAM-PPV was then field tested as a CCC replacement by researchers at the NAWCWD in cooperation with Wright-Patterson Air Force Base (WPAFB) (Figure 3). This EAP was coated on the rear cargo hatch door of the C-5 Galaxy aircraft. BAM-PPV was incorporated into an aerospace coating (non-Cr(VI) epoxy primer and polyurethane topcoat). It was then shown to survive a one-year field test without any evidence of delamination or corrosion. The BAM-PPV military coating was compared to a Cr(VI) aerospace coating (CCC, Cr(VI) epoxy primer and polyurethane topcoat) and had similar corrosion and adhesion performance during Air Force field testing [25].

Based on the successful results from the field tests performed by WPAFB, this study was undertaken to investigate BAM-PPV as a potential alternative to the Cr(VI)-based Army wash primer. The NAWCWD in cooperation with the Army Research Laboratory (ARL) focused on laboratory testing, including accelerated weathering and adhesion tests. In addition to laboratory testing of BAM-PPV's adhesion and corrosion inhibiting properties, toxicology testing was performed on laboratory animals using BAM-PPV powder as the unknown material to determine its potential lethality in the workplace environment. BAM-PPV was tested for its toxicity and found to be a nontoxic, non-dermal irritant and is not a sensitizer [26]. Once these laboratory tests were completed showing that BAM-PPV is a potential Cr(VI)-based Army wash primer alternative and is non-toxic, it was then field tested.

**Figure 3.** Structure of poly(2,5-bis(*N*-methyl-*N*-hexylamino)phenylene vinylene) (BAM-PPV).

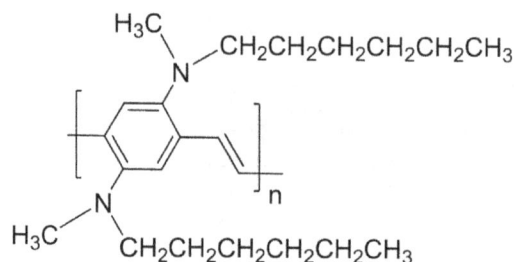

where n >1

## 2. Experimental Section

### 2.1. Materials

The monomer, 2,5-bis(chloromethyl)-4-(hexamethylamino)-phenyl)hexamethylamine dihydrochloride, was prepared according to published work [27–29]. Acetonitrile, methanol, toluene, tetrahydrofuran (THF) and potassium *t*-butoxide (K-*t*-OBu) were obtained from Aldrich Chemical Company and used without further purification. 4130 steel coupons and aluminum alloys (AA 6061-T6 and AA 5083-H116) were purchased from Q-Lab Corporation.

### 2.2. Polymerization of 2,5-Bis(chloromethyl)-4-(hexamethylamino)-phenyl)hexamethylamine Dihydrochloride

The polymerization of the monomer, 2,5-bis(chloromethyl)-4-(hexamethylamino)-phenyl)hexamethylamine dihydrochloride, was carried out in a 5-L reactor with a mechanical stirrer. The reactor was placed inside a large stainless steel secondary container to allow cooling. Toluene (2 L) and THF (2 L) were added to the reactor. Dry ice was added to the secondary container containing acetonitrile to cool the reactor to −45 °C. While the solvent was cooling, 100–150 g of monomer were added, and when the temperature reached −45 °C, 8 molar equivalents of K-*t*-OBu were added. The temperature was kept between −45 °C and −55 °C for 2 h. After 2 h, the solution was then allowed to warm to ambient temperature overnight. After 18 h, the reaction mixture was found to be a very viscous orange, semi-gelatinous material. The polymer was precipitated by pouring this viscous orange material into methanol, filtered through a glass frit and dried under vacuum (1.0 mmHg) at 100 °C overnight.

After drying, the polymer was placed in a Soxhlet thimble and extracted with methanol to remove oligomers and potassium chloride (KCl) salts. The extraction was followed by rinsing with methanol, filtering and drying to constant weight. The crude polymer was mechanically agitated with vigorous stirring using de-ionized water (DI water):methanol (90:10, V/V) as the extractant for one week to remove any remaining KCl salts. The suspension was filtered and the powder dried for one day under vacuum (0.5 mmHg, 25 °C) to constant weight. BAM-PPV was obtained as an orange-yellow powder in 93% yield. The BAM-PPV was characterized via $^1$H NMR, $^{13}$C NMR and FTIR spectroscopy for structure and elemental analysis for chemical composition. The data was consistent with previous reports on BAM-PPV properties, confirming the product [24,28,29].

## 2.3. BAM-PPV Solution Preparation

The BAM-PPV polymer was ground to a fine powder using a commercial coffee grinder. 4-Chlorobenzotrifluoride (Oxsol-100) (VOC-exempt) was used to dissolve the BAM-PPV powder. The BAM-PPV powder was added to the Oxsol-100 solvent with vigorous stirring. The BAM-PPV suspension was heated at 60 °C to dissolve any remaining BAM-PPV powder. After 3 days of mixing, a ~1 wt% solution was obtained. The BAM-PPV solution was filtered through a Buchner funnel fitted with a coarse filter paper (Fisherbrand filter paper, porosity: coarse; diameter: 18.5 cm) to remove trace amounts of impurities present in the solution. The filtrate was then ready for the next step (Section 2.4), which consisted of coating the metal substrates.

## 2.4. BAM-PPV and DOD-P-15328D Wash Primer Coatings Application on Metal Substrates

The BAM-PPV solutions were sprayed onto aluminum alloy (AA) substrates [(AA 6061-T6 (3 × 6 × 0.032"); AA 5083-H116 (3 × 6 × 0.125")] and high strength 4130 steel substrate (3 × 6 × 0.25") using an air brush technique. The BAM-PPV coating was applied to the as-received aluminum alloys. The as-received and grit-blasted 4130 steel substrate were coated with BAM-PPV solution via an air brush spray. The aluminum and steel alloys coated with BAM-PPV were dried under vacuum (1 mmHg, 60 °C) for 12 h, giving an orange-colored film. The dry film thickness was determined to be between 1.9 and 2.2 μm.

For the field demonstration, the AA 2024-T3 metal substrate on the Bradley vehicle headlight cover was used. This aluminum substrate part of the Army's inventory was grit blasted to remove all coatings from the metal surface. BAM-PPV was sprayed via high volume low pressure (HVLP) spray equipment directly onto the metal surface of the Bradley vehicle headlight cover. The BAM-PPV pretreatment coating was dried under ambient conditions to give a dry coating thickness of >2.0 μm. The Army wash primer, DOD-P-15328D, was used as the control and was applied to metal substrates according to the manufacturer's instructions using HVLP coating methods to give a dry film thickness of between 0.3 and 0.5 mils.

## 2.5. Primer(s) and Topcoat(s) Application to the BAM-PPV Pretreatment Coating

The Army primer that was used to coat the BAM-PPV pretreatment coating was MIL-DTL-53022D. A description of the MIL-DTL-53022D primer is found in Table 1. This primer was applied using HVLP coating application methods. The coatings were mixed according to the manufacturer's instructions and applied using a DeVilbiss GTi HVLP Gun with a #413 needle, #100 cap size, a cap fluid pressure setting of 2 psi, a line pressure setting of 40 psi, a hose length of 30 feet and a hose diameter of 3/8". Coatings were applied using a single cross coat to generate an overall dry film coating thickness of between 0.6 and

0.9 mils. The MIL-DTL-53022D primer was applied onto both laboratory coupons and the Army Bradley vehicle headlight cover.

**Table 1.** Army primers and topcoats used for laboratory and field studies.

| Coating | Military Specification | Description |
|---------|----------------------|-------------|
| Primer | MIL-DTL-53022D | chromate and lead-free, solvent-borne, epoxy primer |
| Topcoat | MIL-DTL-53039 | chromate and lead-free, solvent-borne, single-component moisture cure aliphatic polyurethane camouflage CARC coating |
| Topcoat | MIL-DTL-64159 | chromate and lead-free, water-borne, two-component polyurethane CARC coating |

The Army topcoats, MIL-DTL-53039 and MIL-DTL-64159, were mixed according to manufacturer instructions and applied using the above settings, but with two cross coats to generate an overall dry film coating thickness of approximately 1.2 mils. A description of the topcoats, MIL-DTL-53039 and MIL-DTL-64159 are found in Table 1. MIL-DTL-53039 and MIL-DTL-64159 were applied to laboratory coupons, and only MIL-DTL-64159 was applied onto the Army Bradley vehicle headlight cover. The primer and both topcoats were dried using standard drying conditions (ambient conditions ~77 °F and 50% RH for 14 days).

### 2.6. Pull-Off Adhesion ASTM D 4541 (Pneumatic Adhesion Tensile Test Instrument) PATTI

The pneumatic adhesion tensile test instrument (PATTI) pull-off test is designed to give specific information concerning both the inter-coat adhesion and the intra-coat cohesion of organic coating systems [30]. The pull-off adhesion testing of BAM-PPV coatings with primer and topcoats was conducted to assess the effects of the surface preparation at the substrate and the pretreatment. These tests were performed in accordance with ASTM D 4541. An Elcometer Model 108 Hydraulic Adhesion Test Equipment (HATE) was used for this procedure. In addition to being a more quantitative test method, pull-off adhesion is also less prone to human errors in testing, such as variations in pressure applied during scribing, as well as interpretation and perception of results. For the pull-off adhesion test, a loading fixture commonly referred to as a "dolly" was secured normal to the coating surface using an adhesive. The adhesive used was cyanoacrylate. After allowing the adhesive to cure for 24 h at 25 °C in ambient conditions, the attached dolly was inserted into the test apparatus.

The load applied by the apparatus was gradually increased and monitored on the gauge until a plug of coating was detached. The failure value (in psi) was recorded, and the failure mode was characterized. For pull-off data to be valid, the specimen substrate must be of sufficient thickness to ensure that the coaxial load applied during the removal stage does not distort the substrate material and cause a bulging or "trampoline effect". When a thin specimen is used, the resultant bulge causes the coating to radially peel away outwards from the center instead of being uniformly pulled away in pure tension and, thus, results in significantly lower readings than for identically prepared specimens with greater substrate thickness. At 0.25 inches, all of the metallic panels evaluated in the test matrix had adequate thickness for valid pull-off test results. Measurements for each coating system and substrate were obtained by taking 16 measurements on each of the two panels. Any failure measurements due to coating separation between the topcoat surface and the cyanoacrylate adhesive were rejected.

## 2.7. Neutral Salt Spray (NSS) Exposure Testing

Neutral salt spray (NSS) exposure testing was performed to evaluate the ability of the coating systems to withstand a 5 wt% sodium chloride solution, pH-adjusted to a range of 6.5–7.2 [31]. This test was performed on pretreatment, primed and CARC topcoated systems on both steel and aluminum alloy substrates. NSS exposure testing was conducted, in accordance ASTM B 117, Standard Practice for Operating Salt Spray (Fog) Apparatus. All samples subjected to NSS exposure tests were photographed before and after testing to document the performance of the coating. There were three replicates per coating system. The guidance for sample evaluation was taken from MIL-PRF-23377, Performance Specification, Primer Coatings: Epoxy, High Solids. All coupons were primed, topcoated and then scribed with an X-scribe (full size). The 4130 steel coupons were then exposed to NSS for up to 432 h. The AA 6061-T6 and 5083-H116 coupons were also exposed to NSS exposure testing for up to 2016 h.

All samples were checked for blistering, loss of adhesion, undercutting, pitting and corrosion build-up in the scribe. The rating systems for pretreatment, primed and CARC topcoated samples exposed to NSS are given in Table 2. ASTM D 1654 Method A is used for the scribed regions and Method B for the non-scribed regions. Final images were taken upon completion of exposure. BIF stands for blisters in field and means that there was creep from scribe and blistering of the non-scribed regions. Furthermore, where there are two numbers in a rating for the scribed panel, the first number indicates the ASTM D 1654 rating of the scribe and the second is the rating for those areas away from the scribe.

**Table 2.** Performance rating for ASTM 1654 Methods A and B.

| Color Code | Rating | Blister Size |
|---|---|---|
| green | 8–10 (pass) | very small-none |
| yellow | 6–7 (pass) | small |
| orange | 4–5 (marginal) | medium |
| red | 1–3 (failure) | delamination-large |

# 3. Results and Discussion

## 3.1. Adhesion Testing of BAM-PPV Military Coatings

There are several modes of failure when a coating is removed from a substrate using axial tension [32–34]. These modes of failure include: adhesive debonding, cohesive debonding and mixed-mode debonding. Adhesive debonding is the force necessary to separate components of different substances (metal-coating), whereas cohesive debonding is the force necessary to separate components of the same substance (coating-coating) [35]. The results presented in Table 3 are only comparative; they do not provide information regarding whether a coating passes or fails this specific test.

The results of the PATTI adhesion tests (see Table 3) shows that BAM-PPV pretreatment with MIL-DTL-53022 primer and either CARC topcoat MIL-DTL-53039 or MIL-DTL-64159 provided acceptable adhesion. The adhesion results were acceptable for both as-received and grit-blasted 4130 steel coupons. The adhesion was slightly improved using the MIL-DTL-64159 CARC topcoat on grit-blasted 4130 steel coupons, but slightly lower adhesion was observed on grit-blasted BAM-PPV with the MIL-DTL-53039 topcoat. Grit-blasted steel coupons were used to mimic Army depot de-painting operations and to determine if there were any adhesion differences between as-received and grit-blasted surfaces. Overall, the data showed that the adhesion was acceptable for both as-received and grit-blasted steel surfaces.

**Table 3.** Adhesion performance of BAM-PPV with MIL-DTL-53022 and CARC topcoats on 4130 steel coupons.

| Substrate 4130 Steel | Pretreatment | Primer | Topcoat | Mode of Failure | Average psi |
|---|---|---|---|---|---|
| as-received | BAM-PPV | MIL-DTL-53022 | MIL-DTL-53039 | adhesive | 917 |
| grit blasted | BAM-PPV | MIL-DTL-53022 | MIL-DTL-53039 | adhesive | 866 |
| as-received | BAM-PPV | MIL-DTL-53022 | MIL-DTL-64159 | adhesive | 801 |
| grit blasted | BAM-PPV | MIL-DTL-53022 | MIL-DTL-64159 | adhesive | 848 |

### 3.2. NSS Results for BAM-PPV on As-Received and Grit-Blasted 4130 Steel Coupons

NSS exposure testing was performed on grit-blasted steel surfaces to mimic Army depot de-painting and cleaning operations prior to re-application of Army military coatings. Steel coupons were coated with BAM-PPV pretreatment, MIL-DTL-53022 primer and topcoated with MIL-DTL-53039 or MIL-DTL-64159. The failure criterion was an ASTM D 1654 rating of three or less. The majority of the 4130 steel coupon surfaces that were grit-blasted prior to coating with the BAM-PPV pretreatment completed the 432 h of NSS exposure testing. Table 4 provides the results of NSS exposure testing on as-received and grit-blasted 4130 steel coupons coated with pretreatment(s) and MIL-DTL-53022 primer and CARC topcoats (MIL-DTL-53039 or MIL-DTL-64159).

**Table 4.** ASTM D 1654 ratings for as-received and grit-blasted 4130 steel coupons in the neutral salt spray (NSS) chamber. BIF, blisters in field.

| Substrate 4130 Steel | Pretreatment | Primer | Topcoat | Scribe | 124 h | 432 h |
|---|---|---|---|---|---|---|
| as-received | DOD-P-15328D | MIL-DTL-53022 | MIL-DTL-64159 | No | 9 | 7 |
| as-received | DOD-P-15328D | MIL-DTL-53022 | MIL-DTL-64159 | Yes | 9 BIF 8 | 6 BIF 6 |
| as-received | BAM-PPV | MIL-DTL-53022 | MIL-DTL-53039 | No | 9 | 4 |
| as-received | BAM-PPV | MIL-DTL-53022 | MIL-DTL-53039 | Yes | 5 BIF 8 | 4 BIF 2 |
| grit blasted | DOD-P-15328D | MIL-DTL-53022 | MIL-DTL-64159 | No | 7 | 5 |
| grit blasted | DOD-P-15328D | MIL-DTL-53022 | MIL-DTL-64159 | Yes | 9 BIF 7 | 5 BIF 6 |
| grit blasted | BAM-PPV | MIL-DTL-53022 | MIL-DTL-64159 | No | 8 | 5 |
| grit blasted | BAM-PPV | MIL-DTL-53022 | MIL-DTL-64159 | Yes | 8 BIF 7 | 5 BIF 5 |

The DOD-P-15328D Army wash primer military coating (epoxy primer and CARC topcoat) was exposed to 432 h of NSS exposure testing. At 124 h of NSS exposure testing, almost all of the DOD-P-15328D-coated steel coupons (as-received or grit-blasted) showed similar corrosion performance. There was corrosion deterioration of the DOD-P-15328D-coated steel coupons at 432 h of NSS exposure testing.

The BAM-PPV military coating system (epoxy primer and CARC topcoat) provided similar corrosion protection as compared to the control DOD-P-15328D Army military wash primer coating at 124 h of NSS exposure testing. Overall, the best performing BAM-PPV military coating was BAM-PPV with MIL-DTL-53022 epoxy primer and MIL-DTL-64159 topcoat. After 432 h of NSS exposure testing, whether scribed or non-scribed, there was no evidence of delamination of the coating system from the steel coupons.

Figure 4 shows an example of BAM-PPV on as-received and grit-blasted 4130 steel coupon coated with MIL-DTL-53022 primer and topcoated with MIL-DTL-53039 after 432 h of NSS exposure testing. Figure 5 shows an example of the control 4130 steel coupons coated with DOD-P-15328D (Army Cr(VI)-based wash primer), MIL-DTL-53022 epoxy primer and topcoated with MIL-DTL-64159. The Army wash primer was coated onto both (a) as-received and (b) grit-blasted steel coupons. In all cases using either MIL-DTL-53039 or MIL-DTL-64159 topcoats for both the scribed and non-scribed coupons showed evidence of corrosion and blistering. Both the scribed and non-scribed 4130 steel coupons coated with the control and the BAM-PPV military coating showed similar corrosion performance after 432 h of NSS exposure testing.

**Figure 4.** BAM-PPV coated on (**a**) as-received and (**b**) grit-blasted 4130 steel coupons with MIL-DTL-53022 epoxy primer and MIL-DTL-53039 topcoat after 432 h of NSS exposure testing.

(**a**)    (**b**)

**Figure 5.** DOD-P-15328D coated on (**a**) as-received and (**b**) grit-blasted 4130 steel coupons with MIL-DTL-53022 primer and MIL-DTL-64159 topcoat after 432 h of NSS exposure.

(**a**)    (**b**)

*3.3. NSS Results for BAM-PPV on As-Received Aluminum Coupons*

All of the aluminum coupons (AA 6061-T6 and AA 5083-H116) exposed to NSS were as-received without any surface finish. All of the scribed AA 6061-T6 and AA 5083-H116 coupons coated with BAM-PPV pretreatment followed by MIL-DTL-53022 epoxy primer and topcoated with CARC topcoat (MIL-DTL-53039 or MIL-DTL-64159) did not complete the 2016 h NSS exposure testing before extensive corrosion and delamination was evident in the scribed areas. This corrosion performance was consistent for either Army topcoat used during the NSS exposure testing. The data in Table 5 shows that all of the non-scribed panels independent of the topcoat used lasted through 2016 h for both the AA 6061-T6 and AA 5083-H116 coupons. There was no evidence of delamination for the non-scribed BAM-PPV Army coating at 2016 h of NSS exposure testing.

**Table 5.** ASTM D 1654 ratings for as-received AA 6061-T6 and AA 5083-H116 coupons in the NSS chamber.

| Substrate | Pretreatment | Primer | Topcoat | Scribe | 168 h | 504 h | 1248 h | 2016 h |
|-----------|--------------|--------|---------|--------|-------|-------|--------|--------|
| AA 6061-T6 | BAM-PPV | MIL-DTL-53022 | MIL-DTL-53039 | No | 10 | 10 | 10 | 8 |
| AA 6061-T6 | BAM-PPV | MIL-DTL-53022 | MIL-DTL-53039 | Yes | 7 | 4 | <1 | <1 |
| AA 6061-T6 | BAM-PPV | MIL-DTL-53022 | MIL-DTL-64159 | No | 10 | 10 | 7 | 7 |
| AA 6061-T6 | BAM-PPV | MIL-DTL-53022 | MIL-DTL-64159 | Yes | 5 | 5 BIF 7 | 5 BIF 5 | 5 BIF 4 |
| AA 5083-H116 | BAM-PPV | MIL-DTL-53022 | MIL-DTL-53039 | No | 10 | 10 | 10 | 10 |
| AA 5083-H116 | BAM-PPV | MIL-DTL-53022 | MIL-DTL-53039 | Yes | 8 | 7 | 2 | 2 |
| AA 5083-H116 | BAM-PPV | MIL-DTL-53022 | MIL-DTL-64159 | No | 10 | 10 | 10 | 10 |
| AA 5083-H116 | BAM-PPV | MIL-DTL-53022 | MIL-DTL-64159 | Yes | 6 | 6 | 3 | 1 |

### 3.4. Field Test Results for BAM-PPV on Army Bradley Vehicle Aluminum Headlight Cover

The aluminum headlight cover (AA 2024-T3) on the Bradley vehicle was chosen for the Army field testing due to the BAM-PPV military coating showing acceptable laboratory performance results in NSS exposure testing and adhesion studies. The reasons for selecting the headlight cover of the Bradley vehicle are as follows: (1) the Bradley vehicle is an ideal Army platform for assessing the performance of the BAM-PPV coating; and (2) the headlight cover is constructed of an aluminum alloy that provides the EAP a fair approximation of Army inventory.

The location of this vehicle at the Aberdeen Test Center (ATC) provided evaluators with several critical advantages over other locations. These critical advantages included: (1) the vehicle was assigned to the nearby test track, so it received a more aggressive exposure than would be expected from the typical environment of central Maryland; (2) the test track is used to test Army military coatings (wash primer, epoxy primer and CARC topcoat) in an accelerated environment that approximates the GM 9540 cyclic accelerated corrosion test; (3) ATC proximity to the Chesapeake Bay provides raised night time humidity and ameliorates temperature extremes; and (4) the ATC is only 1.5 miles from the personnel who would perform the periodic evaluations.

The Bradley vehicle headlight cover was grit blasted prior to coating with BAM-PPV to remove any Army coatings present on the metal surface. This was done to mimic the current de-painting operations at Army depots. BAM-PPV was sprayed onto the substrate using HVLP spray equipment, coated with MIL-DTL-53022 and topcoated with MIL-DTL-64159 (Figure 6). The vehicle was field tested at the ARL outdoor weathering test track and exposed to rain, snow, sleet, sun, wind, coastal moisture and humidity. Visual inspections for corrosion, adhesion, peeling and blistering were performed every three months. After one year of field testing, no significant corrosion, no blistering near edges and no undercutting of the coating adjacent to chip sites was observed. The chip sites that were present on the headlight cover were a result of wear during the one-year field test. This is considered acceptable performance for ARL field testing. There was slight corrosion present at chip sites where all coatings were removed (epoxy primer and CARC topcoat), including BAM-PPV (Figure 7). When the test method ASTM D 1654 Evaluation of Painted Specimens Subjected to Corrosive Environments there was limited corrosion damage after 12 months to areas where all coatings were removed, including BAM-PPV.

**Figure 6.** Coating procedure employed for de-painting and re-painting using BAM-PPV as the wash primer alternative coating followed by epoxy primer and the CARC topcoat (**a**) grit-blasted headlight cover; (**b**) the headlight cover coated with BAM-PPV; and (**c**) the headlight cover coated with MIL-DTL-53022 and MIL-DTL-64159.

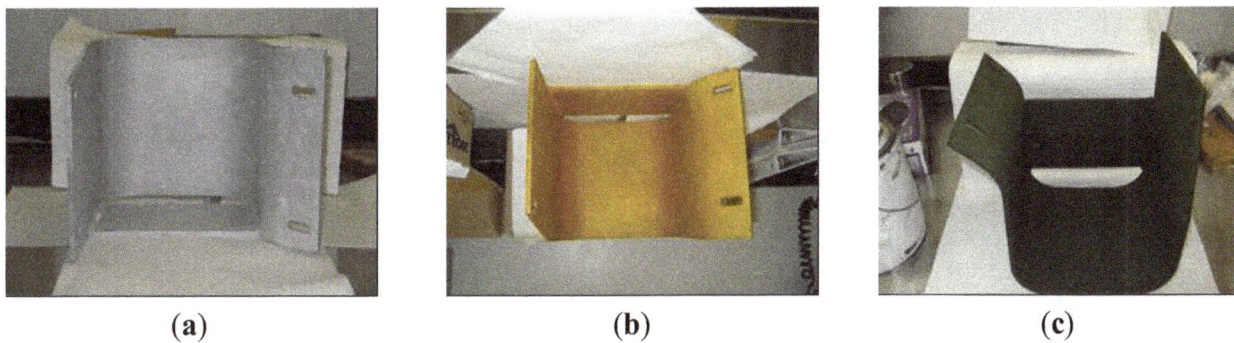

**(a)**                    **(b)**                    **(c)**

**Figure 7.** Wear damage after one-year of field testing the headlight cover: (**a**) wear damage front view; and (**b**) wear damage backside view.

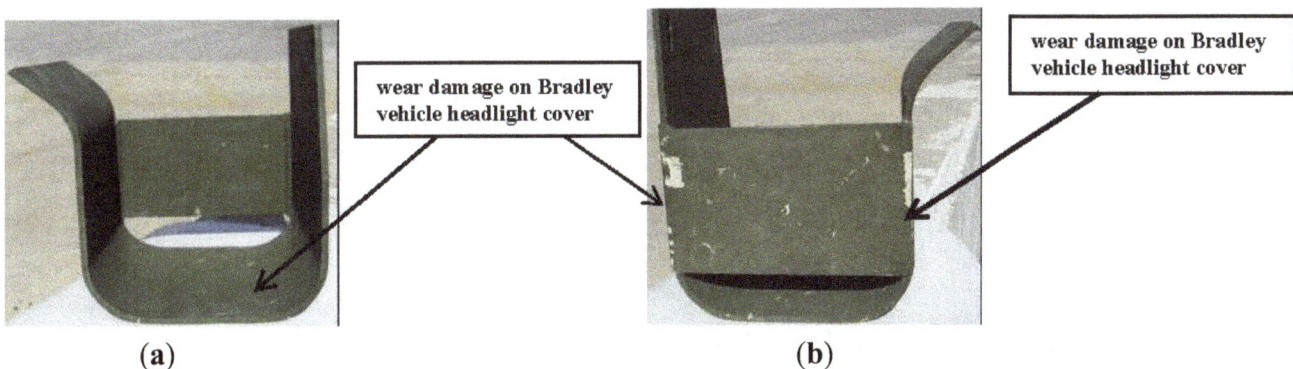

**(a)**                                        **(b)**

## 4. Conclusions

Several conclusions can be made regarding the laboratory and field testing studies incorporating BAM-PPV as a Cr(VI)-based wash primer replacement with full military coatings (epoxy primer and CARC topcoat):

(a) BAM-PPV showed slightly improved adhesion on grit-blasted 4130 steel surfaces as compared to as-received;

(b) BAM-PPV and DOD-P-15328D showed similar corrosion performance in NSS exposure testing;

(c) Results from the ARL field testing showed that BAM-PPV incorporated into Army military coating (epoxy primer and CARC topcoat) provided acceptable field performance; and

(d) BAM-PPV could be a potential alternative to Cr(VI)-based wash primer on AA 2024-T3 substrates for the Army Bradley vehicle.

## Acknowledgments

The financial support of the Department of Defense (DOD) Environmental Security Technology and Certification Program (ESTCP), under the direction of Jeffrey Marqusee and Bruce D. Sartwell, Weapons Systems and Platforms Program Manager, is gratefully acknowledged.

## Author Contributions

Peter Zarras, Cindy Webber, Nicole Anderson and John D. Stenger-Smith were responsible for synthesis, polymerization and coating BAM-PPV onto the as-received aluminum and steel substrates. Christopher E. Miller was responsible for coating BAM-PPV onto grit-blasted coupons and coating with primer and CARC topcoat. Christopher E. Miller was responsible for laboratory and field evaluation studies of BAM-PPV and Army wash primer coating controls.

## Conflicts of Interest

The authors declare no conflict of interest.

## References

1. *Primer (Wash) Pretreatment (Formula No. 117 for Metals) DOD-P-15328D*, Federal Specifications; Department of Defense: Washington, DC, USA, 2001.
2. Smith, P.; Chesonis, K.; Escarsega, J. *Demonstration and Validation of a Replacement Alternative to the Chromate Wash Primer DOD-P-15328D*, ARL-TR-3756; Army Research Laboratory: Adelphi, MD, USA, 2006; pp. 1–18.
3. Daley, R.R.; Hodges, S.A. Corrosion inhibitor for self-etching wash primers. *Polym. Paint Colour J.* **2010**, *200*, 18–20, 22, 24.
4. Chesonis, K.S.; Miller, C.E. *VOC Compliant Wash Primer*, ARL-MR-190; Army Research Laboratory: Adelphi, MD, USA, 1994.
5. Smith, P.; Chesonis, K.; Escarsega, J. *Replacement Alternatives to the Chromate Wash Primer DOD-P-15328D*, ARL-TR-3220; Army Research Laboratory: Adelphi, MD, USA, 2004.
6. Kendig, M.; Jeanjaquet, S.; Addison, R.; Waldrop, J. Role of hexavalent chromium in the inhibition of corrosion of aluminum alloys. *Surf. Coat. Technol.* **2001**, *140*, 58–66.
7. Illevbare, G.O.; Scully, J.R.; Yuan, J.; Kelly, R.G. Inhibition of pitting corrosion on aluminum alloy 2024-T3: Effect of soluble chromate additions *vs.* chromate conversion coating. *Corrosion* **2000**, *56*, 227–242.
8. Kendig, M.W.; Davenport, A.J.; Isaacs, H.S. The mechanism of corrosion inhibition by chromate conversion coatings from X-ray absorption near edge spectroscopy (XANES). *Corros. Sci.* **1993**, *34*, 41–49.
9. Chidambaram, D.; Halada, G.P.; Clayton, C.R. Spectroscopic elucidation of the repassivation of active sites on aluminum by chromate conversion coating. *Electrochem. Solid State Lett.* **2004**, *7*, B31–B33.

10. Xia, L.; Akiyama, E.; Frankel, G.; McCreery, R. Storage and release of soluble hexavalent chromium from chromate conversion coatings equilibrium aspects of CrVI concentration. *J. Electrochem. Soc.* **2000**, *147*, 2256–2262.

11. Ramsey, J.D.; Xia, L.; Kendig, M.W.; McCreery, R.L. Raman spectroscopic analysis of the speciation of dilute chromate solutions. *Corros. Sci.* **2001**, *43*, 1557–1172.

12. Wise, S.S.; Wise, J.P. Chromium and genomic stability. *Mutat. Res. Fundam. Mol. Mech. Mutagen.* **2012**, *733*, 78–82.

13. Nickens, K.P.; Patierno, S.R.; Ceryak, S. Chromium genotoxicity: A double-edged sword. *Chem. Biol. Interact.* **2010**, *188*, 276–288.

14. Katz, S.A.; Ballantyne, B.; Salem, H. The inhalation toxicity of chromium compounds. In *Inhalation Toxicity*, 2nd ed.; Salem, H., Katz, S.A., Eds.; CRC Press: Boca Raton, FL, USA, 2006; pp. 543–564.

15. Sedman, R.M.; Beaumont, J.; McDonald, T.A.; Reynolds, S.; Krowech, G.; Howa, R. Review of the evidence regarding the carcinogenicity of hexavalent chromium in drinking water. *J. Environ. Sci. Health Part C Environ. Carcinogen. Ecotoxicol. Rev.* **2006**, *24*, 155–182.

16. Mengoli, G.; Munari, M.T.; Bianco, P.; Musiana, M.M. Anodic synthesis of polyaniline coatings onto iron sheets. *J. Appl. Polym. Sci.* **1981**, *26*, 4247–4257.

17. DeBerry, D.W. Modification of the electrochemical and corrosion behavior of stainless steels with an electroactive coating. *J. Electrochem. Soc.* **1985**, *132*, 1022–1026.

18. Ahmad, N.; MacDiarmid, A.G. Inhibition of corrosion of steels with the exploitation of conducting polymers. *Synth. Met.* **1996**, *78*, 103–110.

19. Yan, M.C.; Tallman, D.E.; Rasmussen, S.C.; Bierwagen, G.P. Corrosion control coatings for aluminum alloys based on neutral and n-doped conjugated polymers. *J. Electrochem. Soc.* **2009**, *156*, C360–C366.

20. Cecchetto, L.; Delabouglise, D.; Petit, J.-P. On the mechanism of the anodic protection of aluminum alloy AA5182 by emeraldine base coatings evidences of a galvanic coupling. *Electrochim. Acta* **2007**, *52*, 3485–3492.

21. Wessling, B. From conductive polymers to organic metals. *Chem. Innov.* **2001**, *31*, 35–40.

22. Pan, T. Intrinsically conducting polymer-based heavy-duty and environmentally friendly coating system for corrosion protection of structural steels. *Spectrosc. Lett.* **2013**, *46*, 268–276.

23. Sathiyanarayanan, S.; Syed, A.S.; Venkatachari, G. Corrosion protection of galvanized iron by polyaniline containing wash primer coating. *Prog. Org. Coat.* **2009**, *65*, 152–157.

24. Zarras, P.; He, J.; Tallman, D.E.; Anderson, N.; Guenthner, A.; Webber, C.; Stenger-Smith, J.D.; Pentony, J.M.; Hawkins, S.; Baldwin, L. Electroactive Polymer Coatings as Replacements for Chromate Conversion Coatings. In *Smart Coatings, ACS Symposium Series 957*; Provder, A., Baghdachi, J., Eds.; American Chemical Society: Washington, DC, USA, 2007; pp. 135–152.

25. Zarras, P.; Anderson, N.; Webber, C.; Stenger-Smith, J.D.; Spicer, M.; Buhrmaster, D. Electroactive materials as smart corrosion-inhibiting coatings for the replacement of hexavalent chromium. *CoatingsTech* **2011**, *8*, 40–44.

26. Zarras, P.; Stenger-Smith, J.D. Electroactive polymer (EAP) coatings for corrosion protection of metals. In *Handbook of Smart Coatings for Materials Protection*; Makhlouf, A.S.H., Ed.; Woodhead Publishing: Cambridge, UK, 2014; pp. 328–369.

27.  Stenger-Smith, J.D.; Anderson, N.; Webber, C.; Zarras, P. Poly(2,5-bis(*N*-methyl-*N*-hexylamino) phenylene vinylene) as a replacement for chromate conversion coatings. *ACS Polym. Prepr.* **2004**, *45*, 150–151.

28.  Anderson, N.; Irvin, D.J.; Webber, C.; Stenger-Smith, J.D.; Zarras, P. Scale-up and corrosion inhibition of poly(bis(dialkylamino)phenylene vinylenes). *ACS PMSE Prepr.* **2002**, *86*, 6–7.

29.  Irvin, D.J.; Anderson, N.; Webber, C.; Fallis, S.; Zarras, P. New Synthetic Routes to poly(bis(dialkylamino)phenylene vinylenes). *ACS PMSE Prepr.* **2002**, *86*, 61–62.

30.  *ASTM D4541 Standard Test Method for Pull-Off Strength of Coatings Using Portable Adhesion Testers*; ASTM International: West Conshohocken, PA, USA, 2007.

31.  *ASTM B117 Standard Practice for Operating Salt Spray (Fog) Apparatus*; ASTM International: West Conshohocken, PA, USA, 2007.

32.  Mendels, D.A. Adhesion of multilayered coatings by low load scratch test. *Tribol. Mater. Surf. Interfaces* **2008**, *2*, 232–244.

33.  Li, H.Z.; Jia, Y.; Luan, S.; Xiang, W.; Qian, H.; Charles, C.; Mamtimin, G.; Han, Y.; An, L. Influence of inter-fiberspacing and interfacial adhesion on failure of multi-fiber model composites: Experiment and numerical analysis. *Polym. Compos.* **2008**, *29*, 964–971.

34.  Buchheit, R.G. Corrosion resistant coatings and paints. In *Handbook of Environmental Degradation of Materials*; Kutz, M., Ed.; William Andrew Publishing: Norwick, NY, USA, 2005; pp. 367–385.

35.  Packman, D.E. Adhesion-Fundamental and Practical. In *Handbook of Adhesion*; Packman, D.E., Ed.; John Wiley and Sons: West Sussex, UK, 2005; pp. 17–21.

# 7

# Thermal Conductivity Analysis and Lifetime Testing of Suspension Plasma-Sprayed Thermal Barrier Coatings

**Nicholas Curry** [1,*]**, Kent VanEvery** [2]**, Todd Snyder** [2] **and Nicolaie Markocsan** [1]

[1] Department of Engineering Science, University West, Gustava Melins Gata 2, Trollhattan 461 86, Sweden; E-Mail: nicolaie.markocsan@hv.se

[2] Progressive Surface, Grand Rapids, MI 49512, USA;
E-Mails: KVanEvery@progressivesurface.com (K.V.E.); TSnyder@progressivesurface.com (T.S.)

* Author to whom correspondence should be addressed; E-Mail: nicholascurry84@gmail.com

**Abstract:** Suspension plasma spraying (SPS) has become an interesting method for the production of thermal barrier coatings for gas turbine components. The development of the SPS process has led to structures with segmented vertical cracks or column-like structures that can imitate strain-tolerant air plasma spraying (APS) or electron beam physical vapor deposition (EB-PVD) coatings. Additionally, SPS coatings can have lower thermal conductivity than EB-PVD coatings, while also being easier to produce. The combination of similar or improved properties with a potential for lower production costs makes SPS of great interest to the gas turbine industry. This study compares a number of SPS thermal barrier coatings (TBCs) with vertical cracks or column-like structures with the reference of segmented APS coatings. The primary focus has been on lifetime testing of these new coating systems. Samples were tested in thermo-cyclic fatigue at temperatures of 1100 °C for 1 h cycles. Additional testing was performed to assess thermal shock performance and erosion resistance. Thermal conductivity was also assessed for samples in their as-sprayed state, and the microstructures were investigated using SEM.

**Keywords:** thermal barrier coating; suspension plasma spray; thermal shock; thermo-cyclic fatigue; thermal conductivity

# 1. Introduction

The development of the gas turbine for propulsion and power generation is pushing the evolution of new materials and processes to allow for greater performance and efficiency [1]. In the hot section of the engine, demands for higher operating temperatures have led to the need for thermal protection for both rotating and non-rotating parts [1,2]. The thermal barrier coating (TBC) allows the operating temperature of the engine to be increased without increasing the operating temperature for the metallic components [1,3].

Traditionally, TBC ceramics have been applied either by air plasma spraying (APS) or electron beam physical vapor deposition (EB-PVD) [4]. APS has the advantage of lower overall cost, higher process flexibility, lower as-sprayed thermal conductivity and long lifetimes [5,6]. However, APS coatings are highly susceptible to sintering, which degrades the mechanical and thermal properties of the coatings [7–9]. On the other hand, the EB-PVD process can produce columnar structures that can tolerate larger amounts of expansion and contraction [10], but these coatings tend to be more thermally conductive than APS coatings. Additionally, compared to APS, EB-PVD production requires higher costs and involves more technical limitations. Consequently, EB-PVD coatings are preferred for smaller critical components in the turbine section that experiences centrifugal loading and the highest levels of thermal shock; while, because of the superior thermal insulation properties, APS is commonly used to coat combustor liners.

Strain-tolerant microstructures, known as dense vertically cracked (DVC) or segmented coatings, can be produced with APS by controlling the deposition parameters [11–13]. However, achieving this type of microstructure can be more difficult than with the EB-PVD process [11]. Furthermore, DVC coatings typically have a lower porosity content than traditional, non-segmented APS coatings. Therefore, the increased strain tolerance of DVC coatings is accompanied by a decreased thermal insulating ability in comparison to traditional APS coatings.

Suspension plasma spraying has become a promising contender for the production of high performance thermal barrier coating systems, as vertically cracked [14,15], columnar [16] or highly porous coatings [17] are possible to manufacture. Suspension plasma spraying (SPS) coatings can be produced with high levels of porosity that are not normally achievable with APS or EB-PVD coatings [18]. The amount and size of the SPS porosity allows for the reduction in coating thermal conductivity through decreased phonon transport and decreased IR radiation transmittance. As IR radiation is an important contributor to the total heat flux in a combustion environment; the ability to reduce the radiative contribution has an advantage for increased engine performance [19].

This study aims to outline some developments of SPS ceramic coatings in combination with bond coats produced via different thermal spray routes. The lifetime in both thermal fatigue and thermal shock are shown together with the coating thermal properties.

# 2. Theory of Suspension Fragmentation and Deposition

As with conventional powder thermal spray processes, particle treatment history will directly influence coating morphology and properties. SPS is a far more complicated process, as the particle size on coating deposition is not pre-determined, as it is with powder processes. When injecting a suspension as a solid stream into a DC arc plasma gun jet, the fragmentation of the suspension stream occurs roughly two orders of magnitude faster than the vaporization of the solvent [20]. It is therefore important to consider first the influence of the various conditions on the fragmentation of the suspension and the resulting droplet size.

The droplet sizes generated in the SPS process are a function of two opposing forces: suspension surface tension and viscosity, reducing droplet fragmentation, and shear force exerted by the plasma jet, which acts to further fragment the droplets [21,22]. The balance in these effects will determine the final suspension fragment size in the plasma jet. Such a final fragment may contain one or several solid primary particles depending on the powder size distribution and concentration used to manufacture the suspension [23].

After solvent evaporation, the powder mass in the final suspension fragment determines the size of the depositing molten particle. Therefore, controlling deposition particle size in SPS involves adjusting the fundamental suspension properties of surface tension and viscosity that help govern suspension fragmentation in the plasma jet [24]. Surface tension is controlled by the solvent used in the suspension and is difficult to influence. Viscosity is a function of several parameters, such as solids load, powder size distribution and the addition of any dispersants to the suspension.

The depositing particle size plays an important role in SPS, because it determines the overall characteristics of the coating microstructure. Berghaus et al. demonstrated how, in plasma spraying, decreasing the particle mass at a fixed velocity leads to increasing changes in particle velocity by the drag from the plasma jet flow as it interacts with the substrate [25]. Specifically, zirconia particles with diameters <5 μm tend to follow the flow of the plasma as it changes direction and moves parallel to the substrate. These dynamics can be related to the Stokes number of such particles in the boundary layer close to the substrate. Such small particles having a Stokes number smaller than one, the particle trajectory is expected to be strongly influenced of the plasma drag [21]. Under these conditions, deposition occurs when a particle encounters an asperity on the substrate surface. As the particle size reduces, then the trajectory before deposition becomes increasingly parallel to the substrate. VanEvery et al. proposed differing deposition regimes depending on the particle size and, therefore, impact trajectory [16]. When particles are sufficiently small, the shallow angle at which the particle will come into contact with the substrate results in the generation of columnar structures. If the particle size is increased, then the influence in the plasma stream drops, and the trajectory becomes more normal to the substrate. This change results in the intermediate cracked-columnar structure. If the particles are sufficiently large, then their momentum is not influenced significantly by the plasma flow, and deposition occurs in the same fashion as in conventional APS deposition. Therefore, columnar structures suitable for thermal barrier coatings can be generated though the control of both suspension and plasma conditions.

## 3. Experimental Procedures

### 3.1. Suspension Parameters

The suspension used in this study was an 8 wt% yttria-stabilized zirconia (YSZ) material produced by Treibacher Industrie AG (Althofen, Austria). Due to the complexity of the SPS process, a number of factors will be held constant within this study. The suspensions used ethanol as a solvent; therefore, surface tension can be ignored as a factor. Suspensions contained the same solids loading of 25% by weight. Two size distributions were used for this study: a nano-suspension with a median particle size of approximately 50 nm and a sub-micron suspension with a median particle size of 500 nm. Rheology measurements at a shear rate of 1000 (1/s) for the two suspensions show that the nano-suspension had a viscosity of 5.9 mPa·s compared to 1.73 mPa·s for the sub-micron suspension. The nano-suspension displays a viscosity roughly 3-times greater than the sub-micron suspension, due to the smaller particle size in the suspension [22].

### 3.2. Sample Production

In combination with the two suspensions, three bond coat application techniques were used in order to produce a total of 5 experimental SPS TBC systems together with one reference TBC system. The samples in this study are summarized in Table 1, with each row representing one of the 6 coating types. For bond coat application, two well-established methods—high velocity oxy-fuel spraying (HVOF) and air plasma spraying (APS)—were used for producing the first two bond coat types. The third type was produced using the more recently developed high-velocity air-fuel (HVAF) technique; effectively a development of the HVOF process in which compressed air is utilized instead of oxygen gas.

**Table 1.** Experimental coatings and their respective production routes. HVOF, high velocity oxy-fuel spraying; HVAF, high-velocity air-fuel; SPS, suspension plasma spraying; DVC, dense vertically cracked; APS, air plasma spraying.

| Coating ID | Bond coat method | Bond coat feedstock | Bond coat thickness (µm) | Top coat feedstock | Top coat condition | Top coat thickness (µm) |
|---|---|---|---|---|---|---|
| H1 | HVOF | AMDRY 365-1 | 161 | Nano suspension | Type 1 SPS | 339 |
| H2 | HVOF | AMDRY 365-1 | 161 | Nano suspension | Type 2 SPS | 320 |
| P1 | Plasma | AMDRY 365-2 | 186 | Nano suspension | Type 1 SPS | 325 |
| P2 | Plasma | AMDRY 365-2 | 186 | Nano suspension | Type 2 SPS | 275 |
| A3 | HVAF | AMDRY 386-2 | 220 | Sub-micron suspension | Type 3 SPS | 303 |
| DVC | Plasma | AMDRY 386-4 | 168 | Powder | DVC-APS | 444 |

The HVOF samples were sprayed using the gas-fuelled DJ-2600 hybrid gun (Sulzer Metco, Wohlen, Switzerland), and the air plasma spray coatings were produced with a F4-MB gun (Sulzer Metco, Wohlen, Switzerland). The HVOF and APS bond coats were both produced using a proprietary NiCoCrAlY alloy powder, AMDRY 365 (Sulzer Metco, Wohlen, Switzerland), with powder cuts selected appropriately for deposition technique. These samples are labeled "H" and "P", respectively. The HVAF bond coat type was sprayed using the Uniquecoat M3 gun (Uniquecoat, Richmond, VA, USA). The powder used in this case was the proprietary NiCoCrAlY, AMDRY 386-2 powder. AMDRY 386 differs by having additions of <1% Hf and <0.7% Si. These samples are labeled "A".

As a reference sample for the study, DVC coatings were also produced via APS using the 100HE plasma spray system (Progressive Surface, Grand Rapids, MI, USA). Bond coats were produced using a proprietary NiCoCrAlY AMDRY 386-4 (Sulzer Metco, Wohlen, Switzerland) material. This powder differs from AMDRY 386-2 only in the powder size distribution chosen to be more suitable for plasma spraying. Top coats were produced using a proprietary 8 wt% YSZ powder; SPM-2000 (Sulzer Metco, Wohlen, Switzerland). These DVC coatings represent the state-of-the-art when discussing strain-tolerant APS coatings.

The substrates used for thermo-cyclic fatigue testing samples in this study were Haynes 230 alloy with a size of 50 mm × 30 mm × 5 mm. All other samples used Hastelloy-X as a substrate material. The plates for microstructural analysis were 25 mm × 25 mm × 1.6 mm. Test buttons for thermal shock testing were 25 mm in diameter and 6 mm thick.

### 3.3. SPS-Specific Coating Deposition Conditions

SPS coatings were deposited using the 100HE Plasma system and the LiquifeederHE suspension feed system (Progressive Surface, Grand Rapids, MI, USA). For illustration purposes, a close up image of the spray process is shown in Figure 1. The suspension stream can be observed entering the plasma jet, after which, it is atomized in the jet. For all SPS coatings in this study, the plasma gun conditions were the same, consisting of a power level of 105 kW and a stand-off distance of 70 mm. The suspension was fed into the plasma jet using a solid stream nozzle mounted orthogonal to the plasma flow at a flow rate of 45 mL/min.

**Figure 1.** Suspension plasma spraying using the 100HE plasma system.

In order to produce different coating structures, the surface speed of the samples relative to the torch was varied by increasing the rotational speed of the sample holder during spray deposition. Type 1 coatings were produced with a surface speed of 380 mm/s. Type 2 coatings were produced with a surface speed of 600 mm/s. Both Type 1 and 2 samples utilize the nano-suspension, as shown in Table 1. Type 3 coatings were produced using the sub-micron suspension and a surface speed of 1015 mm/s. For each coating type, the top coat was applied to all sample substrates during a single SPS experiment, e.g., the H1 and P1 samples were coated together. Again, the summary of coatings can be seen in Table 1.

Specimens for SEM analysis were sectioned using a diamond cutting blade and mounted in low viscosity epoxy-based resin using a vacuum impregnation technique. Samples were subsequently polished using well-established methods for TBC specimens. Gold sputtering was used to allow the ceramic layers to be observed in the SEM.

## 3.4. Thermo-Cyclic Fatigue

Thermo-cyclic fatigue (TCF) testing was performed primarily to study the ability of a coating to resist high temperature oxidation and the stress of oxide growth at high temperature. While failure in TCF testing is driven by bond coat oxidation, the ability of the top coat to survive thermal shock and the stress of oxide growth will also determine the final lifetime of the coating [5]. The TCF test is a simplification of a real exposure test in that once the sample has passed its transient heating period, there is no thermal gradient present across the coating system. As such, the choice of testing temperature is dictated by the upper operating limit for the substrate and NiCoCrAlY bond coat used in the coating system.

For each coating group, four samples were prepared for testing. TCF testing involved placing samples in an automated cycling furnace at 1100 °C for a period of 1 h. After the heating period, the samples moved from the hot zone to a cool zone, where they were immediately photographed. The samples were then force-cooled with compressed air at 25 °C. Cooling resulted in a sample temperature of no more than 100 °C within 10 min of the cooling cycle start. After the cooling cycle was completed, the samples were returned to the hot zone for another heating cycle. Failure was considered to have occurred when 20% of the ceramic surface de-bonded. Failed TCF samples were mounted in low viscosity epoxy-based resin and prepared for microscopic analysis to assess the coating microstructure at failure.

## 3.5. Thermal Shock Testing

Thermal shock testing investigates the ability of coatings to survive very rapid heating and cooling events. Such testing primarily checks the ability of the coating to cope with the stress of thermal expansion mismatch, sintering and thermal gradients. Unlike TCF testing, the influence of the oxidation of the bond coat is of less importance, as the exposure time at elevated temperatures is short and the bond coat/ceramic interface temperature is some 100 °C lower.

Thermal shock samples were prepared from 25 mm-diameter, 6 mm-thick buttons coated with the complete TBC system. Before testing, overspray was ground away from the edge of the sample and the button mounted to a carrier plate using a single spot weld. Thermal shock testing was conducted using a burner rig at GKN Aerospace (Trollhättan, Sweden) [26]. Samples were subjected to 75-s cycles with heating to surface temperatures of 1200 °C, bond coat temperatures of approximately 1000 °C and rear face temperatures of between 960 and 980 °C. Samples were preheated before the test start to 600 °C from the rear face of the sample with hot air guns. The fuel gas used in the combustion burners was propane. Samples were monitored once per revolution of the test fixture using a video recording system and pyrometer measurements. Failure was deemed to have occurred when 20% of the ceramic surface had spalled from the test sample. The results are accurate to within the nearest 4 cycles of failure.

## 3.6. Thermal Conductivity Analysis

The thermal conductivity of the coating systems was determined using the laser flash method that has been long accepted for the analysis of coating thermal properties and has been discussed in detail by Taylor [27]. Ten millimeter-diameter samples for thermal property evaluation were water-jet cut from coated plates of 25 mm × 25 mm. For each coating group, a minimum of four samples were measured. The thermal diffusivity for the complete coating system was measured using a Netzsch LFA 247 (Netzsch Gerätebau GmbH, Selb, Germany). During thermal diffusivity measurements, a laser pulse of known pulse width is fired at the sample rear face. An infra-red detector above the front face of the sample records the temperature increase on the front face due to the laser energy. In order to minimize the radiation effects and maximize heat absorption, the sample is commonly placed with the substrate face towards the laser. Because a YSZ coating is transparent to IR radiation, a thin film of carbon or gold was applied to the top coat surface; this layer prevents direct detection of heat radiating from the bond coat. Measurements at room temperature were performed in an air atmosphere. High temperature measurements (>30 °C) were performed under a dynamic argon atmosphere at atmospheric pressure. Argon gas is used to prevent oxidation of the sample at high temperatures.

The thermal diffusivity of the sample can be calculated using the formula:

$$\alpha = (0.1388 \times L^2)/(t\,(0.5)) \tag{1}$$

where $\alpha$ is the thermal diffusivity ($m^2 \cdot s^{-1}$); L is the thickness of the sample; and $t\,(0.5)$ is the half time taken for the total temperature rise. For this study, the various coatings were measured as a complete TBC system (ceramic, bond coat and substrate). In this case, the ceramic layer is treated as an unknown within a three-layer system, where the thickness, thermal diffusivity, density and specific heat capacity for the substrate and bond coat have been measured previously.

The specific heat capacity of the YSZ material was measured using a DSC 404C differential scanning calorimeter (Netzsch Gerätebau GmbH, Selb, Germany). Measurements of both the bond coat and substrate materials had been completed previously [9]. The density of the ceramic coatings was measured using Archimedes displacement. The thickness of the layers within the coating system was evaluated from microstructural cross-sections using an average of more than 25 measurements of layer thickness along the cross-section of the sample. Finally, calculation of the thermal conductivity was performed using the Proteus thermal analysis software (Netzsch Gerätebau GmbH, Selb, Germany) using the three-layer model according to Clark-Taylor, which accounts for heat loss with pulse correction.

## 3.7. Erosion Testing

Erosion testing was performed at GKN Aerospace Engine Systems according to a standard erosion test procedure. Erosion testing involves feeding of alumina powder though a blast feeding nozzle at 60 degrees to the sample surface at a distance of 10 cm. The test uses alumina grit-blast media of a median particle size 50 μm. The calibration of the erosion rate is first accomplished using a polycarbonate reference piece. The test is run for 40 s, after which the sample is removed. The depth of the erosion pit generated during the test is then measured using a micrometer gauge. The erosion test is repeated on a minimum of 3 samples in order to evaluate erosion resistance.

# 4. Results and Discussion

## 4.1. Microstructure

When discussing the microstructures of the SPS layers produced in this study, it is important to mention that differences in microstructure relate to three possible changes in the experimental set-up:

- Bond coat surface roughness;
- Suspension properties;
- Surface speed during coating deposition.

Example microstructures for Type 1 and 2 coatings are displayed for HVOF and APS bond coats in Figures 2 and 3, respectively. In the case of the Type 1 and 2 coatings, a nano-suspension was used for spraying. Due to its higher viscosity, the nano-suspension leads to larger fragmented suspension droplets, generating larger spray particles that results in what shall be referred to as a cracked-columnar structure.

**Figure 2.** HVOF bond coat with Type 1 SPS coating (H1) (**A**); HVOF bond coat with Type 2 SPS coating (H2) (**B**).

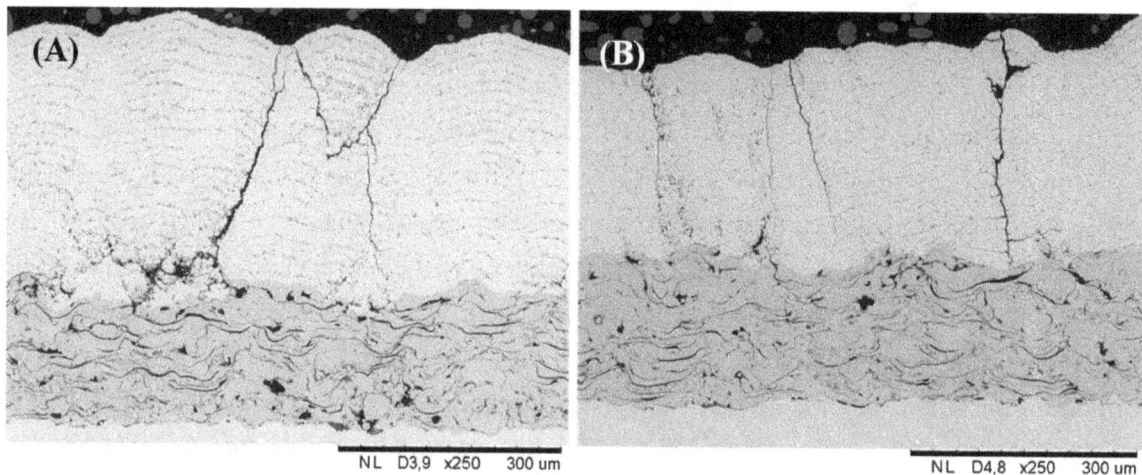

**Figure 3.** Plasma bond coat with Type 1 SPS coating, (P1) (**A**); plasma bond coat with Type 2 SPS coating (P2) (**B**).

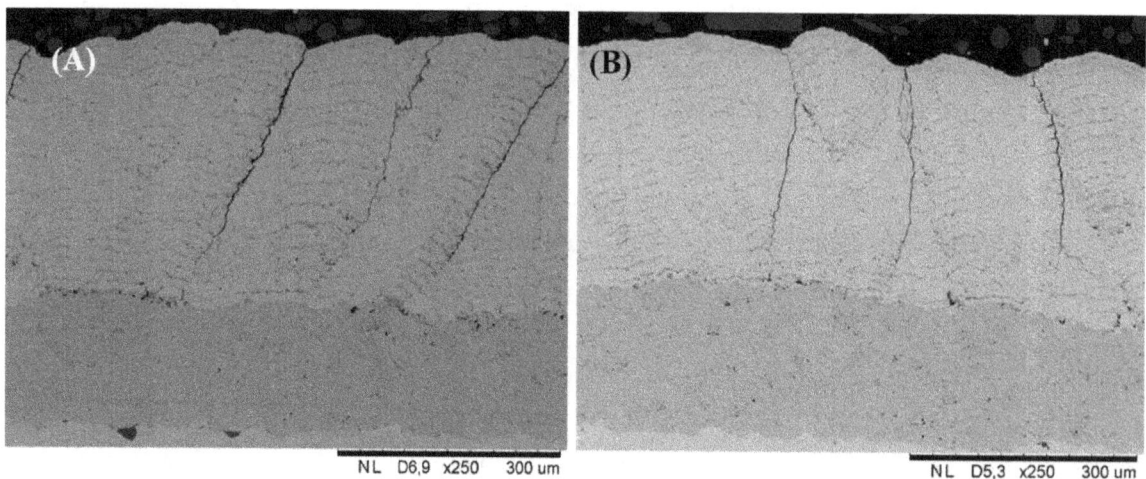

The differences in microstructure between coatings deposited on HVOF bond coats (see Figure 2) *versus* APS bond coats (see Figure 3) are due to the surface roughness differences of the bond coats. As coating build-up is influenced by the asperities on the substrate, coatings (P1 or P2) that formed on a rough surface, such as a plasma-sprayed bond coat, will form wider and less uniform columns than those (H1 and H2) on a smoother surface, such as an HVOF bond coat. It can be observed that the Type 1 coatings (H1 and P1) have more pronounced layers within the cross-section that are separated by zones of higher porosity, which are attributed to the overspray between deposited layers. This overspray results from the deposition of particles treated in the plasma periphery that land ahead of and behind the material within the plume core as the plasma moves over the substrate. Because the particles in the plume periphery experience less heating from the plasma and more cooling from the atmosphere, they tend to deposit in a partially-molten or solid state, which reduces spreading during impact and produces higher porosity zones. Hence, the lower surface speed during deposition is primarily responsible for the presence of more distinct inter-pass porosity layers in the Type 1 coatings, as compared to the Type 2 coatings, which were produced with roughly twice the surface speed.

The microstructure of a Type 3 (A3) coating produced with the sub-micron suspension is shown in Figure 4. Compared to the nano-suspension coatings, the samples were composed of significantly narrower columns with a higher degree of branching along their length. The sub-micron suspension displays a lower viscosity than the nano-suspension used for the Type 1 and 2 samples. This fact results in smaller droplets generated during suspension break-up. The resulting smaller spray particles are more susceptible to plasma flow changes close to the substrate surface, which enhances the formation of columnar structures in the coating. Furthermore, like the Type 1 and 2 coatings, the A3 coatings were influenced by the surface topography of the substrate due to the build-up of material on asperities. The HVAF bond coat used for the A3 coatings has a lower roughness than either the HVOF or APS bond coats, meaning that the surface asperities are more closely spaced, which results in more columns per unit surface area relative to the other coatings.

**Figure 4.** HVAF bond coat with Type 3 SPS coating (A3).

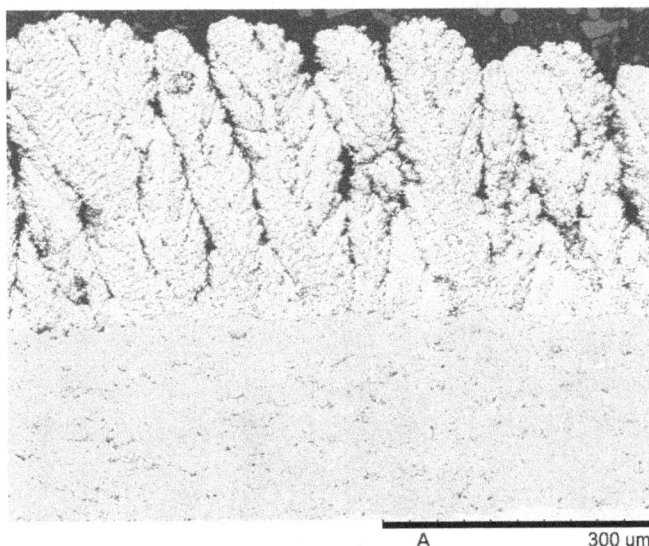

The reference APS DVC microstructure can be seen in Figure 5. As is common within these coatings, vertical cracks (indicated by arrows) are shown along the length within the coating cross-section. Additionally, the area between the cracks is characterized by isolated pores and regions of dense microstructure.

**Figure 5.** Dense vertically cracked APS coating with vertical cracks arrowed.

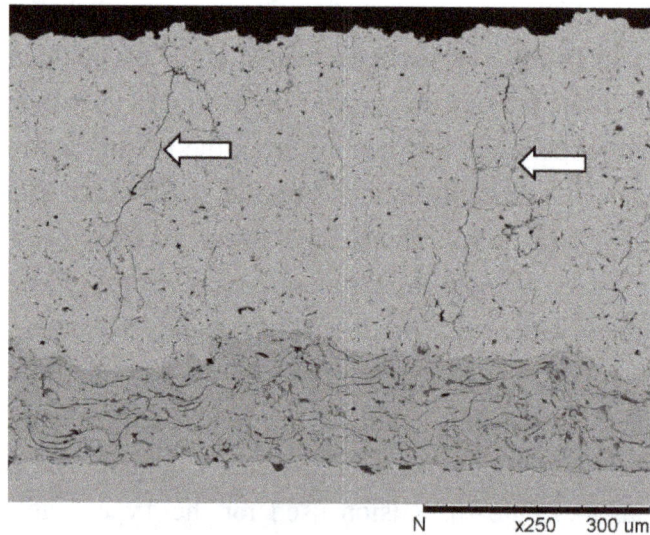

## 4.2. Thermal Conductivity

Thermal conductivity results are shown in Figure 6. Trends in the coating thermal conductivity can be related to the microstructures shown in Figures 2–5. In comparison with the more porous SPS coatings, the APS DVC coating displays a higher thermal conductivity in the as-sprayed state. While a porous APS coating may have an as-produced thermal conductivity in the region 0.5–1.0 ($W \cdot m^{-1} \cdot K^{-1}$) depending on the microstructure [5], the thermal conductivity of a DVC coating is 2–4-times higher due to its more dense structure. This conductivity increase displays the trade-off required in conventional APS spraying in order to produce a strain-tolerant coating.

**Figure 6.** Thermal conductivity of the coating systems.

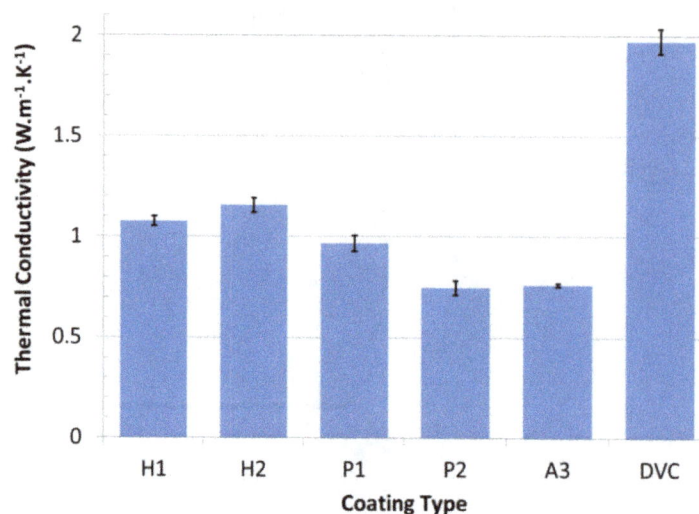

The difference in thermal properties amongst the SPS samples with different bond coats can also be connected to expected and observable microstructural differences. The coatings on plasma-sprayed bond coat (P1 and P2) show lower thermal conductivity than their HVOF counterparts (H1 and H2). This difference is consistent with the more irregular structure of the SPS YSZ on the rougher plasma-sprayed bond coat surface, increasing the coating porosity. However, because the nano-suspension was used for both the H and P coatings, the depositing particle sizes were comparable, which resulted in a similar porosity orientation between these sample sets (Figures 2 and 3). The thermal conductivity of the P2 samples was lower than that of the P1 samples. The P2 samples were comprised of more layers than the P1 samples, due to the 1.6-times faster surface speed relative to that used for the P1 samples. Therefore, the P2 samples contained more overspray porosity zones, which were oriented roughly perpendicular to the heat flow during the laser flash testing, which would act as barriers to heat transport. The same trend is not observed in the H1 and H2 samples, though there is no statistically significant difference in thermal conductivity between the two coating types. The lack of expected variation may again be related to the bond coat influence on the microstructure. The smoother HVOF interface produces less variability in coating porosity due to the different surface speeds used for spraying.

The sub-micron suspension used for the A3 samples produced a porosity distribution that differed from that of the Type 1 and 2 samples. The A3 samples exhibited wider inter-columnar gaps and a higher number of these gaps per cross-sectional area. This porosity, which is orientated parallel to the primary direction of heat flow, does not lower the thermal conductivity as significantly as that which is oriented perpendicular to the heat flow. Consequently, although the A3 coatings contained a significantly higher amount of porosity, these samples display the same level of thermal conductivity as the P2 coatings.

Thermal conductivity *versus* temperature data is displayed in Figure 7 for the A3 sample. The full line represents the as-sprayed thermal properties; whereas the dashed line represents the thermal properties after several hours at high temperature. It can be noted that, as with an APS coating, there is some increase in thermal conductivity after high temperature exposure, due to sintering of the microstructure after several hours at high temperature [9]. The degree of microstructural change is expected to be consistent with Stage 1 sintering, as defined by Cernuschi *et al.* [7], which occurs during the first hours of exposure at high temperatures. During this time, crack healing and bridging of the microstructure occurs, resulting in improved thermal conduction through the coating.

*4.3. Thermo-Cyclic Fatigue Lifetime and Failure Microstructures*

The results of TCF sample testing are shown in Figure 8. The standard deviation of each sample set is shown by the error bars. It can be seen that large differences in thermal fatigue life were achieved with different bond coat systems. Samples with HVOF bond coats (H1 and H2) exhibited less than half the lifetime of samples with the same top coat type on air plasma-sprayed bond coats (P1 and P2). Similar results for APS *versus* HVOF bond coats have been reported previously with conventional APS top coats [6].

**Figure 7.** Thermal conductivity *versus* temperature for coating A3 in the as-produced state and after one heat cycle.

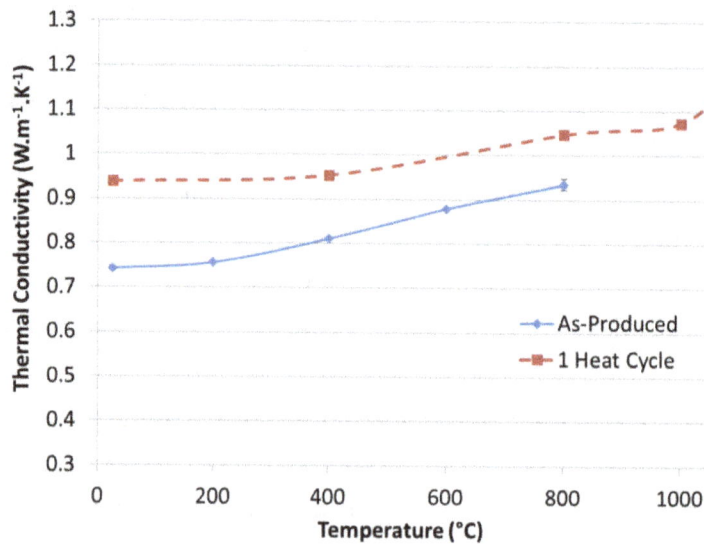

**Figure 8.** Thermo-cyclic fatigue lifetime of experimental coatings.

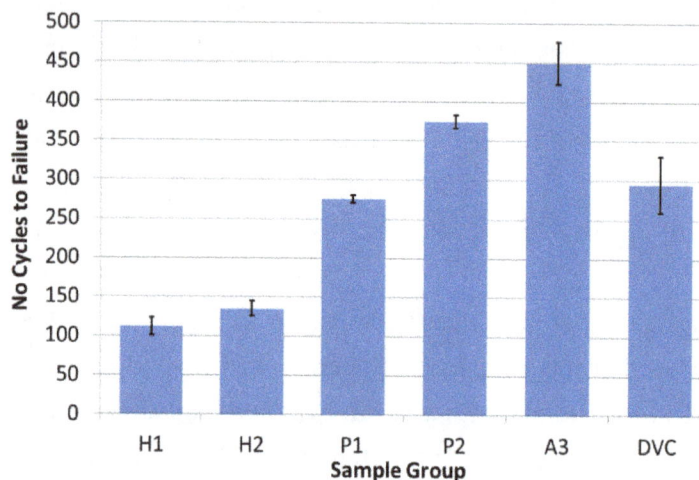

For all of the coating combinations tested, failure resulted from transverse crack propagation that could be connected to stress produced by a thermally-grown oxide (TGO) layer at the bond coat-top coat interface [28,29]. This layer is an unavoidable by-product during the high temperature exposure of YSZ-MCrAlY TBCs in atmospheric conditions and results from oxygen diffusing to and reacting with the bond coat material. Thus, the lifetime of a YSZ-MCrAlY system in TCF testing was determined by a combination of the oxygen diffusion rate to the bond coat, the reaction rate at the bond coat, the top coat strain tolerance and the ease of crack propagation along the top coat-TGO interface.

Initially, the rate of oxygen reaching the bond coat is determined by gas infiltration into the cracks and pores of and lattice diffusion through the YSZ top coat. As the TGO develops, oxygen must then diffuse through this layer to react with the bond coat; thus, the TGO composition can affect bond coat oxidation. For example, a reaction between the diffusing oxygen and the aluminum of the bond coat can form an alumina layer that produces a slower growing oxide layer than reactions yielding mixed oxides consisting of nickel, chromia spinels, which are considered highly detrimental to coating lifetime [28]. Hence, the bond coat chemistry also influences TGO development by determining which oxides form. For instance, a β Ni-Al phase in the bond coat acts as an aluminum reservoir that promotes the formation of a passivating alumina layer, so that the beta phase content can be tracked and used to determine the degree of bond coat oxidation and the remaining chemical life of the bond coat [28]. A slower growing, passivating oxide is preferred, because the formation of any TGO introduces stress into the TBC system. The greater the capacity of the top coat to accommodate this stress, the better the TCF performance of the TBC tends to be [6]. Once the TGO-induced stress exceeds a critical level, cracks will be initiated and propagate along the lowest energy pathway. The planarity of the interface may influence the rate of propagation; less planar interfaces generate higher total stresses, but make crack propagation more difficult.

Considering the above factors, the lifetime disparity between H and P samples was likely to have resulted mainly from bond coat topography differences. As mentioned previously, the P samples contained more porosity than the H samples, but, given the similarities in the coating cross-sections, the higher porosity in the P samples would not be expected to improve the strain tolerance enough to account for the majority of the ~2.5 increase in the average TCF lifetime when comparing the H samples to the P samples. Therefore, while porosity may have contributed to the TCF lifetime differences between the H and P samples, the dominant factor was most probably that the TGO-induced cracking required more energy to propagate along the interface structure produced by the plasma-sprayed bond coat (P samples), because it was significantly rougher than the interface structure producing the HVOF bond coat (H samples).

The plasma-sprayed bond coat data also show a larger difference in lifetime between the Type 1 and Type 2 SPS top coats. This can be explained firstly by the greater difference in microstructure generated by changing surface speed when depositing coatings on an APS bond coat. Additionally, the P2 coatings have exhibited lower thermal properties than P1 coatings; this could insulate the P2 samples further during thermal transient stages in the TCF cycle. The lack of a significant lifetime difference between the H1 and H2 samples is again related to the influence of the bond coat on the coating microstructure. As the HVOF bond coat induces little variability in microstructure between Type 1 and 2 coatings, there is similarly little difference in their TCF lifetimes.

The TCF lifetime of A3 samples with a HVAF bond coat are also shown on the right of the graph in Figure 8. The A3 samples featured a different powder composition (AMDRY 386) for the HVAF bond coat than that used for the HVOF and plasma spray bond coats (AMDRY 365). The suspension used for the A3 top coat utilized a different particle size distribution than that sprayed to produce the H and P samples. Both of these changes likely helped to improve the TCF lifetime. As stated above, the AMDRY 386 bond coat chemistry may reduce the oxidation rate and formation of spinel oxides relative to the AMDRY 365 bond coats. Additionally, the sub-micron suspension used for the A3 top coat produced a more porous columnar and, thus, strain-tolerant microstructure than that resulting from the nano-suspension used to generate the H and P samples.

The TCF results for the reference DVC APS samples are shown on the far right of the graph in Figure 8. The DVC samples used the same AMDRY 386 bond coat composition as was used for the A3 samples, but the bond coat was applied via plasma spray instead of HVAF. The TCF lifetime of the DVC samples was on par with that of the P1 samples and gives a useful benchmark for the TCF performance of current industry standard strain-tolerant plasma spray coatings.

The cross-section of a failed H1 sample is shown in Figure 9. This figure shows that failure was connected to cracks (labeled A) propagating along the interface between the thermally-grown oxide (TGO) layer and the bases of the columnar structures forming the top coat. In addition to this cracking, propagation also occurred through the inter-pass porosity at some locations. Due to the higher porosity of the inter-pass layers, they represent a low energy pathway for crack propagation.

The TGO layer consists of a dark alumina oxide layer (labeled B) at the bond coat interface that acts to prevent further oxidation. There is, however, a large amount of lighter colored mixed oxide regions above the alumina layer (labeled C). The mixed oxide consists of nickel, chromia spinels, and due to its faster growth, it is considered highly detrimental to coating lifetime [28]. Investigation of the HVOF bond coat after 100–140 cycles reveals that a substantial amount of the beta phase ($\beta$ Ni-Al) remained within the bond coat. This phase can be seen in Figure 9 as the darker regions within the bond coat layer indicated by the dashed lines and double arrow.

**Figure 9.** Microstructure of an H1 sample after approximately 120 thermo-cyclic fatigue cycles. A: Interface cracking; B: alumina thermally-grown oxide; C: mixed oxide.

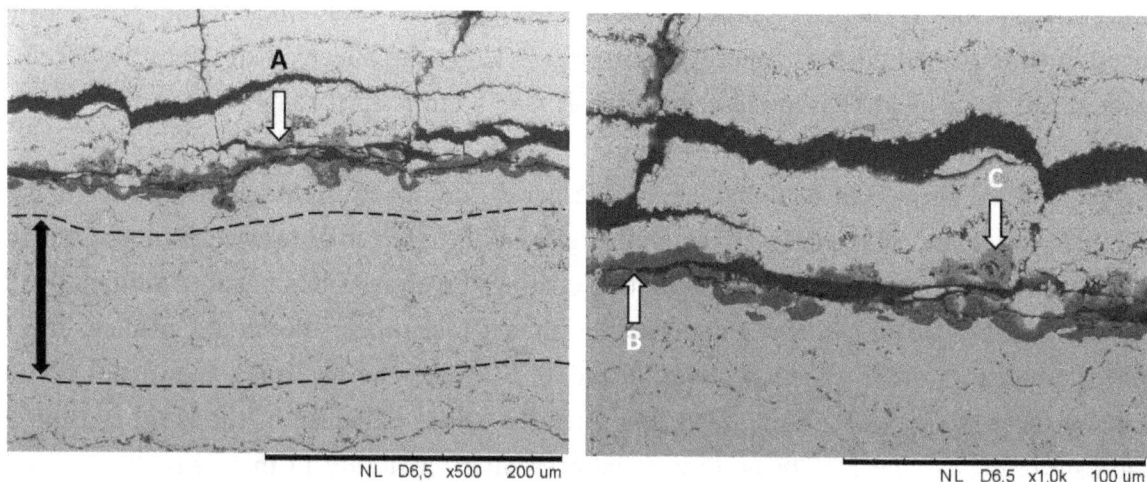

A cross-sectioned HVAF bond coat sample after failure is displayed in Figure 10. Like the HVOF bond coat samples, failure resulted from cracking at the TGO-top coat interface; though cracks propagating through inter-pass porosity were not observed, cracking within the ceramic layer above the TGO is indicated by the white arrow (labeled A). The remaining beta phase region is denoted by a double-headed arrow, with an approximate width of the region being 45 $\mu$m. As expected, there is less beta phase remaining compared to the HVOF bond coat sample; this decreased beta content is due to the four-times longer exposure time at high temperatures. Unlike the TGO layer in the HVOF coating, the HVAF TGO layer is denser in structure and contains very little of the mixed oxides found within the HVOF or plasma bond coats. In some locations, rumpling of the oxide layer has been observed (labeled B); this may contribute to crack initiation and failure [28]. The improved oxidation performance in TCF testing explains the longer cyclic lifetime. The reason for the improved performance may be related to the HVAF deposition itself or the different chemistry of the starting powder.

**Figure 10.** Failure microstructure of an A3 coating system after approximately 450 thermo-cyclic fatigue cycles. A: Interface cracking; B: Oxide rumpling.

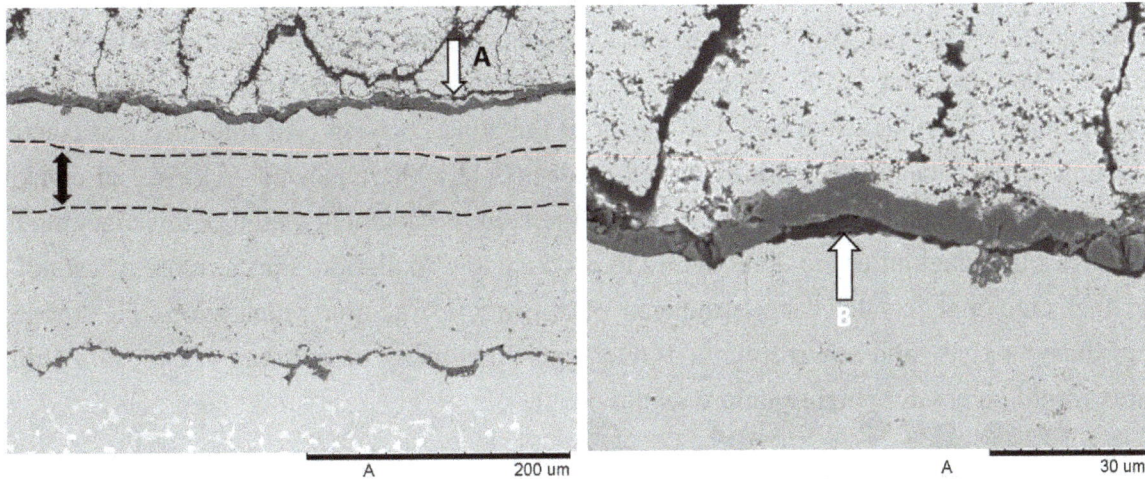

## 4.4. Thermal Shock Testing

The results for the thermal shock testing of the coatings are shown in Figure 11. It should be noted that in the case of all SPS coatings, the thermal shock test was discontinued with no observed failures, due to limitations on testing time. Previous testing experience with this thermal shock rig over several decades has shown an "average" lifetime for thin (~300 μm) TBC to be 1800 cycles. The maximum lifetime achieved has been 3000 cycles, though without the ability to replicate the result. As Figure 11 shows, the SPS coatings of this study have significantly greater thermal shock life than exhibited by APS systems tested on this rig. Long thermal shock lifetimes are corroborated by Guignard *et al.* [15] for vertically cracked SPS coatings at similar temperature levels. The results indicate that the structures produced by SPS praying are highly strain tolerant. Further testing against EB-PVD coatings would be required to show their ultimate performance.

**Figure 11.** Thermal shock testing data. The red dashed line indicates the number of cycles at which the test was halted.

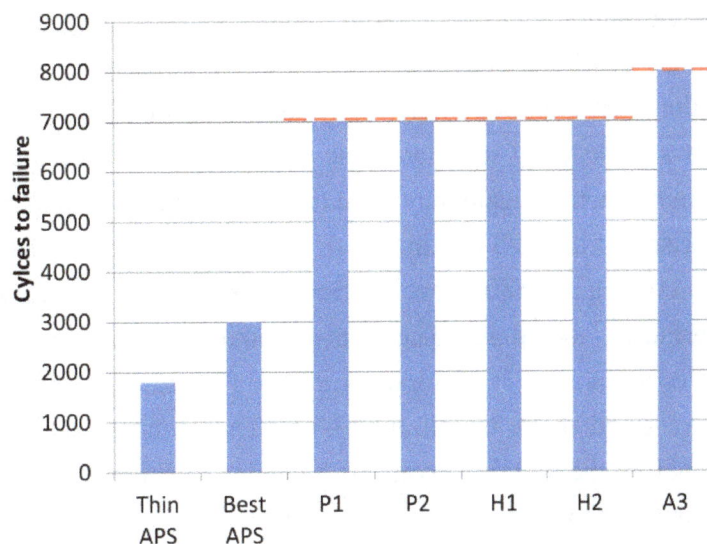

After testing was discontinued, exposed samples were sectioned and prepared for microstructural investigation. A cross-section of an H1 sample is displayed in Figure 12. The H1 microstructure can be considered representative of the H2 sample after the same amount of cyclic exposure. Deposits labeled A in the micrograph are not part of the coating and are in fact transferred iron-chrome oxides from the test rig fixture. The bond coat can be observed to contain two phases: the darker beta-phase and the lighter gamma phase. There is also a thin TGO layer at the bond coat-top coat interface that contains both an alumina layer (dark) and regions of mixed oxides (light). There is some evidence of cracking (labeled B) close to the TGO layer and at the edges of the columns/segments, though this cracking had yet to be significant enough after the 7000 cycles to produce any spallation. Furthermore, it cannot be discounted that cracking may have been induced during microstructural preparation. The minimal depletion of the beta-phase and relatively thin TGO suggest that many more hours of exposure at high temperatures would be possible before failure would occur.

**Figure 12.** H1 coating microstructure after 7000 thermal shock cycles. Deposits from A: the combustor burner; B: interface cracking.

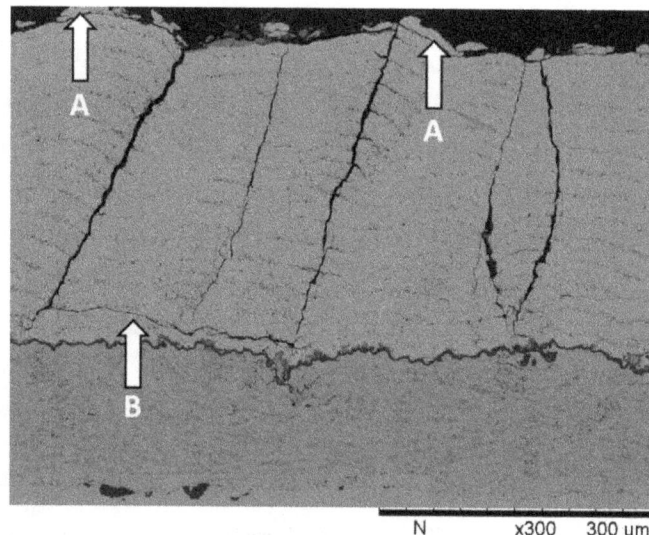

Figure 13 displays the microstructure of a P2 sample after approximately 7000 cycles. The sample is also representative of the P1 samples after the same number of test cycles. The growth of a thin TGO layer at the interface between bond coat and top coat can be observed. Additionally there has been growth of TGO along the inter-splat boundaries or delaminations internally within the bond coat. This P2 sample likewise shows evidence of some cracking (arrowed) in the inter-pass porosity layers at the edges of the columns/segments, which again may have occurred during metallographic preparation.

Figure 14 shows a comparison of the SPS coating close to the bond coat interface (left) and close to the top surface (right) for A3 coating after 8000 cycles. Beta phase zones are present within the HVAF bond coat close to the interface, (labeled A). The thin alumina layer can be seen at the interface between bond coat and top coat (labeled B). Unlike in in A3 TCF testing samples, the presence of mixed oxides (labeled C) was found above the alumina layer. These mixed oxides likely grow during early cycling of the coating before the alumina layer has a chance to completely form. Comparing the left and right images of Figure 14 shows that a difference in morphology exists between SPS coating porosity near the bond coat and that near the top surface. Pores close to the bond coat interface are irregularly shaped, while those close to the interface show a more rounded shape. This difference results from the temperatures close to the top surface being high enough to enable sintering. However, sintering in SPS coatings results in only minor changes in pore structure, primarily spheroidization of the pores and healing of boundaries and cracks. As the microstructure changes are not uniform throughout the coating, it can be expected that the impact of this sintering will be lower than that suggested in the thermal conductivity data of Figure 7, which corresponds to an isothermally exposed sample. The lower thermal conductivity increase for samples subjected to thermal gradient conditions indicates that the degradation in performance for real applications within a gas turbine will not be as severe as isothermal testing would suggest.

**Figure 13.** P2 coating after 7000 thermal shock cycles. Interface cracking arrowed.

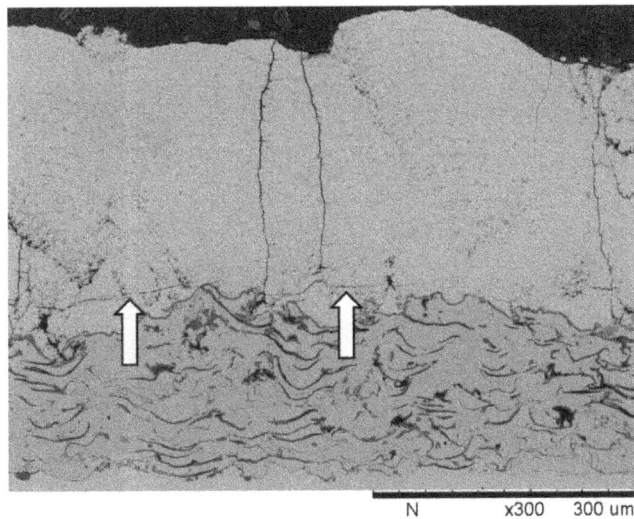

**Figure 14.** A3 coating microstructure after 8000 thermal shock cycles. (**A**) The bond coat interface is shown with: A: beta phase; B: alumina oxide TGO; C: mixed oxide; (**B**) The microstructure close to the surface is shown.

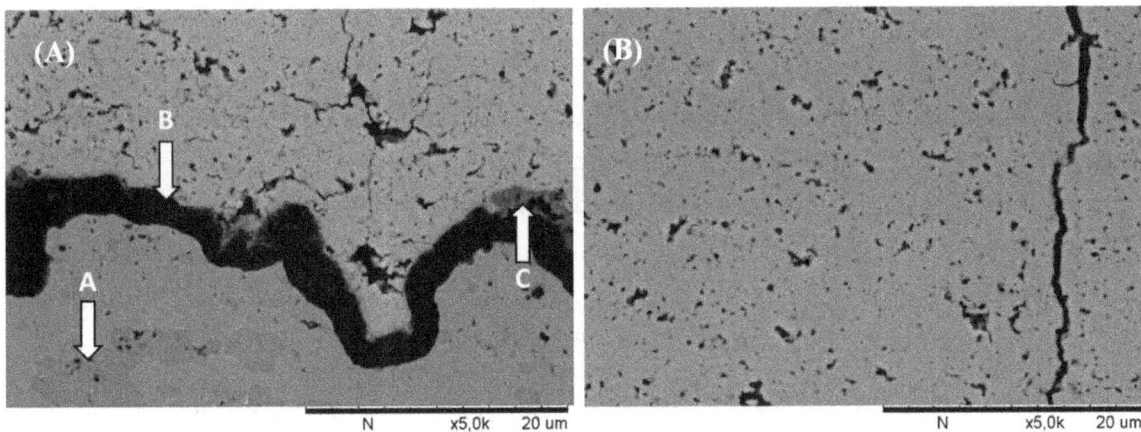

*4.5. Erosion Test Results*

Erosion test data is shown in Figure 15. Due to a limited number of samples, only A3 specimens were tested from amongst the SPS samples in this study. As a reference, the erosion rate of a porous

APS coating is shown along with the test data for an APS DVC sample. It can be seen that the SPS sample A3 erosion rate is over four-times lower than that of the porous APS coating. The erosion rate of the DVC coating is lower still, at approximately 60% of the A3 value. The microstructure of the coating A3 suggests it to be the most porous of the SPS coatings investigated here; thus, it is feasible that a denser SPS coating could have an erosion rate on par with a DVC coating. Cernuschi *et al.* [30] conducted a more detailed study comparing DVC structured APS coatings with EB-PVD coatings. Their work suggests that the erosion rates of EB-PVD and DVC structures are similar.

**Figure 15.** Erosion rate of coating A3 compared to dense vertically cracked and conventional porous APS ceramic thermal barrier coating (TBC) systems.

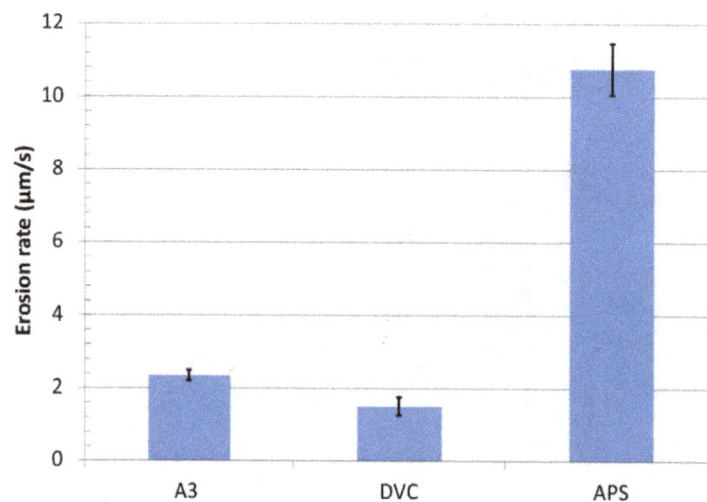

## 5. Conclusions

Dense vertically cracked coatings, as the existing "state-of-the-art" in strain-tolerant TBCs that can be applied under atmospheric conditions, were compared to YSZ samples produced by the emerging atmospheric coating technology: suspension plasma spray. The suspension plasma spray YSZ coatings demonstrated a number of important improvements over established plasma spray coatings in the tests of this study:

- The thermal conductivity of SPS coatings is in a similar range as that of conventional porous APS coatings and much lower than the competing DVC coatings;
- The thermo-cyclic fatigue lifetime for SPS coatings is greater than for conventional strain-tolerant coatings, and their thermal shock lifetime is an order of magnitude greater than conventional APS;
- The erosion resistance of the best performing SPS coating was only 40% lower than a DVC coating, bringing SPS into the same erosion rate as EB-PVD coating.

Overall, the performance of the SPS YSZ coatings shows promise as top coats in the thermal barrier systems for gas turbines. However the coatings presented in this article are presently still at the experimental stage. Further development is still required to gain better properties for SPS TBC as a whole system. In particular, more work is necessary to determine the appropriate bond coat for these SPS top coats. Furthermore, the correct balance of microstructural features (column morphology, column uniformity, intra-columnar porosity, *etc.*) in SPS coatings is not presently known. Lastly, confirmation of their viability as replacements to current TBC top coats requires directly comparing SPS TBC coatings to systems with EB-PVD ceramic layers.

## Acknowledgments

Special thanks are given to Stefan Björklund for assistance with the HVAF bond coat production. Thanks are also given to Lars Östergren and Nicholas Erb at GKN Aerospace Engine Systems for help with the thermal shock rig and access to the HVOF production equipment. Further thanks are given to Wyszomir Janikowski for assisting with sample testing and evaluation. Additional thanks are given to Toni Bogdanoff, Jönköping University, for assistance with the thermal property evaluation.

## Author Contributions

Nicholas Curry and Nicolaie Markocsan designed the experiments. Bond coat deposition was supervised by Nicholas Curry. Kent VanEvery and Todd Snyder performed the SPS coating experiments. The microstructure preparation, thermo-cyclic fatigue testing, SEM analysis, thermal diffusivity measurement and data analysis was carried out by Nicholas Curry. The paper was written by Nicholas Curry and Kent VanEvery.

## Conflicts of Interest

The authors declare no conflict of interest.

## References

1.  Soechting, F.O. A design perspective on thermal barrier coatings. *J. Therm. Spray Technol.* **1999**, *8*, 505–511.
2.  Mutasim, Z.; Brentnall, W. Thermal barrier coatings for industrial gas turbine applications: An industrial note. *J. Therm. Spray Technol.* **1997**, *6*, 105–108.
3.  Miller, R. Thermal barrier coatings for aircraft engines: History and directions. *J. Therm. Spray Technol.* **1997**, *6*, 35–42.
4.  Schulz, U.; Leyens, C.; Fritscher, K.; Peters, M.; Saruhan-Brings, B.; Lavigne, O.; Dorvaux, J.-M.; Poulain, M.; Mévrel, R.; Caliez, M. Some recent trends in research and technology of advanced thermal barrier coatings. *Aerosp. Sci. Technol.* **2003**, *7*, 73–80.
5.  Curry, N.; Markocsan, N.; Li, X.-H.; Tricoire, A.; Dorfman, M. Next generation thermal barrier coatings for the gas turbine industry. *J. Therm. Spray Technol.* **2010**, *20*, 108–115.
6.  Curry, N.; Markocsan, N.; Östergren, L.; Li, X.-H.; Dorfman, M. Evaluation of the lifetime and thermal conductivity of dysprosia-stabilized thermal barrier coating systems. *J. Therm. Spray Technol.* **2013**, *22*, 864–872.
7.  Cernuschi, F.; Lorenzoni, L.; Ahmaniemi, S.; Vuoristo, P.; Mäntylä, T. Studies of the sintering kinetics of thick thermal barrier coatings by thermal diffusivity measurements. *J. Eur. Ceram. Soc.* **2005**, *25*, 393–400.
8.  Golosnoy, I.; Cipitria, A.; Clyne, T. Heat transfer through plasma-sprayed thermal barrier coatings in gas turbines: a review of recent work. *J. Therm. Spray Technol.* **2009**, *18*, 809–821.

9.  Curry, N.; Donoghue, J. Evolution of thermal conductivity of dysprosia stabilised thermal barrier coating systems during heat treatment. *Surf. Coat. Technol.* **2012**, *209*, 38–43.

10. Schulz, U.; Bernardi, O.; Ebach-Stahl, A.; Vaßen, R.; Sebold, D. Improvement of EB-PVD thermal barrier coatings by treatments of a vacuum plasma-sprayed bond coat. *Surf. Coat. Technol.* **2008**, *203*, 160–170.

11. Karger, M.; Vaßen, R.; Stöver, D. Atmospheric plasma sprayed thermal barrier coatings with high segmentation crack densities: Spraying process, microstructure and thermal cycling behavior. *Surf. Coat. Technol.* **2011**, *206*, 16–23.

12. Bengtsson, P.; Ericsson, T.; Wigren, J. Thermal shock testing of burner cans coated with a thick thermal barrier coating. *J. Therm. Spray Technol.* **1998**, *7*, 340–348.

13. Guo, H.B.; Kuroda, S.; Murakami, H. Segmented thermal barrier coatings produced by atmospheric plasma spraying hollow powders. *Thin Solid Films* **2006**, *506–507*, 136–139.

14. Kaßner, H.; Siegert, R.; Hathiramani, D.; Vaßen, R.; Stoever, D. Application of Suspension Plasma Spraying (SPS) for manufacture of ceramic coatings. *J. Therm. Spray Technol.* **2007**, *17*, 115–123.

15. Guignard, A.; Mauer, G.; Vaßen, R.; Stöver, D. Deposition and characteristics of submicrometer-structured thermal barrier coatings by suspension plasma spraying. *J. Therm. Spray Technol.* **2012**, *21*, 416–424.

16. VanEvery, K.; Krane, M.; Trice, R.; Wang, H.; Porter, W.; Besser, M.; Sordelet, D.; Ilavsky, J.; Almer, J. Column formation in suspension plasma-sprayed coatings and resultant thermal properties. *J. Therm. Spray Technol.* **2011**, *20*, 817–828.

17. Killinger, A.; Gadow, R.; Mauer, G.; Guignard, A.; Vaßen, R.; Stöver, D. Review of new developments in suspension and solution precursor thermal spray processes. *J. Therm. Spray Technol.* **2011**, *20*, 677–695.

18. Bacciochini, A.; Ilavsky, J.; Montavon, G.; Denoirjean, A.; F, B.; Valette, S.; Fauchais, P.; Wittmann-teneze, K. Quantification of void network architectures of suspension plasma-sprayed (SPS) yttria-stabilized zirconia (YSZ) coatings using Ultra-small-angle X-ray scattering (USAXS). *Mater. Sci. Eng. A* **2010**, *528*, 91–102.

19. Stuke, A.; Kaßner, H.; Marqués, J.-L.; Vaßen, R.; Stöver, D.; Carius, R. Suspension and air plasma-sprayed ceramic thermal barrier coatings with high infrared reflectance. *Int. J. Appl. Ceram. Technol.* **2012**, *9*, 561–574.

20. Fazilleau, J.; Delbos, C.; Rat, V.; Coudert, J.F.; Fauchais, P.; Pateyron, B. Phenomena involved in suspension plasma spraying part 1: Suspension injection and behavior. *Plasma Chem. Plasma Process.* **2006**, *26*, 371–391.

21. Fauchais, P.; Rat, V.; Coudert, J.-F.; Etchart-Salas, R.; Montavon, G. Operating parameters for suspension and solution plasma-spray coatings. *Surf. Coat. Technol.* **2008**, *202*, 4309–4317.

22. Pawlowski, L. Suspension and solution thermal spray coatings. *Surf. Coat. Technol.* **2009**, *203*, 2807–2829.

23. Delbos, C.; Fazilleau, J.; Rat, V.; Coudert, J.F.; Fauchais, P.; Pateyron, B. Phenomena involved in suspension plasma spraying. Part 2: Zirconia particle treatment and coating formation. *Plasma Chem. Plasma Process.* **2006**, *26*, 393–414.

24. Rampon, R.; Marchand, O.; Filiatre, C.; Bertrand, G. Influence of suspension characteristics on coatings microstructure obtained by suspension plasma spraying. *Surf. Coat. Technol.* **2008**, *202*, 4337–4342.

25. Berghaus, J.O.; Bouaricha, S.; Legoux, J.-G.; Moreau, C.; Chráska, T. Suspension Plasma Spraying of Nano-Ceramics using and Axial Injection Torch. In Proceedings of the International Thermal Spray Conference, Basel, Switzerland, 2–4 May 2005.

26. Vaßen, R.; Cernuschi, F.; Rizzi, G.; Scrivani, A.; Markocsan, N.; Östergren, L.; Kloosterman, A.; Mevrel, R.; Feist, J.; Nicholls, J. Recent activities in the field of thermal barrier coatings including burner rig testing in the European Union. *Adv. Eng. Mater.* **2008**, *10*, 907–921.

27. Taylor, R.E. Thermal conductivity determinations of thermal barrier coatings. *Mater. Sci. Eng. A* **1998**, *245*, 160–167.

28. Evans, H.E. Oxidation-induced stresses in thermal barrier coating systems. In *Advanced Ceramic Coatings and Interfaces V: Ceramic Engineering and Science Proceedings*; Zhu, D., Lin, H.-T., Mathur, S., Ohji, T., Eds.; John Wiley & Sons, Inc.: Hoboken, NJ, USA, 2010; Volume 31, doi:10.1002/9780470943960.ch3.

29. Hille, T.S.; Turteltaub, S.; Suiker, A.S.J. Oxide growth and damage evolution in thermal barrier coatings. *Eng. Fract. Mech.* **2011**, *78*, 2139–2152.

30. Cernuschi, F.; Lorenzoni, L.; Capelli, S.; Guardamagna, C.; Karger, M.; Vaßen, R.; von Niessen, K.; Markocsan, N.; Menuey, J.; Giolli, C. Solid particle erosion of thermal spray and physical vapour deposition thermal barrier coatings. *Wear* **2011**, *271*, 2909–2918.

# 8

# Noise Reduction Properties of an Experimental Bituminous Slurry with Crumb Rubber Incorporated by the Dry Process

**Moisés Bueno** [1,2,*]**, Jeanne Luong** [1,3]**, Fernando Terán** [1]**, Urbano Viñuela** [1]**, Víctor F. Vázquez** [1] **and Santiago E. Paje** [1]

[1]  Laboratory of Acoustics Applied to Civil Engineering (LA[2]IC), Universidad de Castilla-La Mancha, Avda. Camilo José Cela s/n, 13071 Ciudad Real, Spain; E-Mails: fernando.teran@uclm.es (F.T.); urbano.vinuela@uclm.es (U.V.); victoriano.fernandez@uclm.es (V.F.V.); santiago.exposito@uclm.es (S.E.P.)

[2]  Road Engineering Laboratory, Empa, Swiss Federal Laboratories for Material Science and Technology, Ueberladstr. 129, CH-8600 Duebendorf, Switzerland

[3]  Environmental Sciences and Technologies Department, University of Liège, Passage des Déportés 2, 5030 Gembloux, Belgium; E-Mail: jeanne.luong@ulg.ac.be

*  Author to whom correspondence should be addressed; E-Mail: moises.bueno@empa.ch

**Abstract:** Nowadays, cold technology for asphalt pavement in the field of road construction is considered as an alternative solution to conventional procedures from both an economic and environmental point of view. Among these techniques, bituminous slurry surfacing is obtaining an important role due to the properties of the obtained wearing course. The functional performance of this type of surfaces is directly related to its rough texture. Nevertheless, this parameter has a significant influence on the tire/road noise generation. To reduce this undesirable effect on the sound performance, new designs of elastic bituminous slurries have been developed. Within the FENIX project, this work presents the acoustical characterization of an experimental bituminous slurry with crumb rubber from wasted automobile tires incorporated by the dry process. The obtained results show that, under controlled operational parameters, the close proximity sound levels associated to the experimental slurry are considerably lower than those emitted by a conventional slurry wearing course. However, after one year of supporting traffic loads and different weather conditions, the evaluated bituminous slurry, although it conserves the original noise reduction properties in relation to the conventional one, noticeably increases the generated sound emission. Therefore, it is required to continue improving the design of experimental surfaces in order to enhance its long-term performance.

**Keywords:** bituminous slurry surfacing; acoustic characterization; crumb rubber; dry process; noise reduction properties; tire/road close proximity; close proximity (CPX) method

## 1. Introduction

Cold microsurfacing and slurry surfacing are considered as one of the efficient pavement technology from both economic and ecological points of view, mainly due to their single application at low temperature. They consist of a mixture of high-quality mineral aggregate and bitumen emulsion stabilized by the addition of an emulsifier which is uniformly spread in a thin layer over a properly prepared surface. The difference between the two designs is that whereas slurry surfacing uses slightly smaller aggregates, microsurfacing can incorporate polymer modified bitumen in order to improve the cohesion of a mixture with aggregates up to 10 mm [1]. Traditionally, bituminous slurry has been widely used as a maintenance and rehabilitation treatments for road surfaces without structural defects to prolong the service life of the pavements, filling cracks and sealing against atmospheric agents preventing water intrusion, with a fast application process and minimal influence on the traffic. Furthermore, after improvements over several years, this cold technology is currently even used as a wearing course on roads with lower traffic requirements, offering worthy technical properties such as a high skid-resistance owing to its rough texture [2–4]. Nevertheless, the texture of a slurry wearing course is turning into a problem presenting a negative effect on the acoustic behavior, increasing impacts and vibrational mechanisms generated in the contact between tire and surface and, therefore, the emitted noise [5–8]. One solution which is being researched in order to reduce the so-called tire/road noise, is the development of more elastic surfaces to minimize the impact effect through the incorporation of crumb rubber (CR) from wasted automobile tires. In the last decades, various technical and scientific studies have reported that the addition of CR and other polymers improves mechanical properties of the road surface, such as permanent deformation, stiffness or reflective and thermal cracking [9–13]. Usually, the incorporation of CR as modifier of the bituminous binder, via the wet process, has reflected an improvement of the acoustic performance of the asphalt roads, obtaining a decreasing of the generated sound levels in relation to the percentage of rubber employed in the binder and/or the maximum aggregate size (macrotexture) used in the mixture [14–17]. Another procedure to incorporate the CR is via the dry process, partially replacing the mineral aggregates in the grading and mixing prior to the addition of the bituminous binder [11,16,18]. Different designs, such as Rubit™ in Sweden and Plusride™ in the USA, have presented promising results reflecting a good sound behavior [19,20]. Although the dry process presents some advantages in relation to the wet process, concerning the cost involved and the higher amount of rubber to be used [21,22], the research has been concentrated mainly on the wet process. This choice could be explained by the irregular performance along its service life of some experimental sections with CR introduced by the dry process [23].

The present work summarizes the main results of the acoustical characterization of a test section with a new experimental design of a bituminous slurry with CR incorporated by the dry process. The study has been carried out by the Laboratory of Acoustics Applied to Civil Engineering (LA$^2$IC) of the University of Castilla-La Mancha (UCLM) through close proximity (CPX) measurements with the semi-anechoic chamber Tiresonic Mk4-LA$^2$IC [16] over the test section during its first year of service in order to evaluate its acoustical properties.

This experimental work was performed in the area of safety and comfort in asphalt roads within the FENIX Project framework ("Strategic Research on Safer and More Sustainable Road") that has been undertaken in Spain [24]. The FENIX Project is the greatest effort in the research and development of road paving made in Europe to date. It has been structured around the following main research lines: warm mixtures, perpetual pavements, recycling (cold and hot), by-products, safety and comfort, nanomaterials, low energy consumption plants and fluidized bed.

## 2. Experimental Set up

### 2.1. Test Section and Slurry Surface

After different previous laboratory works and analysis of the obtained results, ELSAN Company layered a conventional slurry wearing course of 6 mm maximum aggregate size [1] from siliceous nature with a bituminous emulsion type C60BP5 MIC (CE denomination), between the KP 54+700 and the KP 70+100 of the road CM-4106 in the region of Ciudad Real (Spain). In this road, the same slurry design with CR from waste tires as an experimental section of 400 m long around the KP 68+000 was layered. The GPS coordinates of the test track are shown in Figure 1. This experimental surface incorporates 7% of CR by weight of aggregates by the dry process. The particles of rubber in the experimental design present a size inside the range of 2/4 mm and they are incorporated at ambient temperature, so none chemical reaction happens, performing only as a different fraction of the mineral aggregates. Figure 2 shows representative pictures of the surface texture of the different mixtures at the test section.

**Figure 1.** Location (GPS coordinates) of the test section at the road CM-4106 (Spain) with the different road surfaces evaluated.

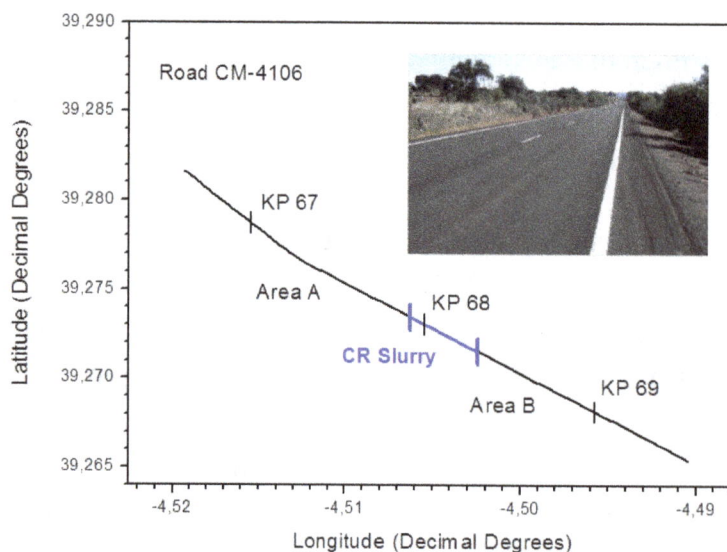

**Figure 2.** Representative photographs of the bituminous slurry wearing courses conventional (**a**) and experimental with crumb rubber (**b**).

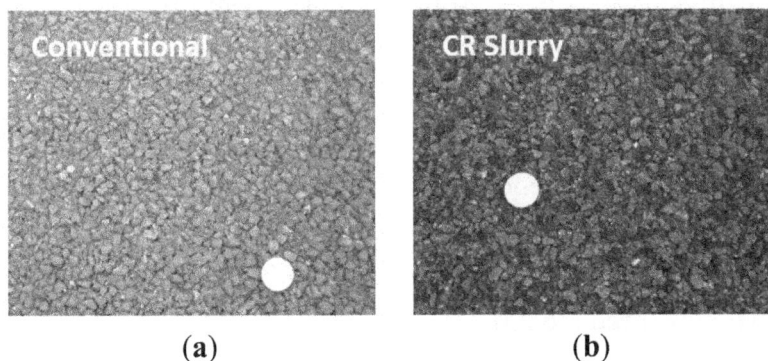

(**a**)                    (**b**)

The first acoustic geo-auscultation of the experimental track with CR slurry was carried out two months after its construction and opening to traffic. Moreover, in order to assess the effect of the supported traffic on the conservation of the experimental CR surface, and thus on its acoustic behavior, another characterization was carried out one year later. Regarding this analysis it is essential to remark that the test road supports a low-traffic volume of approximately 710 vehicles per day, with 5% of these being heavy vehicles (*Spanish General Road Office's database*).

In acoustic characterizations of different road surfaces, the pavement temperature is one important parameter to take into account to be able to compare their noise levels. The obtained sound performance of a same surface can vary in a ratio of 0.06 dB(A)/°C due to the influence of the surface temperature [25]. Hence, it is necessary to accomplish the acoustic characterizations at the same surface temperature or introduce a correction temperature factor for a proper comparison of the behaviors of different pavements or maintenance states. In this aspect, in the present work the two acoustic geo-auscultations were carried out at the same temperature, within a range of 17–18 °C; thus, the influence of the temperature on the analysis should be considered negligible.

## 2.2. Measurement Equipment

The measurements of the tire/road noise were carried out following the CPX methodology used in previous acoustic characterizations of road surfaces [16,25]. Trailer Tiresonic Mk4-LA$^2$IC (Figure 3) is made up of a semi-anechoic chamber which isolates tire/pavement sound from the external traffic or wind noises, in a frequency range of 300 to 4000 Hz where the traffic noise emission is focused. The CPX method provides a measure of tire/road sound levels ($L_{CPtr}$) in close proximity to the contact patch, and should provide insight into the acoustical characterization of asphalt pavements. The A-weighted pressure levels emitted by the rolling of a reference tire are continuously measured every 0.4 s by two BSWA MP201 1/2 in. microphones located close to the tire in the frequency range from 200 Hz to 16 kHz. The microphones placed inside the semi-anechoic chamber are located at a horizontal distance of 20 cm from the plane of the nearest tire sidewall and at a height of 10 cm above the road pavement surface. Front (FM) and rear (RM) microphones are positioned at angles of 45 and 135 deg to the rolling direction, respectively. A portable NI Compact Rio control and acquisition system with a four channel module and a cRio mobile module for global position determination are used to geo-register continuously the close proximity sound levels. In this study, the reference tire was a Pirelli P6000 205/55 R16 with an inflated pressure in cold conditions of 240 kPa. Before test measurements, the reference tire was warmed up by driving for more than 20 min and the sensitivity of the whole acoustic measurement set up was checked with an acoustic calibrator 4231 B&K before and after the measurements over the tested road surface. An optical tachometer is used during the test to measure the instantaneous vehicle speed. For this evaluation, the measurements were carried out at a reference speed of 80 km/h. Knowing that the rolling speed has a high influence on the emitted sound levels, the vehicle cruise control system was used to maintain automatically the steady reference speed. Moreover, in this case, as consequence of the operational difficulties to develop the specific measurements in order to obtain an associated speed constant factor (B) for the test surfaces, a general constant B = 30 has been taken to correct the possible speed deviations of the reference speed [16,25].

**Figure 3.** Tiresonic Mk4-LA$^2$IC for acoustic characterization of road surfaces and detail of the microphone positions in the close proximity of a reference tire inside the semi-anechoic chamber.

After one year, a second test of measurements was carried out in order to evaluate the evolution of the pavement. All the operational parameters were kept constant using in this analysis the same reference tire (with less than 100 km of working). In this second set of measurement the CPX sound levels were measured every 0.2 s.

As the texture is one of the most important parameters in the sound emission during the interaction between the tire and the surface of the road, geo-referenced measurements of macrotexture (wavelengths ranging from 0.5 mm to 50 mm) were also carried out to correlate them with the close proximity sound levels. The evaluation of the surface macrotexture was accomplished with a laser texture scanner system produced by Ames Engineering. The portable system is designed to measure *in situ* sections of pavement of 7 cm × 10 cm. The texture scanner is a stand-alone unit that can be placed on the surface on three point contact feet. This laser scanner allows the representative parameter MPD (Mean Profile Depth) to be measured [26] through the multiple lines scanning for the patch of surface directly under the scanner. To achieve a representative value of MPD, different measurements were carried out along the test segments, whereby global position coordinates were logged by means of an integrated GPS receiver.

## 3. Results and Discussion

### 3.1. Surface Texture

As previously it has been said, the macrotexture is a determinant factor that has a significant effect on tire/road sound emission. In this sense, this parameter mainly affects the vibrational mechanisms, increasing or decreasing vibrations generated by the contact between tire and surface, and indirectly it could influence on aerodynamic mechanisms throughout sound dispersion phenomena. Originally, this property is related to the maximum aggregate size used in the design of the asphalt mixture. However, the lay-down of the mixture and its compaction define the final profile of the surface. Afterwards, the supported traffic as well as the weather conditions will influence on its ageing and state of conservation. Therefore, before presenting any sound result, it is important to characterize the surface texture in order to compare the different slurry surfaces studied in this work. Table 1 presents the different MPD indexes associated to the surface with CR and the conventional surfaces (A and B) placed just before and after the experimental section. Besides, the parameters obtained after one year for these dense surfaces are also shown.

**Table 1.** Mean Profile Depth (MPD) index obtained from macrotexture measurements at different points on the analyzed test sections.

| MPD (mm) | Conventional Slurry A | CR Slurry | Conventional Slurry B |
|---|---|---|---|
| Initial | 1.25 | 2.70 | 1.50 |
| After one year | 1.20 | 2.80 | – |

Observing the results, it is important to remark that the experimental surface presents an unusually high value compared to a conventional bituminous slurry surfaces. This fact can be due to the randomly spreading and orientation of the rubber particles in and along the surface. Attending to the evolution of the texture after one year of service life, it is also unexpected that the different evaluated surfaces conserve their macrotextures, presenting quite constant MPD results what reflects the optimal performance of this type of surfaces under traffic requirements.

### 3.2. CPX Sound Levels

In this section, the results of the measurements carried out to evaluate the influence of the CR as part of the aggregate on the acoustic behavior of the experimental bituminous slurry are presented. First of all, the evolution of the instantaneous close proximity sound levels throughout the kilometric points of the test track at a reference speed of 80 km/h ($L_{CPtr,80 km/h}$ ($t$)) associated with the conventional and experimental segments are shown in Figure 4. The resulted sound profile of the slurry surface with CR can be compared with the profiles of the surfaces layered just before (A) and after (B) the experimental area. Regardless of the registered acoustical variability [27], it can be observed that the experimental CR slurry presents lower close proximity sound levels that the conventional design. Attending to the surfaces placed before and after the experimental section, it is reflected that despite being the same bituminous design, the emitted sound levels seem to be different. Usually, this difference can be due to slight differences in construction or state of conservation. In this case, it is relevant to remark that the section after the rubberized experimental area coincides with the access to the residence area of the closer town what implies a reduction on the driving speed. These deceleration zones have to support different traffic loads and thus, they can be worn in a different scale. Considering this fact, in this study the conventional surface before the experimental section (A) has been selected as the reference one.

**Figure 4.** Tiresonic Mk4-LA$^2$IC for acoustic characterization of road surfaces and detail of the microphone positions in the close proximity of a reference tire inside the semi-anechoic chamber.

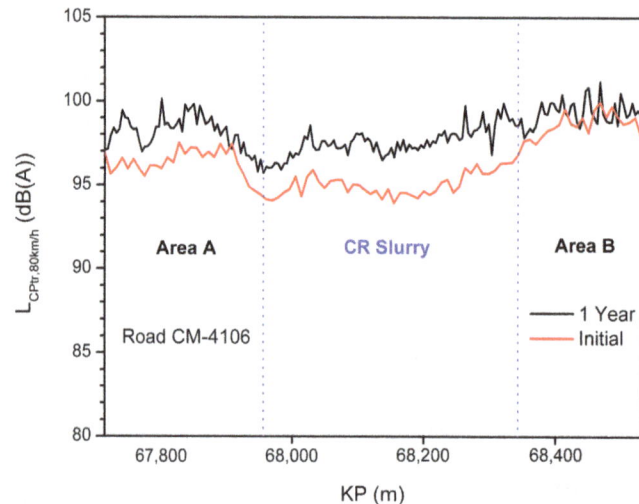

To make easier a general analysis, Table 2 summarizes the characteristic close proximity level ($L_{CPtr,80\ km/h}$), the acoustical variability ($\sigma$) and the total range of variation ($\Delta$) of each evaluated segment. Whereas the profiles shown in Figure 4 correspond to single measurements, these characteristics parameters are obtained as average from different measurements.

**Table 2.** Characteristic close proximity level ($L_{CPtr,80\ km/h}$), acoustical variability ($\sigma$) and total range of variation ($\Delta$) associated to the studied surfaces.

| $v_{ref}$ **80 km/h** | **Initial** | | | **After One Year** | | |
|---|---|---|---|---|---|---|
| | $L_{CPtr}$ **dB(A)** | $\sigma$ **dB(A)** | $\Delta$ **dB(A)** | $L_{CPtr}$ **dB(A)** | $\sigma$ **dB(A)** | $\Delta$ **dB(A)** |
| Conventional slurry A (reference) | 96.6 | 0.6 | 2.0 | 98.3 | 0.8 | 3.1 |
| CR Slurry | 95.2 | 0.8 | 2.6 | 97.3 | 0.8 | 3.0 |
| Conventional slurry B | 98.2 | 0.9 | 3.3 | 99.4 | 0.7 | 3.1 |

Analyzing the characteristic acoustic parameters, it can be seen that, two months after its construction, the experimental surface with CR as part of aggregates generates on average 1.4 dB(A) less than a conventional slurry design. This difference can reach values higher than 2.5 dB(A) in some evaluated points. Although a similar effect was already obtained for gap-graded asphalt mixtures [16], now this fact confirms that slurry surfaces, which usually present poor acoustical performance due to its rough texture, can reduce the sound emission incorporating crumb rubber particles, even when the measured texture is higher.

The acoustical variability ($\sigma$) and the total range of variation ($\Delta$) are aspects to be considered in an acoustic assessment of pavement surfaces [16,27]. The standard deviations around the mean and the difference between the lowest and the highest close proximity sound levels quantify the longitudinal surface homogeneity remarking punctual differences that can be found along the test sections in order to indicate the suitability of the surface and its construction technique from acoustical standpoint. On the other hand, considering the temporal evolution, this information allows the identification of different states of maintenance of pavement surfaces. In this line, the experimental and the conventional surface present a variability less than 1.0 dB(A) and a total range of variation around 3 dB(A), even after one year of service. Attending to the results presented by Paje *et al.* [16], where different asphalt surfaces with CR were also studied, it can be observed that the evaluated bituminous slurries present a good acoustical homogeneity. However, the punctual differences along the test sections found in both studies show that the technique of incorporation of CR in asphalt mixtures and bituminous slurries by the dry process has to be improved.

Regularly, slurries as rolling surface are placed on roads which are not required to support heavy traffic conditions due to its special design. For this reason, a second set of measurements was carried out after one year in order to analyze the evolution of the studied surfaces and their acoustical properties. Figure 4 and Table 2 show the close proximity sound levels generated and the characteristic acoustical parameters associated to the different surfaces, respectively. The results registered after one year since the first acoustic characterization reveal that the different characteristic close proximity levels have significantly increased and the experimental rubberized slurry has lost sound reduction effect compared to the conventional surface. The original difference of 1.4 dB(A) has been reduced to 1.0 dB(A). The increase of the close proximity sound levels associated to the bituminous slurry with crumb rubber could be related to the loss of rubber particles from the surface [22], decreasing the damping effect over the impacts and vibrations. Nonetheless, the increase of the sound levels generated by the conventional surface and the conservation of the initial textures could be informing about the alteration of other surface properties which also influence on the acoustic performance. This fact is checked in the next section where the analysis of the sound spectra can give relevant information about the different generation mechanisms involved in the tire/road noise.

### 3.3. CPX Sound Spectra

In this section, In order to understand the mechanisms related to the interaction between the reference tire and the evaluated surfaces, the sound emission during the rolling process at 80 km/h was analyzed in 1/3-octave band between 200 Hz and 16 kHz. Analyzing these sound spectra, the influence of the two main groups of mechanisms on the sound generation can be determined, since impacts and vibration mechanisms are related to low and medium frequencies (below 1 kHz) and aerodynamic mechanisms are associated to higher frequencies (above 1 kHz). In this sense, Figure 5 presents the characteristic close proximity sound spectra associated to the experimental and the conventional surfaces. It can be observed that the incorporation of CR have no significant effect on sound levels at frequencies above 1 kHz, usually associated to air-pumping and other mechanisms related to the flow of air in and around the tread grooves of the tires, what was expected since both of them are dense surfaces. Nevertheless, noticeable differences are found in the range below 2 kHz, such as those found in other asphalt surfaces with CR added by the dry process [16]. In this area, related to vibrational mechanisms, the CPX sound levels emitted by the rubberized slurry are lower than those emitted by the conventional surface. Therefore, it can be concluded that the reduction of the impacts effect, due to crumb rubber incorporation, can be the main reason for the decrease in the sound emission generated over the experimental surface [28].

**Figure 5.** Comparison of the representative close proximity noise spectra of the different test sections for the reference speed ($v_{ref} = 80$ km/h).

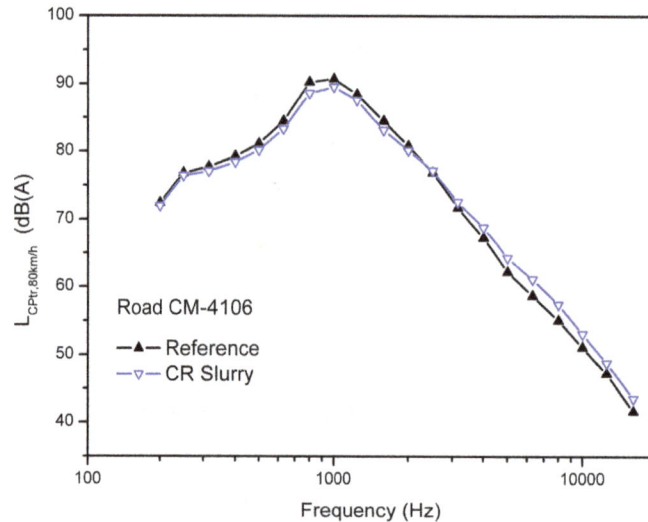

The evolution of the generation mechanisms after one year of service has been also analyzed. In this sense, in Figure 6 the first obtained close proximity sound spectrum associated to the surface with CR is compared to the spectrum obtained from the second set of measurements. Analyzing the figure, it can be observed that the shape of the spectra is quite similar, but after one year, the intensity of the close proximity levels has increased along the evaluated range of frequencies. This means that the ageing has effect on the different generation and propagation mechanisms. Previous studies [13] have reported that the addition of CR to asphalt mixes, especially by the dry process, increases their stiffness modulus. In this line, the obtained results indicate that it is possible that a higher stiffness due to ageing can be the surface property with more influence on the acoustic performance of the slurry surfaces evaluated in this study.

**Figure 6.** Time evolution of tire/road noise. Spectra at a reference speed of 80 km/h associated to the experimental bituminous slurry with crumb rubber.

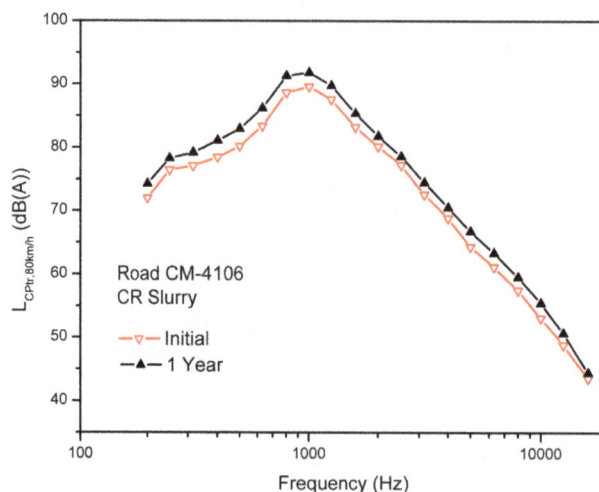

## 4. Conclusions

The analysis of the results obtained from the acoustic characterization of a new design of bituminous slurry, which incorporates crumb rubber by the dry process, as experimental wearing course has been presented in this work. Two different sets of measurements were carried out with the trailer Tiresonic Mk4-LA$^2$IC in order to evaluate the close proximity sound generated by the interaction between the reference tire and the rubberized slurry and its temporal evolution after one year of traffic service. The results show that the experimental slurry emits lower close proximity sound levels at 80 km/h compared to the conventional surface. Analyzing the mechanisms of generation by means of the associated sound spectra, it can be observed that, despite presenting a rougher surface macrotexture, the studied surface with crumb rubber has a considerable influence on the vibration mechanisms reducing the effect of the impacts. This aspect reflects that the improvement of the acoustic performance is related to the incorporation of crumb rubber since a more elastic wearing course is obtained. The second set of measurements carried out after one year indicates that the noise reduction properties of the rubberized slurry in relation to the conventional surface are kept quite constant, decreasing the original difference from 1.4 dB(A) to 1.0 dB(A). Nevertheless, it seems important to remark that the close proximity sound levels registered have significantly increased. Considering that the shapes of the associated sound generation spectra are quite similar and the macrotexture presents constant values, and that the different measurements were accomplished under controlled operational parameters (measurement equipment and temperature), it is possible that the increase of the sound levels is related to an increase of the surface stiffness due to an early ageing.

Thus, once the feasibility of the new design with crumb rubber as alternative to the conventional slurry surfacing has been checked, it is necessary to study in more detail its long-term performance and where to place these special surfaces. Furthermore, new experimental designs with varying size and content of crumb rubber, bituminous emulsions or even subgrades layers are required to investigate their influence in order to enhance the acoustic performance.

## Acknowledgments

The authors would like to thank to the Spanish Centre for the Development of Industrial Technology (CDTI) for its support in the development of the FENIX Project "Strategic Research on Safer and More Sustainable Roads" within the framework of the Ingenio 2010 program (www.proyectofenix.es) as well as to the Spanish Ministry of Economy and Competiveness for the project BIA 2012-32177 within the framework of the National Plan for Scientific Research through European Regional Development Funds.

## Author Contributions

Experimental measurements, analysis and interpretation of the results as well as conclusions have been conducted by all the co-authors. The manuscript has been written by M.B. with revision and approval by the others co-authors.

## Conflicts of Interest

The authors declare no conflict of interest.

# References

1. *UNE-EN 12273:2009 Slurry Surfacing Requirements*; European Committee for Standardization: Brussels, Belgium, 2009.

2. Smith, R.E.; Beatty, C.K. Microsurfacing usage guidelines. *Transp. Res. Rec.* **1999**, *1680*, 13–17.

3. Nebrada-Rodrigo, F.J.; Santos, J. Slurry seal and microsurfacing (Lechadas bituminosas y microaglomerados en frío). *Carreteras* **2005**, *4*, 78–87. (In Spanish)

4. Van Kirk, J. Multi-layer pavement preservation strategies using slurry surfacing and chip seals. In Proceedings of the 7th ISSA World Congress, Lyon, France, October 2010.

5. Sandberg, U.; Ejsmont, J.A. *Tyre/Road Noise Reference Book*; Informex: Kisa, Sweden, 2002.

6. Mun, S.; Cho, D.S.; Choi, T.M. Influence of pavement surface noise: The Korea highway corporation test road. *Can. J. Civil Eng.* **2007**, *34*, 809–816.

7. Ahammed, A.M.; Tighe, S.L. Pavement surface friction and noise: Integration into the pavement management system. *Can. J. Civil Eng.* **2010**, *37*, 1331–1340.

8. Wang, G.; Smith, G.; Shores, R. Pavement noise investigation on North Carolina highways: An on-board sound intensity approach. *Can. J. Civil Eng.* **2012**, *39*, 878–886.

9. Takallou, H.B.; Rayner, G.L.; Takallou, M.B. Development of design and construction guidelines for rubberized emulsified asphalt slurry seal. In Proceedings of the 4th ISSA World Congress, Paris, France, March 1997.

10. MacLeod, D.; Ho, S.; Wirth, R.; Zanzotto, L. Study of crumb rubber materials as paving asphalt modifiers. *Can. J. Civil Eng.* **2007**, *34*, 1276–1288.

11. Hernández-Olivares, F.; Witoszek-Schultz, B.; Alonso-Fernández, M.; Benito-Moro, C. Rubber-modified hot-mix asphalt pavement by dry process. *Int. J. Pavement Eng.* **2009**, *10*, 277–288.

12. Pasquini, E.; Canestrari, F.; Cardone, F.; Santagata, F.A. Performance evaluation of gap graded asphalt rubber mixtures. *Constr. Build. Mater.* **2011**, *25*, 2014–2022.

13. Moreno, F.; Rubio, M.C. Influence of crumb rubber on the indirect tensile strength and stiffness modulus of hot bituminous mixes. *J. Mater. Civil Eng.* **2012**, *24*, 715–724.

14. Miró, R.; Pérez-Jiménez, F.; Martínez, A.H.; Reyes, O.; Paje, S.E.; Bueno, M. Effect of using crumb rubber bituminous mixes on functional characteristics of road pavements. *Transp. Res. Rec.* **2009**, *2126*, 83–90.

15. Ahammed, A.M.; Tighe, S.L.; Klement, T. Quiet and durable pavements: Findings from an Ontario study. *Can. J. Civil Eng.* **2010**, *37*, 1035–1044.

16. Paje, S.E.; Bueno, M.; Terán, F.; Miró, R.; Pérez-Jiménez, F.; Martínez, A.H. Acoustic field evaluation of asphalt mixtures with crumb rubber. *Appl. Acoust.* **2010**, *71*, 578–582.

17. Freitas, E.F. The effect of time on the contribution of asphalt rubber mixtures to noise abatement. *Noise Control Eng. J.* **2012**, *60*, 1–8.

18. Pasetto, M.; Spinoglio, S. The use of a synthetic aggregate in non-skid and noise abatement microsurfacings for road maintenance. Performance improvements and environmental impact. In Proceedings of the 4th ISSA World Congress, Paris, France, March 1997.

19. Esch, D.C. Construction and benefits of rubber modified-asphalt pavements. *Transp. Res. Rec.* **1982**, *1680*, 13–17.

20. Amirkhanian, S.N.; Arnold, L.C. *A Laboratory and Field Investigation of Rubberized Asphalt Concretes Mixtures (Pelham road)*; Final Report No. FHWA-SC-93-02; South Carolina Department of Highways and Public Transportation: South Carolina, SC, USA, 1993.

21. Cao, W. Study on properties of recycled tire rubber modified asphalt mixtures using dry process. *Constr. Build. Mater.* **2007**, *21*, 1011–1015.

22. Moreno, F.; Rubio, M.C.; Martinez-Echevarría, M.J. The mechanical performance of dry-process crumb rubber modified hot bituminous mixes: The influence of digestion time and crumb rubber percentage. *Constr. Build. Mater.* **2012**, *26*, 466–474.

23. Emery, J. Evaluation of rubber asphalt demonstration projects. *Transp. Res. Rec.* **1995**, *1515*, 37–46.

24. Centre for the Development of Industrial Technology (CDTI). 2009. FENIX Project. Strategic Research on Safer and More Sustainable Roads. Available online: http://www.proyectofenix.es/ (accessed on 4 August 2014).

25. Bueno, M.; Luong, J.; Viñuela, U.; Terán, F.; Paje, S.E. Pavement temperature influence on close proximity tire/road noise. *Appl. Acoust.* **2011**, *72*, 829–835.

26. *EN ISO 13473–1:2004 Characterization of Pavement Texture by Use of Surface Profiles. Part 1: Determination of Mean Profile Depth*; European Committee for Standardization: Brussels, Belgium, 2004.

27. Paje, S.E.; Bueno, M.; Terán, F.; Viñuela, U. Monitoring road surfaces by close proximity noise of the tire/road interaction. *J. Acoust. Soc. Am.* **2007**, *122*, 2636–2641.

28. Biligiri, K.P.; Kalman, B.; Samuelsson, A. Understanding the fundamental material properties of low-noise poroelastic road surfaces. *Int. J. Pavement Eng.* **2013**, *14*, 12–23.

# 9

# Performance Assessment of Low-Noise Road Surfaces in the Leopoldo Project: Comparison and Validation of Different Measurement Methods

**Gaetano Licitra** [1,2,*], **Mauro Cerchiai** [3], **Luca Teti** [2,4], **Elena Ascari** [4,5], **Francesco Bianco** [4] and **Marco Chetoni** [2]

[1] Department of Lucca, Via A.Vallisneri n.6, Agenzia Regionale per la Protezione Ambientale della Toscana, ARPAT, Lucca I-55100, Italy

[2] Consiglio Nazionale delle Ricerche CNR, IPCF, Via G.Moruzzi n.1, Pisa I-56124, Italy; E-Mails: teti.luca@gmail.com (L.T.); ing.marcochetoni@gmail.com (M.C.)

[3] Agenzia Regionale per la Protezione Ambientale della Toscana, ARPAT, Area Vasta Costa - Settore Agenti Fisici, Via V.Veneto n.27, Pisa I-56127, Italy; E-Mail: mauro.cerchiai@arpat.toscana.it

[4] Physics Department, University of Siena, Via Roma n.56, Siena I-53100, Italy; E-Mails: elena.ascari@unisi.it (E.A.); sephir@gmail.com (F.B.)

[5] Consiglio Nazionale delle Ricerche CNR, IDASC, Via Fosso del Cavaliere n.100, Roma I-00133, Italy

* Author to whom correspondence should be addressed; E-Mail: gaetano.licitra@arpat.toscana.it

**Abstract:** In almost all urban contexts and in many extra-urban conurbations, where road traffic is the main noise pollution source, the use of barriers is not allowed. In these cases, low-noise road surfaces are the most used mitigation action together with traffic flow reduction. Selecting the optimal surface is only the first problem that the public administration has to face. In the second place, it has to consider the issue of assessing the efficacy of the mitigation action. The purpose of the LEOPOLDO project was to improve the knowledge in the design and the characterization of low-noise road surfaces, producing guidelines helpful to the public administrations. Several experimental road surfaces were tested. Moreover, several measurement methods were implemented aiming to select those that are suitable for a correct assessment of the pavement performances laid as mitigation planning. In this paper, the experience gained in the LEOPOLDO project will be described, focusing on both the measurement methods adopted to assess the performance of a low-noise road surface and the criteria by which the experimental results have to be evaluated, presenting a comparison of the obtained results and their monitoring along time.

**Keywords:** tire/road noise; noise mitigation action; Close Proximity Method (CPX); Statistical Pass-By (SPB)

**PACS classifications:** 43.50 Rq; 43.50 Lj; 43.20 Ye

---

## 1. Introduction

Transportation noise is an environmental stressor that causes sleep disturbance and annoyance. The latter is the most frequently ascertained effects of noise for people living in urban areas. The reduction of the urban road traffic noise pollution and of the population noise exposure has become mandatory. A great role in the noise generation mechanism of road infrastructures is played by the road pavement, through the interaction with the rolling tire, which often constitutes the primary source of traffic noise at high speeds [1]. Therefore, the use of road surfaces with low noise emission characteristics is one of the most applied actions all over the world, especially when the source emission must be considered. In fact, the use of noise mitigation solutions based on barriers (e.g., involving only the propagation path) cannot be the only satisfactory solution. There are many cases where a barrier cannot solve the problem at all (e.g., a road in a valley and houses on the surrounding hills) or where it is socially opposed (e.g., in urban contexts). When low-noise road surfaces are used as mitigation actions combined with other actions, as for example traffic planning, it is necessary to have suitable methods to assess the effectiveness of the road surfaces. These methods have to be applicable in all contexts, even where the surrounding conditions are very different from those requested by standards. During last years a great effort was made by some international research projects (such as SILVIA—"Silenda VIA—Sustainable Road Surfaces for Traffic Noise Control"—EU Fifth Framework [2],—HARMONOISE—"Harmonized Accurate and Reliable Methods for the EU Directive on the Assessment and Management Of Environmental Noise" [3]—and IMAGINE—"Improved Methods for the Assessment of the Generic Impact of Noise in the Environment" [4]) to study the traffic road noise sources, developing methods, protocols and models. In Italy, a specific project, called LEOPOLDO, on the evaluation of low noise emission surfaces, started in 2006. In this paper, after a brief description of the LEOPOLDO project objectives, methods and protocols developed are described in detail. Then, results obtained through the time monitoring of the experimental surfaces are reported. Finally, discussions deal with suitability of developed methods, aiming to underline pros and cons of each one, relatively to what should be necessary to evaluate the effectiveness of a road surface laid as mitigation action.

## 2. The LEOPOLDO Project

In Tuscany, the LEOPOLDO project [5] was planned in order to develop innovative noise mitigation techniques to be used in action plans for road infrastructures, based on a new type of pavement layers: the project implementation is part of the required environmental policies to mitigate road noise pollution implemented by the Tuscany Region and other European Community member states, in accordance with the Directive 2002/49/EC [6]. The project participants were the Tuscany Region, the ten provinces of the Tuscany Region, ARPAT—the regional environmental protection agency of Tuscany—and the Civil Engineering Department of the University of Pisa. This project was a contact point between public administrators, which have to face noise problems due to road traffic; the environmental agency, which executes noise controls in order to verify the respect of noise limits provided by regulations; and research institutes, which are able to find innovative procedures and suitable mitigation solutions.

*Objective*

The aim of project was firstly to study some new kinds of road surfaces. Besides the low-noise aim, the LEOPOLDO project took into account the environmental compatibility of the road surfaces, in terms of raw material, industrial and productive processes, and the safety factors requested by ordinary roads in urban and extra-urban contexts. In addition, economic costs were taken into account, mainly in terms of the proposed solution suitability, which depends on the time-stability of the road surface performances. For example, traffic and weather conditions can cause surface degradation which in turn can cause a decay in noise performance, which leads to the necessity of a new laying. Thus, the LEOPOLDO projects aimed also to find the best surface criteria, based on the surrounding conditions of the whole context in which surfaces were laid, in order to choose the most suitable surface to use (e.g., plain, or hill, or mountain; the presence of ice or snow during winter, or of an amount of water for a long time, and so on).

The secondary aim of the LEOPOLDO project was to develop measurement protocols useful to assess the road surface effectiveness and time stability, in terms of both acoustical and safety characteristics.

Within the LEOPOLDO project, different experimental road surfaces have been laid on six sites, all along regional road infrastructures. Then, they have been characterized using several techniques and monitored during time. Moreover, all sites have been equipped with an instrument able to monitor traffic (amount, speed and type of vehicles), asphalt conditions (temperature, humidity, vertical pressure and strain gauge) and meteorological conditions (air temperature, wind and rain).

A multi-year acoustical monitoring was carried out by means of SPB and CPX methods and results are shown in this paper, and the availability of the experimental road surfaces are a good opportunity to perform further side researches on collected data (for example, see [7–12]), including analysis on vibrations whose elaborations are still ongoing.

Results obtained in the LEOPOLDO project recently lead to the Tuscany Region Guidelines, useful to local administrators for choosing and engineering the most suitable surface to be used as mitigation action on local urban and extra-urban roads [13]. Methods and protocols, developed with the aim to assess the effectiveness of a low-noise road surface laid as mitigation action, have been adopted by the Tuscany Region and their application is requested in all actions based on regional funds [14].

## Experimental Surfaces

The six experimental road surfaces are detailed in Table 1. All installations are about 200 m long and placed on extra-urban roads not in densely urbanized areas. For more details on composition, volumetric characteristics, aggregate grading and fractal dimensions of the mixtures see [11].

**Table 1.** Experimental road surfaces details. For more details see Tables 1 and 2 in [11].

| Id | Site | Technology | Bitumen | Depth | Speed Limit |
|----|------|-----------|---------|-------|-------------|
| 1 | Arezzo | Micro-draining open grade 0/10 | Hard 4.8% | 3 cm | 90 km/h |
| 2 | Firenze | SMA optimized texture gap grade 0/8 | Hard 6.8% | 3 cm | 50 km/h |
| 3 | Lucca | ISO10844 optimized texture dense grade 0/8 | Hard 5.0% | 3 cm | 70 km/h |
| 4 | Pisa | Dense grade 0/6 with expanded clay | Hard 8.5% | 4 cm | 70 km/h |
| 5 | Massa | Micro-draining open grade 0/6 | Hard 4.5% | 4 cm | 50 km/h |
| 6 | Pistoia | Asphalt rubber (wet process) gap grade 0/8 | AR 8.7% | 3 cm | 90 km/h |

It is necessary to underline that the experimental surfaces have been laid in contexts which present different local surrounding conditions (lane width, roadside ground, guard-rail, slope, exposure to sun, traffic density and typology *etc.*). All these differences surely influence the wear and in some cases could invalidate roadside measure comparisons.

## 3. Acoustical Analysis: Methods and Developed Protocols

The acoustical effectiveness of a low-noise road surface is given mainly by the reduction of the tire/road noise, because it is the main noise source of a passing vehicle. Noise reduction can be obtained working on one of the many sound generation phenomena or through absorption across the road surface. Depending on which phenomenon is treated, the noise spectra will turn out different [1]. The main way to reduce the tire/road noise is by lowering the tire vibrations excited by the road texture profile, turning down the whole noise emission spectrum. Another approach is to shift the noise emission peak towards the low frequencies, in order to take advantage from the A-weighting. On the downside, to obtain the shift towards the low frequencies it is necessary a porous surface, and this is not always a well-suitable solution because porosity needs high traffic density and high speed to maintain performances along time. Moreover, the low-frequencies are not taken into account to evaluate noise limits, but they play a significant role in causing annoyance of people [10].

Road surface can also absorb energy along the fist part of the propagation path, providing a further reduction in the roadside levels. Obviously, the aim of a low-noise road surface is to provide lower levels at the roadside (e.g., at buildings façades). Therefore, it is important to evaluate both the tire/road and the roadside noise.

The standard ISO 11819 provides two method to measure the influence of the road surface on the tire/road noise: the Statistical Pass By method (SPB) in the part 1 [15] and the Close Proximity method (CPX) in the part 2 [16]. They have been applied within the LEOPOLDO project and new protocols for measurement and for data post-processing have been developed. In addition, the acoustical impedance have been measured by means of the Impedance tube method, described in the UNI EN ISO 10534 [17,18].

*3.1. SPB*

### 3.1.1. The Method

The SPB method is described by the ISO 11819-1 and it involves measuring the noise levels from vehicles cruising-by at a constant speed and with the engine operating at the usual condition for that speed. The method relies on a great number of vehicles from normal traffic, without any constraint on tire or vehicle. The measured physical quantity is the maximum A-weighted level $L_{A,max}$ reached at the microphone positioned 7.5 m far from the center of road lane, at 1.2 m height. Data are related to the vehicle speed and the best fit estimates $L_{A,max}$ value at the reference speed.

Since for every vehicle the levels reach the $L_{A,max}$ at the microphone when passing over almost the same part of the road, the propagation path is assumed equivalent for all. Therefore, the data dispersion is only due to vehicle models variety, driving behaviors and mainly tire variety [Phillips-Abbot]. To further improve the data accuracy, vehicles are classified in some categories depending on the weight and the number of tires/axles. Then, the SPB indexes are calculated as a linear combination of the $L_{A,max}$ values at the reference speed obtained for different categories. The standard prescribes that microphone has to be positioned without reflecting obstacles neither behind nor laterally. Moreover, vehicles should pass at constant speed and between microphone position and the center of the road lane there should be the same surface. These surrounding conditions often avoid the applicability of the SPB method in urban contexts.

### 3.1.2. The Modified Protocol

The protocol applied by ARPAT within the LEOPOLDO project combines the technical international standard with the guidelines provided by HARMONOISE project [19]. HARMONOISE introduces a second measurement position, at 3.0 m height, at 7.5 m far from the center of road lane, to improve the evaluation of the influence of local context, avoiding the roadside ground influence not negligible in case of the 1.2 m height position. Moreover, the applied procedure is based on measuring the acoustical energy of the passing vehicle, using the sound exposure level (SEL) in place of the $L_{A,max}$ prescribed by the ISO. The SEL is calculated in according to the ISO 1996-2 [20], which defines the pass-by event as the signal part in which level exceeds the background noise more than 10 dB(A). Thus, during the measurement session, pass-by sound pressure signal and related speed are registered.

In the post processing analysis, the statistical sample of many single passages constitutes the data-set for a logarithmic regression (1), between the measured speed $v$ and the SEL for each microphone to estimate the level at the reference speed $v_0$, in accordance with the HARMONOISE project.

$$\text{SEL} = A + B\log\left(\frac{v}{v_0}\right) \qquad (1)$$

where $A$ is the SEL at reference speed $v_0$ (tipically 50 km/h) and $B$ is a speed-related correction.

Due to the traffic densities and typologies that characterize experimental sites, only the light vehicles category is significantly populated. Thus, only the SPB index for light vehicles, named $L_1$ level, is calculated.

However, the SPB procedure fails when speed data gather around a specific value (commonly the speed limit): in fact, when the speeds are almost the same, the variability due to driver behavior and to vehicle characteristics dominates and data constitute a cloud. In this case, data outside the cloud influence the fit algorithm. To avoid the influence of possible outliers outside the cloud, the binning technique was applied to the whole data-set and a minimum chi-square fit of central values with their uncertainties is performed. That is: data are grouped in velocity classes—called bins—about 10 km/h wide (the actual width is chosen in order to minimize the total chi square of the final fit); mean and standard deviations of data $L_{EQ}(A)$ in each class are computed with the hypothesis of a Poissonian distribution of data and each class is represented by the tern: center velocity of the class; mean $L_{EQ}(A)$ and standard deviation of data contained in each class. The terns are used in the best fit between $L_{EQ}(A)$ and velocity. In physical statistics this technique is usually called binning.

In this way, the information of data spread out and the numerousness in each speed bin is taken into account by means of the uncertainty associated with the central values. In Figure 1 an example of data binning is provided.

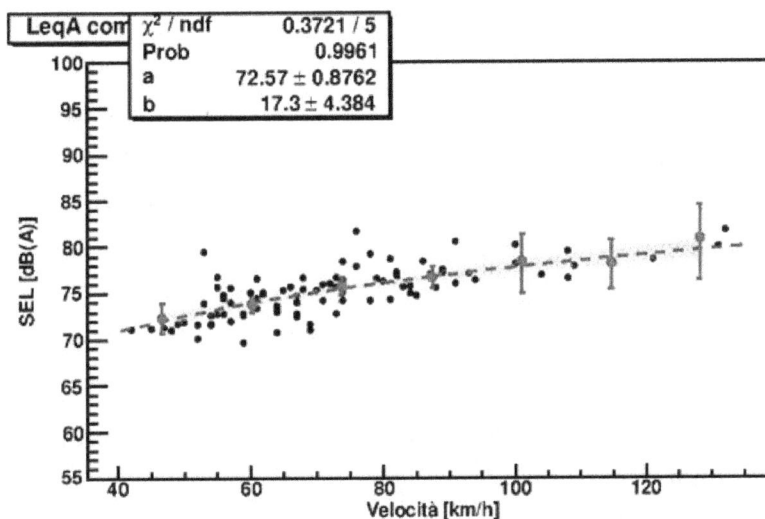

**Figure 1.** Example of fit performed with binning technique.

### 3.2. CPX

#### 3.2.1. The Method

The CPX method is described in the standard draft ISO/DIS 11819-2 (2011) [16]. The CPX measurements can be carried out using a trailer towed by a separated vehicle or a self-powered vehicle. The method uses two microphones placed at 0.2 m from the axis of the wheel and 0.1 m above the road surface. The microphones position is chosen in proximity to the tire/road contact, aiming to evaluate only the road-tire noise without the engine and exhaust system of the car.

The use of four reference tires was requested by the 2002 ISO 11819-2 release, but in the actual one number and technical specifications are committed to the third part actually unwritten. For standard application some runs (at least three) on both wheel tracks are requested. For particular applications, only one single run, carried out on the wheel track closest to the edge of the road and using only one reference tire is allowed. The measurement protocol requires that the sound signal over 20 m long road segments together with the corresponding vehicle speeds are recorded. Then, for each road segment, the A-weighted equivalent third-octave-band level from 315 Hz to 5000 Hz is determined at each microphone position. The energy-based average sound level for the two microphones, normalized at the reference speed by a simple correction procedure or by a logarithmic regression between levels and speed data, is called "Tire/road Sound Level"($L_{CPX}$). Finally, the road surface, preferably longer than 100 m, is characterized averaging all the 20 m long segments, while the standard deviation around the mean is an indication of the homogeneity.

### 3.2.2. The Modified Protocol

An adapted protocol for measurement and data post-processing has been developed to improve the suitability of the CPX method within the LEOPOLDO project [21]. In the present work, results are shown in terms of tire/road noise levels, without strictly referring to CPX indexes; however, for the sake of simplicity, they are hereafter referred as $L_{CPX}$ values. The set-up is based on the measurement system mounted on a self-powered vehicle, as described in [22–24], using the Green Michelin Energy XH1 185/65/R15 as reference tire. In the post-processing step, data analysis is based on the spatial resolution of 5.84 m long segments and the sound pressure level $L_{p_i}$ associated to the $i$-th segment is estimated by fitting experimental data by the well-known bi-logarithmic relationship with speed data. The fit is calculated for each segment, for each third octave band level in the frequency range of 315 to 5000 Hz. It is computed using a minimum chi-squared based iterative algorithm, taking into account the asymmetry of the uncertainties derived from the logarithmic conversion. Finally, the overall A-weighted equivalent sound pressure level, at the reference speed, associated to the $i$-th segment, $L_{CPX_i}$, is obtained through the A-weighted energy-based sum of the third octave bands estimated levels, as required by the ISO.

The $L_{CPX_i}$, levels versus distance are used to characterize the road surface installation through its homogeneity and the averaged noise levels on all segments.

The last improvement of the modified protocol prescribes that during the same measurement session runs have to be extended over a second road surface, typically a DAC 0/12 or a SMA 0/12, as suggested in [21], close to the test one as much as possible. The selected second surface then becomes the "reference" and the evaluation of the acoustical performances of the test one is carried comparing it to the reference one. This reference surface could be as equal as possible to the pre-existing, *ante-operam* one (e.g., long aged and possibly acoustically stable in time), or, alternatively, a road surface coeval to the test one. This choice depends on the purpose of the measurement or the aim of the test surface laying. This procedure, called "the differential criterion", came from the necessity to avoid influence by measurement conditions (especially meteo-climatic ones). Their effect would depend on the particular configuration tire/road and in real scenarios it is nearly impossible to find the appropriate correction for each surface surveyed. Moreover, despite in a single measurement session most of these error sources affect systematically the measurements, they can be assumed as random in case of several measurement sessions carried out in different days and/or with different set-ups or instrumental chains. Thus, comparing absolute values obtained could be not significant.

## 4. Experimental Results

In each site, the SPB method was applied firstly on the pre-existing surface (*ante-operam* one), before the laying of the experimental surface, carrying out the measurement session according to the developed protocol above explained. Thus, results obtained later for the experimental surface were compared to the *ante-operam* one. This allows to evaluate over time the effectiveness of the experimental road surface relative to the pre-existing status (e.g., considering the new laying as a mitigation action).

In case of the CPX method, the *differential criterion* above described, allows to avoid the use of results obtained in measurement sessions carried out before the experimental laying.

Concerning the uncertainties evaluation, it has to be considered that: for the SPB results, data variability derives exclusively from the road traffic, because the road spatial unhomogeneity influences in the same way each pass-by. Thus, the uncertainty associated to results, calculated through the SPB fit algorithm, depends on the sample and it can be considered as a measurement uncertainty; on the contrary, the uncertainty associated to the CPX results derives mainly from data dispersion (e.g., the spatial unhomogeneity of the road surfaces). Thus, CPX results obtained in different measurement sessions carried out on the same road surface, shall show almost equivalent uncertainty, unless unhomogeneity was increasing. These observations are important to better evaluate results reported below.

### 4.1. Site 1—The Micro-Draining Open Grade 0/10

The road is characterized by high traffic density, with a significant percentage of heavy vehicles and an average speed of about 80 km/h. Lanes are wide with good visibility on the test stretch, which is surrounded by containment walls. The experimental road surface is porous and it shows an absorbing peak around 1100 Hz, as shown in Figure 2.

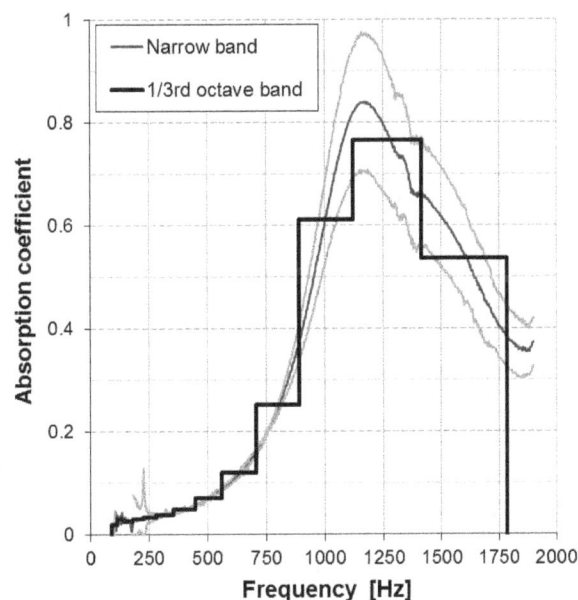

**Figure 2.** Site 1: Absorbing coefficient *vs.* frequency measured using impedance tube on some samples extracted from the surface. In grey the dispersion of data from narrow band analysis and in black the resulting 1/3rd octave band result. The main absorption peak can be found at about 1100 Hz.

The CPX results along time are shown in Figure 3. In Figure 3a absolute $L_{CPX}$ values for reference and experimental surfaces are plotted, while in Figure 3b their difference is plotted for each lane.

In this site, the reference surface is coeval with the experimental one. Looking at the absolute values, the reference surface can be probably considered acoustically settled after the second year. On the experimental road surface, the acoustical characteristics of a *lane 1*, (continuous line in Figure 3b) are getting worse before the other one *lane 2*, (dashed line in figure).

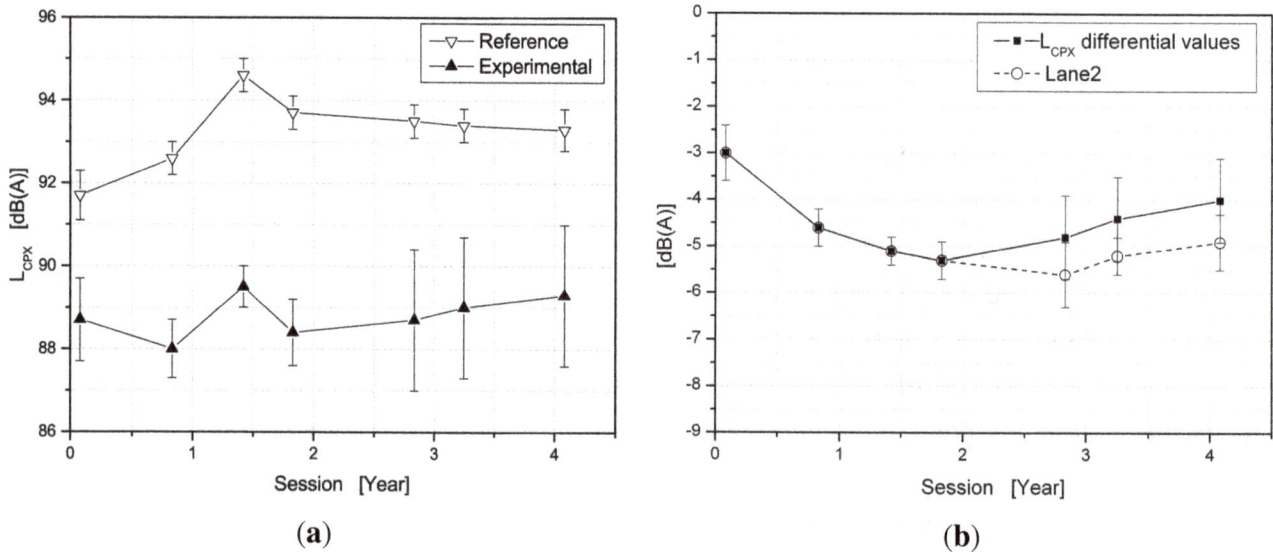

**Figure 3.** Absolute (**a**) and differential (**b**) values of $L_{CPX}$ along measurement sessions (e.g., time, in years) for the micro-draining surface laid in Site 1.

This can be figured out in Figure 4, where the spatial distributions of $L_{CPX}$ levels of both lanes are plotted, pointing out differences between the first and the last measurement session carried out on the experimental surface. Since the first session *lane 1* shows a higher spatial unhomogeneity than the *lane 2*. The hypothesis that the laying of the *lane 1* suffered some troubles (in the temperature, in the mix or other) is clearly confirmed by results obtained in the last measurement session, where it is worsened in terms of both spatial homogeneity and absolute values. This is also highlighted in the $L_{CPX}$ absolute values graph, where the uncertainty for the experimental road surface is increasing in the last measurement sessions, and in the differential values graph, where also results obtained considering only the *lane 2* are plotted.

The SPB measurements were carried out on the *lane 1* and in its first part(referring to the Figure 4, the stretch centered on the 150 m in the plot), so the increasing roadside level is justified. Moreover, comparing absolute values it can be easily noticed that the third measurement session (see Figure 5), about one year and an half after the laying, has been affected by some bias evidently due to surrounding condition (given that CPX and SPB measurement are carried out through two different instrumental chains and by two different couples of operators), probably the not perfectly dry road surface because the day before the measurement session it was a rainy day. It is remarkable as the differential criterion is powerful to minimize the influence of this bias on results.

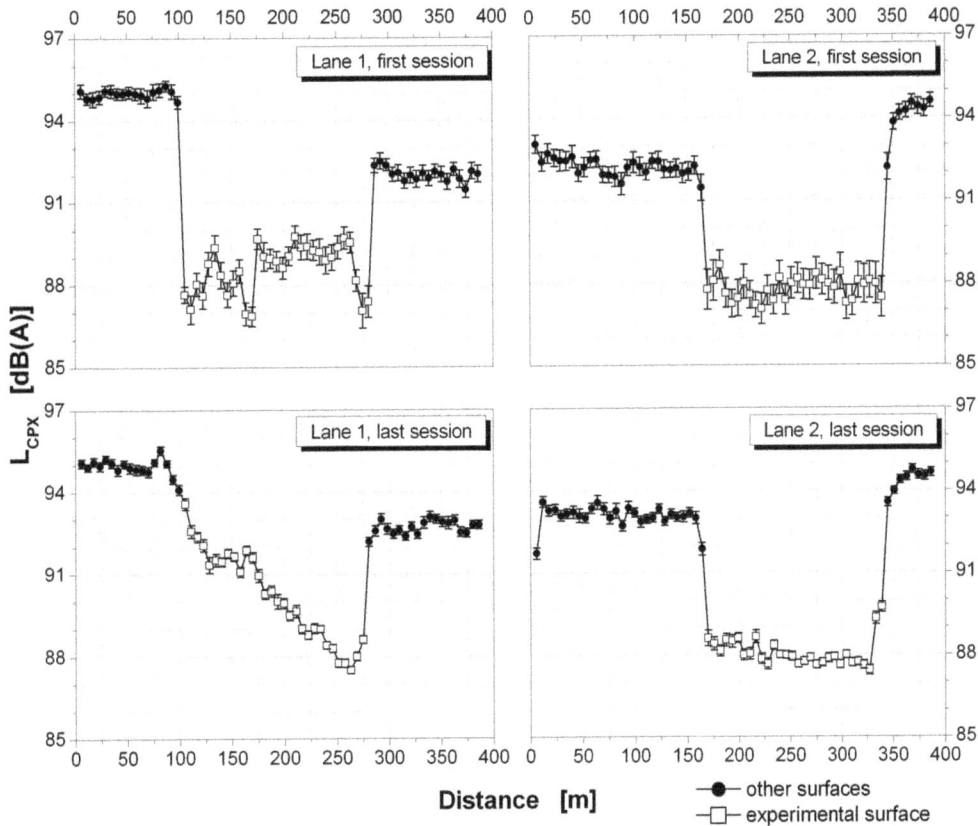

**Figure 4.** CPX data spatial distribution obtained during first session (upper figures) of measurements and during last one (lower figures) at Site 1. *Lane 1* results are represented on the left plots while *lane 2* ones are on the right plots. Experimental stretches data are marked with white squares.

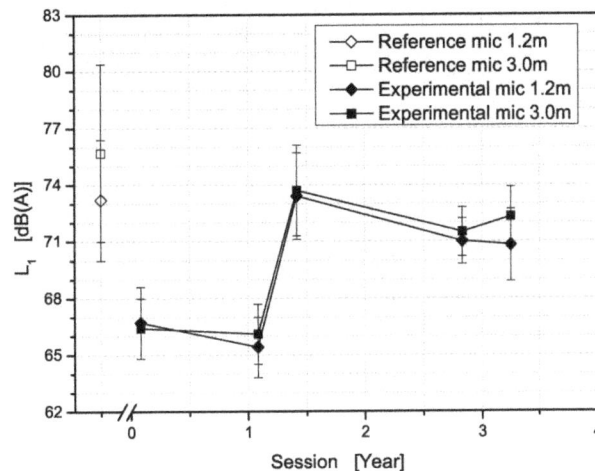

**Figure 5.** Time evolution of SPB values from *lane1*, Site 1. In white *ante-operam* values.

## 4.2. Site 2—The SMA Optimized Texture Gap Grade 0/8

The road is characterized by low traffic density, with limited percentage of heavy vehicles and with an average speed of about 70 km/h. There is no good visibility along the site, going from a bend to a nearby small slope. A hill is present on one side of the road, while the other one has a descending slope with some buildings.

The CPX results are shown in Figure 6, the SPB results are shown in Figure 7. In this case, results show that, after the first year characterized by an initial settling, the acoustical characteristics of the road surface are almost stable around a 3 dB(A) lowering of tire/road noise emission. It is clearly highlighted by the $L_{CPX}$ differential values and confirmed by the SPB ones. Moreover, it can be easily noticed that the 2 year aged measurement session was affected by some bias due to surrounding condition, confirming that the differential criterion is powerful to minimize external influence on results.

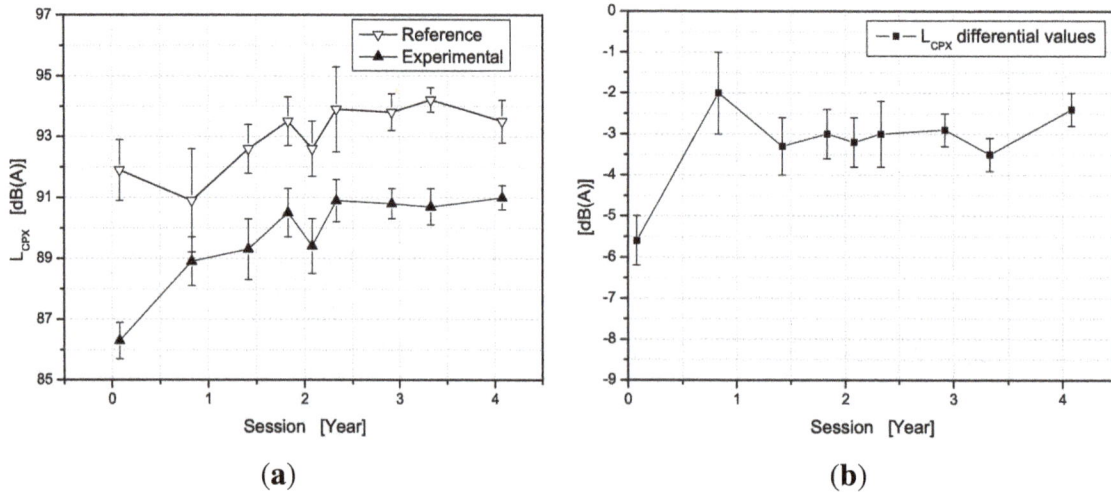

**Figure 6.** Absolute (**a**) and differential (**b**) values of $L_{CPX}$ along measurement sessions (e.g., time, in years) for the SMA optimized texture surface laid in Site 2.

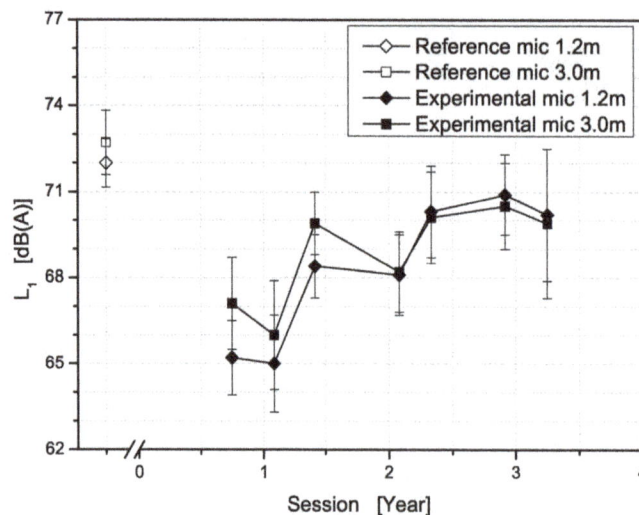

**Figure 7.** Time evolution of SPB values in Site 2. In white *ante-operam* values.

In Figure 8 the CPX data spatial distribution of both lanes is shown, pointing out differences between the first and the last session on the experimental surface. Probably, the *lane 2* in the first session was still settling and loosing the outward bitumen in excess, showing an unhomogeneity unascertainable in the last measurement session. There are not large differences between the two lanes, and both show an equal increasing of levels.

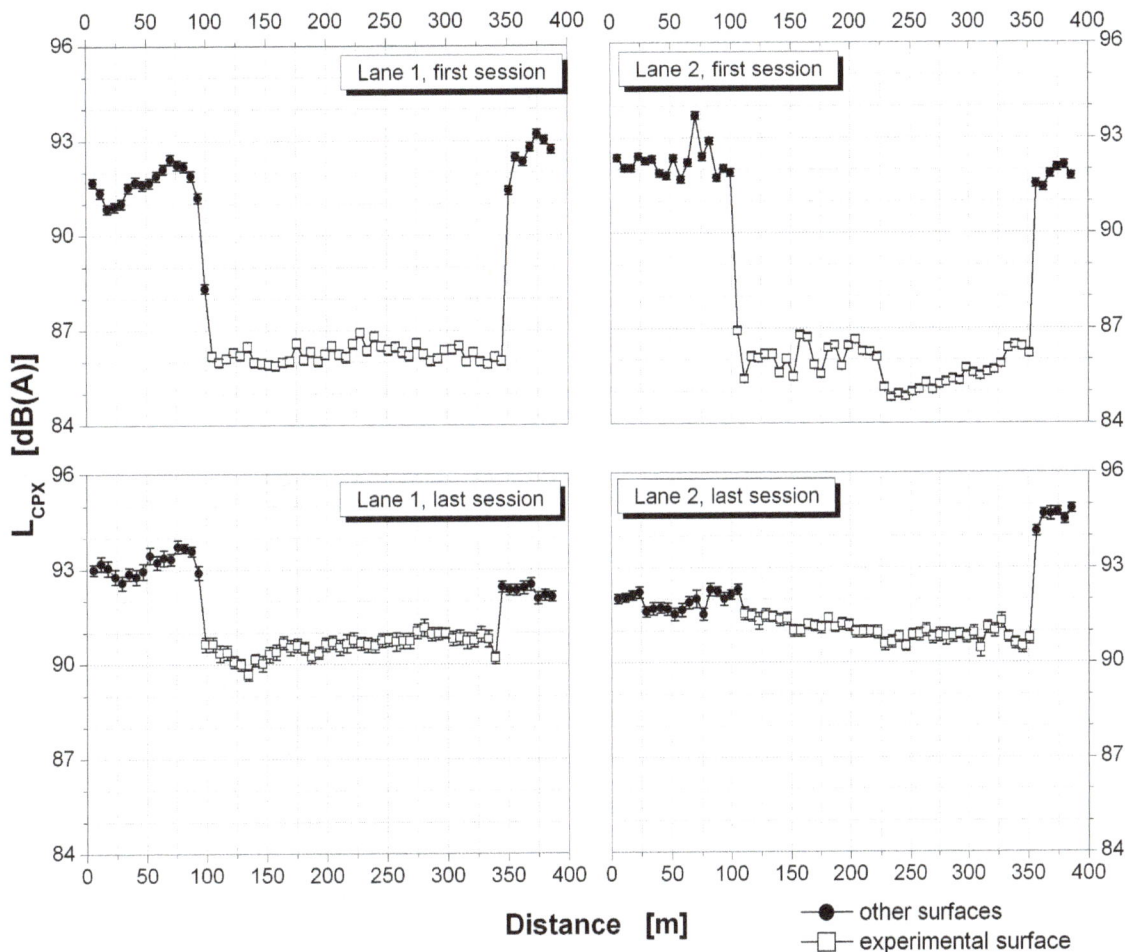

**Figure 8.** CPX data spatial distribution obtained during first session (upper figures) of measurements and during last one (lower figures) at Site 2. *Lane 1* results are represented on the left plots while *lane 2* ones are on the right plots. Experimental stretches data are marked with white squares.

## 4.3. Site 3—The ISO10844 Optimized Texture Dense Grade 0/8

This site shows a not high traffic density, with limited percentage of heavy vehicles and an average speed of about 50 km/h. There is a good visibility along the site and the road has a small slope. On one side of the road there is a steep slope side of a hill, whilst on the other side there is a stream 5 m further down after the small flat lawn where the SPB instrumentation is placed. The CPX results are shown in Figure 9. In this case, results show that in the first 18 months the road surface has been characterized by an initial settling. On the other side, since the second year from the laying, the acoustical characteristics of the road surface are almost stable around a 5 dB(A) lowering of tire/road noise emission.

The CPX spatial distribution, shown in Figure 10, points out its homogeneity and its stability in time: in fact, since the first measurement session and till the last one lanes show equivalent levels.

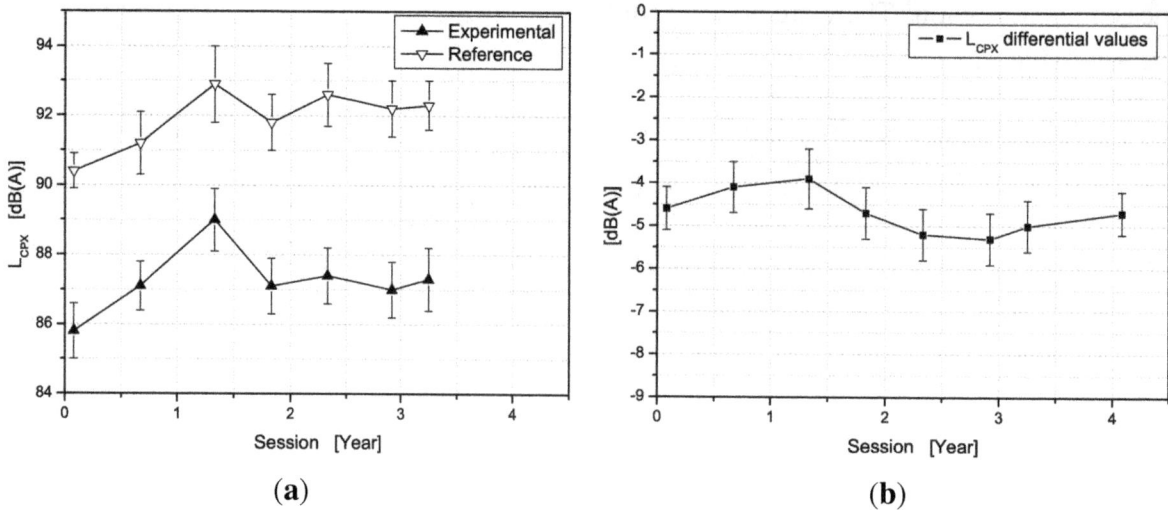

**Figure 9.** Absolute (**a**) and differential (**b**) values of $L_{CPX}$ along measurement sessions (e.g., time, in years) for the ISO10844 optimized texture surface laid in Site 3.

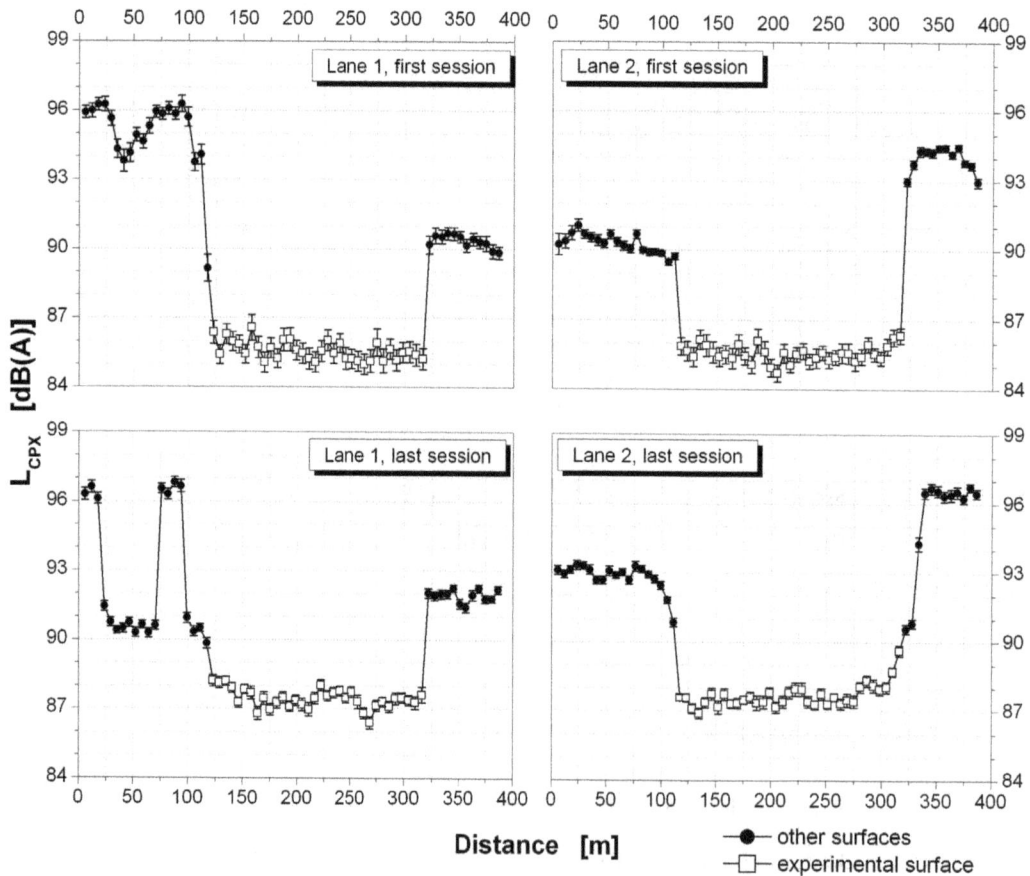

**Figure 10.** CPX data spatial distribution obtained during first session (upper figures) of measurements and during last one (lower figures) at Site 3. *Lane 1* results are represented on the left plots while *lane 2* ones are on the right plots. Experimental stretches data are marked with white squares.

Even if CPX and SPB do not show a clear relationship, the SPB results shown in Figure 11 confirm this good performance. The difference of about 2–3 dB(A) between the two SPB microphones is probably due to the absorbing lawn influence.

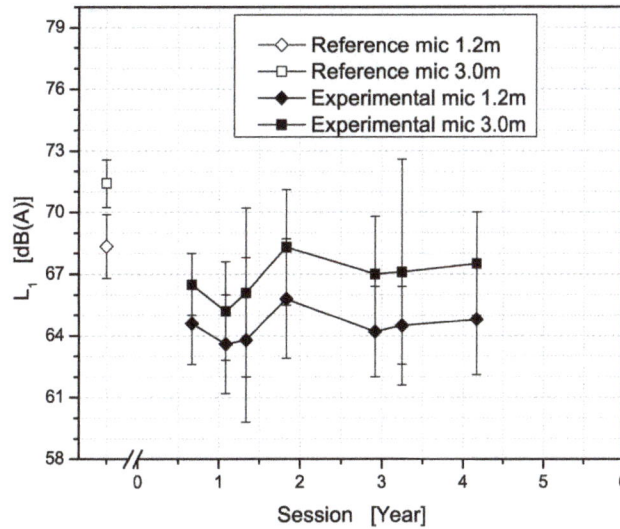

**Figure 11.** Time evolution of SPB values in Site 3. In white *ante-operam* values.

## 4.4. Site 4—The Dense Grade 0/6 with Expanded Clay

This straight site presents a high traffic density with high percentage of heavy vehicles and an average speed of about 80 km/h. No slope is present over this road, with open plane fields on one side and a slight ascending slope of the ground on the other.

The CPX results are shown in Figure 12, the SPB results are shown in Figure 13. In this case, results are highly correlated among the two methods and they show clearly a decaying of the acoustical characteristics of the road surface. According CPX method results decaying might be flattening at around 3 dB(A) of tire/road noise lowering. Despite the high data correlation shown with the CPX data with, SPB results are not able to assess this lowering by means of the comparison with the *ante-operam* value.

The CPX data spatial distributions shown in Figure 14 point out the homogeneity stability since the first measurement session and till last one, showing the clearly increasing levels, similar for both lanes.

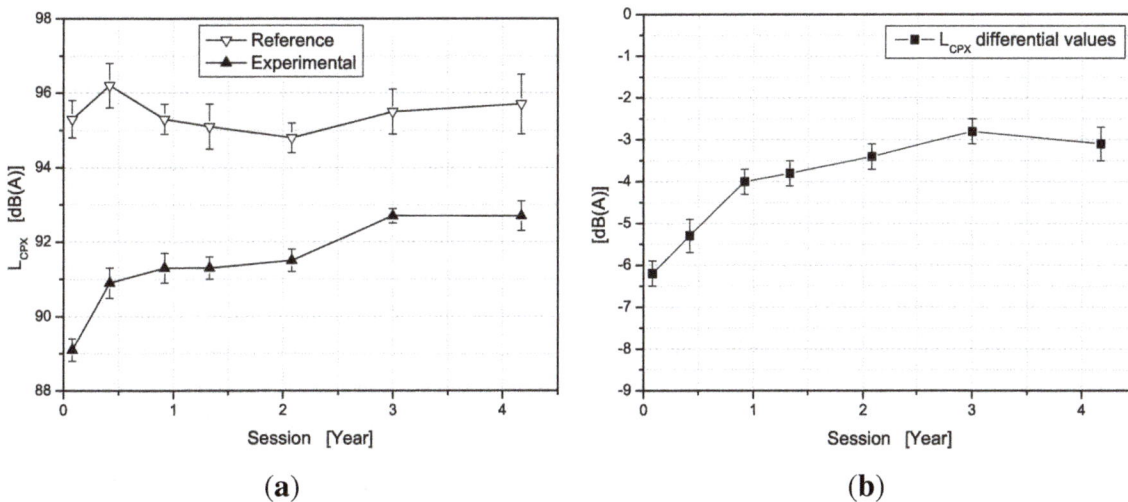

(a)

(b)

**Figure 12.** Absolute (**a**) and differential (**b**) values of $L_{CPX}$ along measurement sessions (e.g., time, in years) for the Dense grade 0/6 with expanded clay laid in Site 4.

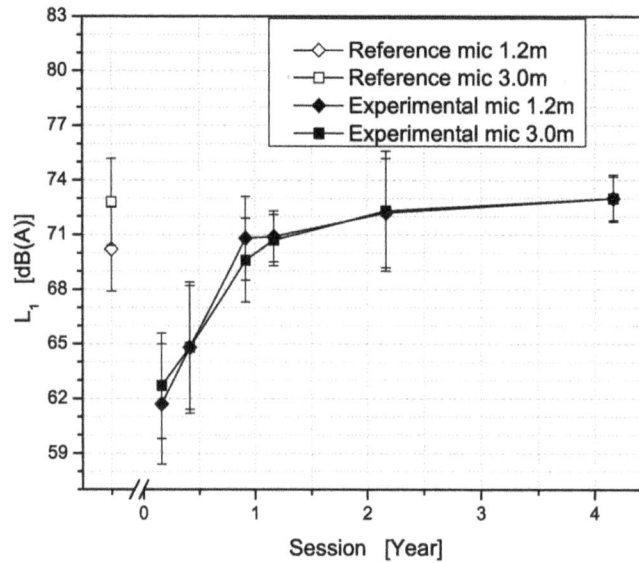

**Figure 13.** Time evolution of SPB values in Site 4. In white *ante-operam* values.

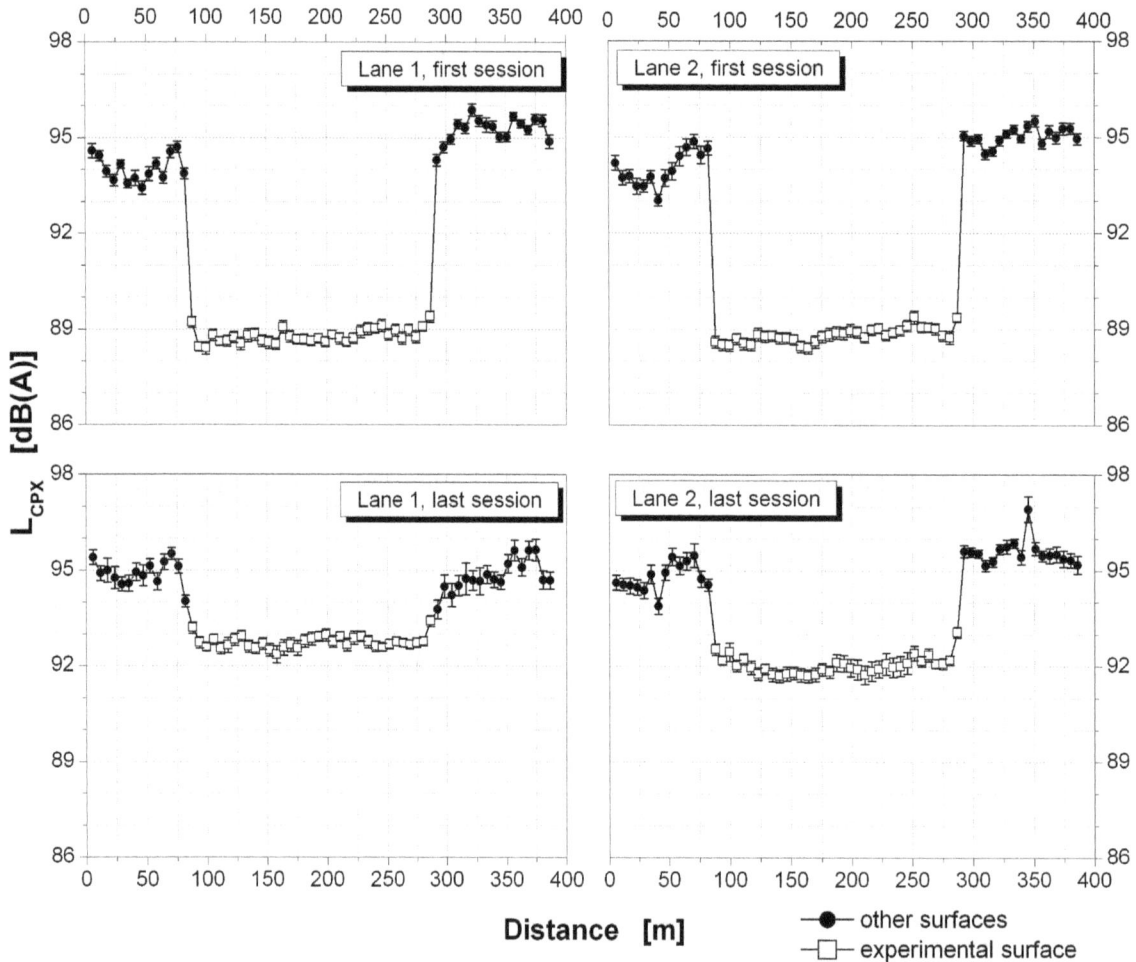

**Figure 14.** CPX data spatial distribution obtained during first session (upper figures) of measurements and during last one (lower figures) at Site 4. *Lane 1* results are represented on the left plots while *lane 2* ones are on the right plots. Experimental stretches data are marked with white squares.

### 4.5. Site 5—The Micro-Draining Open Grade 0/6

This road has very low traffic density, but with a significant percentage of heavy vehicles during daytime. Bends are present with a slow slope and lanes are less than 3.5 m width, and the average speed is lower than 50 km/h. The ground on one side of the road shows a depression, while SPB instrumentation is placed on a wide parking lot on the other side.

The experimental road surface is porous and it shows an absorbing peak around 800 Hz, as shown in Figure 15.

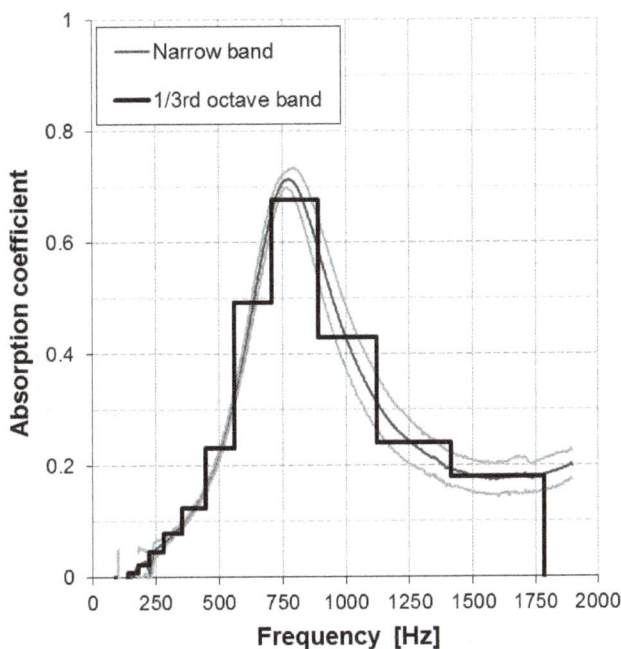

**Figure 15.** Site 5: Absorbing coefficient *vs.* frequency measured using impedance tube on some samples extracted from the surface. In gray the dispersion of data from narrow band analysis and in black the resulting 1/3rd octave band result. The main absorption peak can be found at about 800 Hz.

The traffic density and average speed are too low to maintain the porosity, and therefore the acoustical characteristics are steeply decaying, even though from an optimal initial 8 dB(A) lowering of tire/road noise emission, as well shown by the CPX results in Figure 16.

In Figure 17 the data spatial distribution of both lanes is shown, pointing out high unhomogeneity since the first measurement session, probably due to troubles in the laying. In particular, *lane 2* shows a down-step shape still detectable in the last session. On *lane 1* higher homogeneity can be found except at borders. It is worth noting, however, that the road surface shows a stretch per lane (*lane 1* between 180 and 320 m, *lane 2* between 280 and 350 m) with very low levels even after four years.

On the SPB results (see Figure 18), it has to be pointed out that sample data were subjected to strong statistical fluctuations, due to the low traffic density, which imply both a strong variability due to driving behavior and a gathering at the same speed. As well described above, these sample fluctuations highly influence the fit algorithm and the SPB results do not show the relationship with CPX ones.

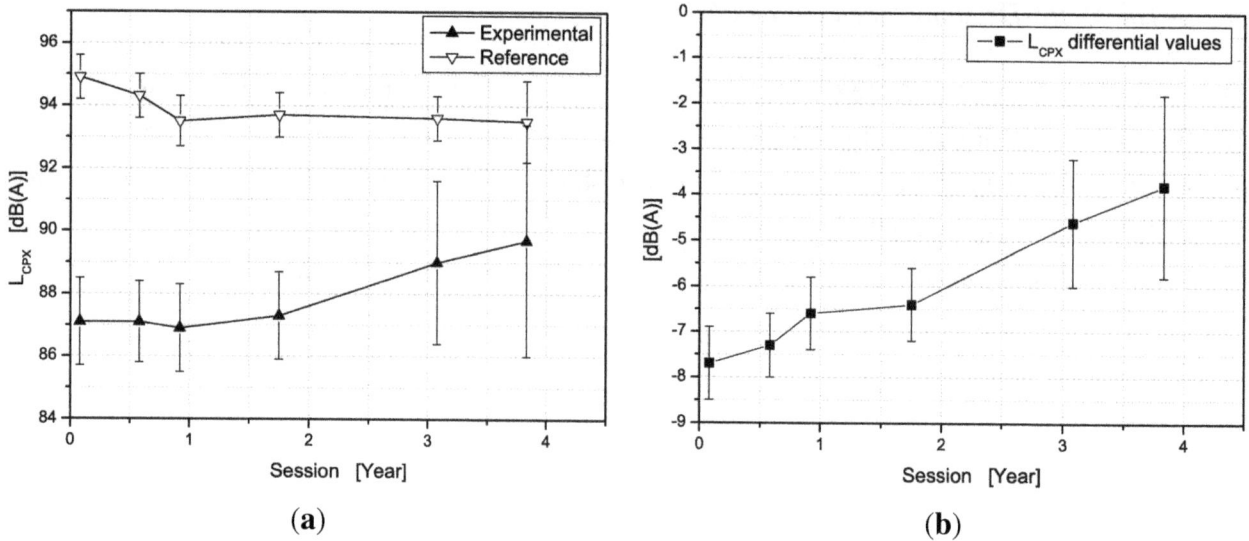

**Figure 16.** Absolute (**a**) and differential (**b**) values of $L_{CPX}$ along measurement sessions (e.g., time, in years) for Micro-draining open grade surface laid in Site 5.

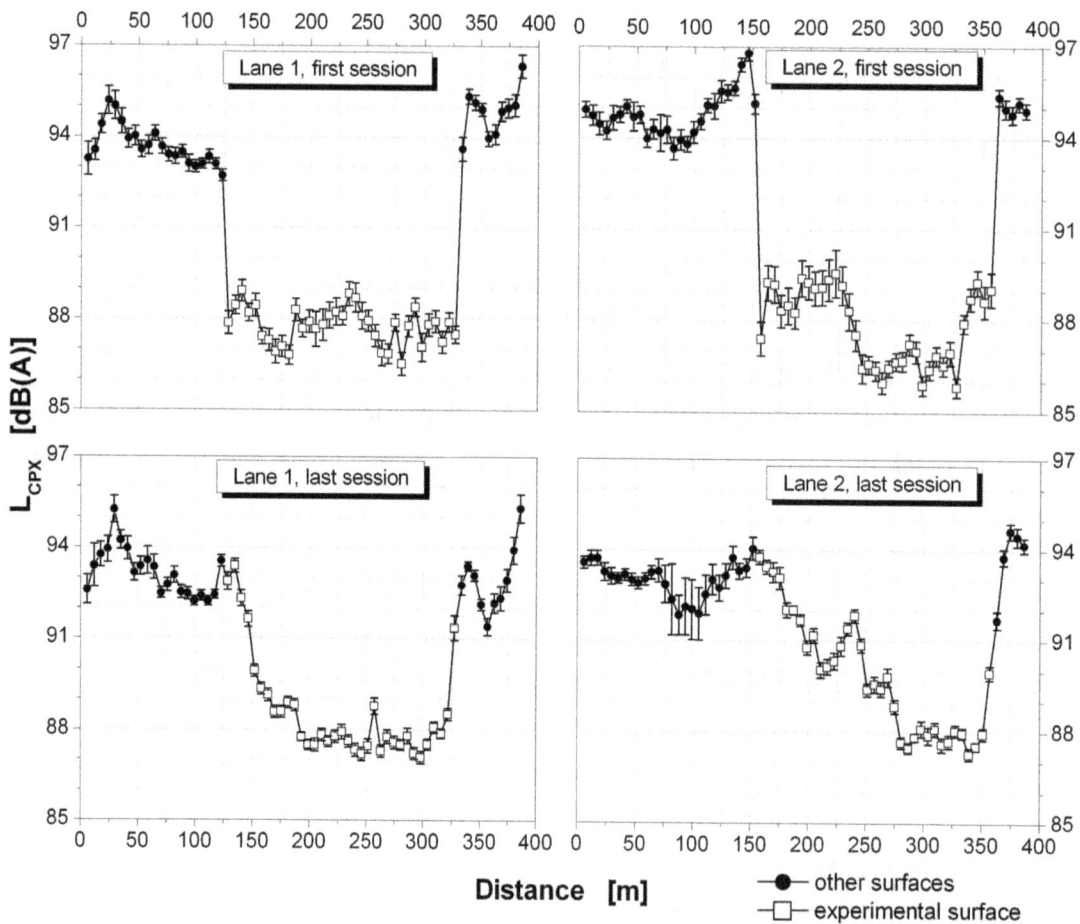

**Figure 17.** CPX data spatial distribution obtained during first session (upper figures) of measurements and during last one (lower figures) at Site 5. *Lane 1* results are represented on the left plots while *lane 2* ones are on the right plots. Experimental stretches data are marked with white squares.

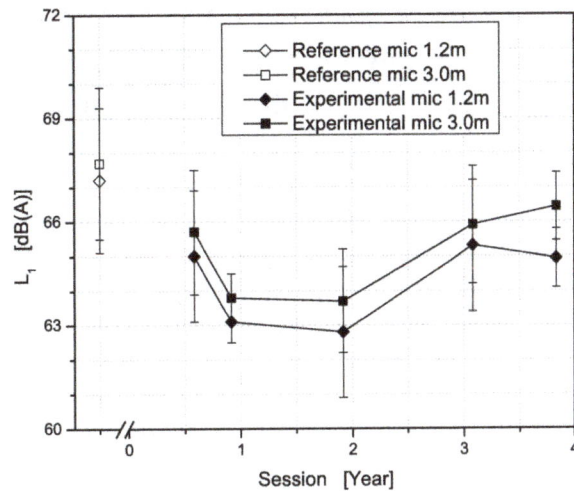

**Figure 18.** Time evolution of SPB values in Site 5. In white *ante-operam* values.

### 4.6. Site 6—The Asphalt Rubber (Wet Process) Gap Grade 0/8

Medium traffic density, with a significant percentage of heavy vehicles and an average speed of about 70 km/h characterize this site, whose wide lanes allow good visibility. The ground around the road is a lawn almost flat on the side where the SPB instrumentation is placed, whilst the other side is a steep descending slope.

The CPX results are shown in Figure 19. In this case, the reference surface used to apply the differential value is coeval with the experimental one (results obtained using a per-existing surface as reference are shown in [9]). Looking at the absolute values, the reference surface probably cannot yet be considered acoustically settled, while the experimental road surface shows a strong time-stability. Therefore, the $L_{CPX}$ differential values are increasing and the lowering of tire/road noise emission is clearly improving.

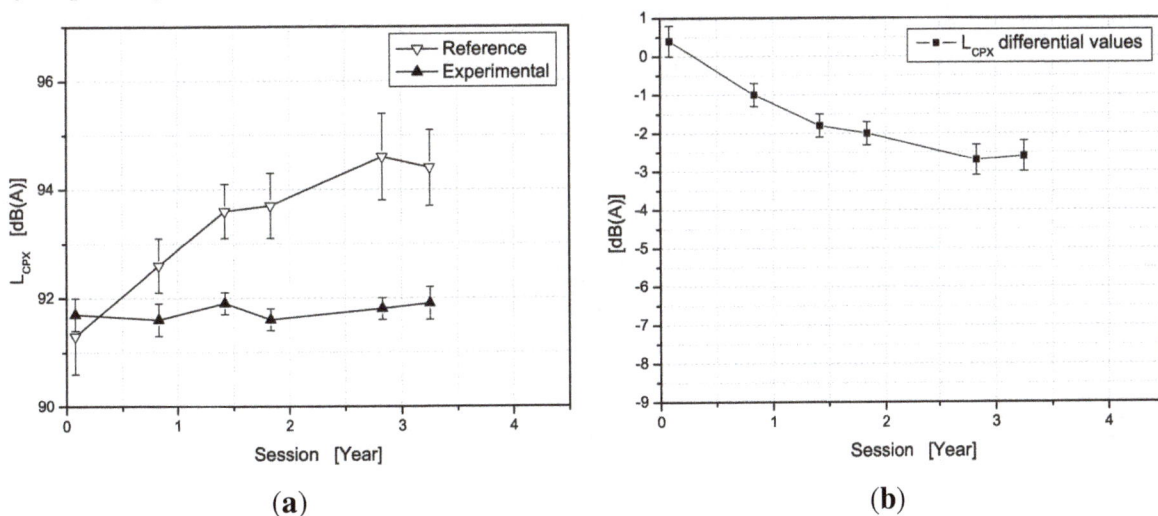

(a)                                                           (b)

**Figure 19.** Absolute (**a**) and differential (**b**) values of $L_{CPX}$ along measurement sessions (e.g., time, in years) for Asphalt rubber (wet process) surface laid in Site 6.

The CPX spatial distribution, shown in Figure 20, points out its homogeneity and its stability in time: since the first measurement session and till the last one, lanes show equivalent levels.

The SPB results (see Figure 21) confirm the good stability of the experimental road surface. The difference of about 2–3 dB(A) between the two SPB microphones is probably due to the absorbing lawn influence, as in case of site 3. It has to be underlined that for the SPB the reference surface was the per-existing one.

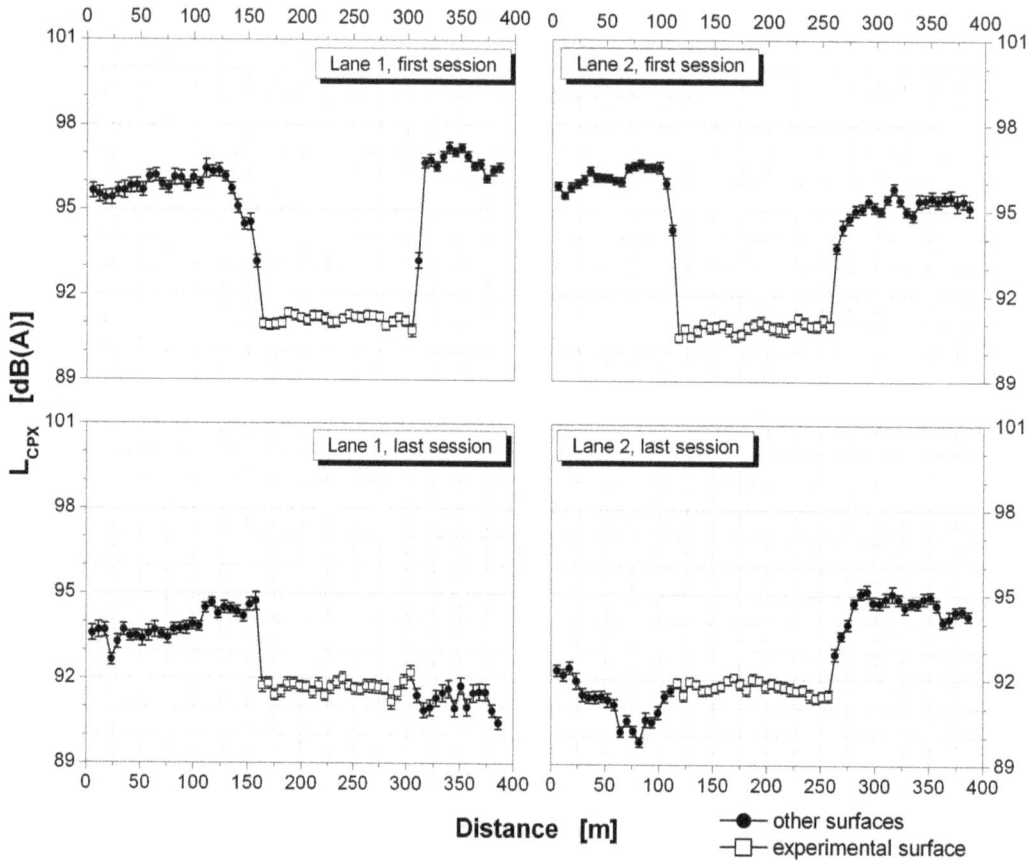

**Figure 20.** CPX data spatial distribution obtained during first session (upper figures) of measurements and during last one (lower figures) at Site 6. *Lane 1* results are represented on the left plots while *lane 2* ones are on the right plots. Experimental stretches data are marked with white squares.

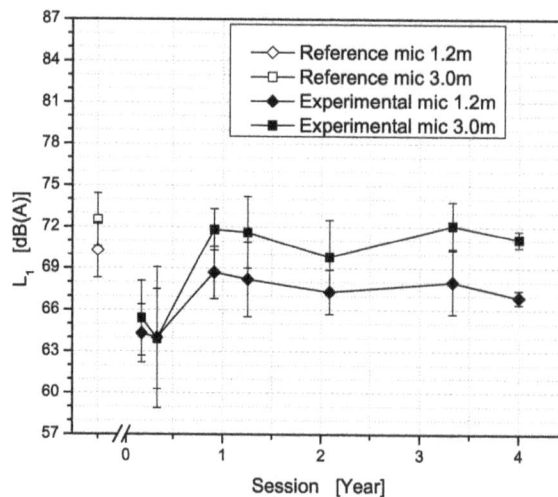

**Figure 21.** Time evolution of SPB values in Site 6. In white *ante-operam* values.

## 5. Discussion

The most challenging task for the LEOPOLDO project was studying and developing methods and protocols useful to assess the effectiveness of the mitigation action planned by means of road surfaces. Indeed, when somewhere there is an excess of noise limit a mitigation action is needed.

The mitigation action should be a solution stable in time as much as possible, to avoid recurring of the noise limit excess and incurring further costs. The main critical issue of the verifications requested by laws are that they are carried out just in few points, or even in only one, and that the time monitoring is not provided for. It means that regulation criteria are not able to leave spatial homogeneity and time stability out of consideration. On the contrary, it is well-known that every road surface is weathered and exposed to traffic: resulting wear and tear corresponds to a worsening of acoustical performances. Moreover, many road surface technologies prescribe specific production methods, that if unattended leads to inhomogeneity in case of very long laying. Thus, choosing the right road surface to be used in the mitigation action planning phase does not guarantee the expected outcome, neither in terms of spatial regularity nor in terms of time durability.

A second critical issue is the ability of identifying which action has failed when the limit excess is assessed after the realization of a mitigation project planned with several actions combined together. In fact, the verifications requested by laws are not able to provide the answer. Similarly, when the excess of noise limit recurs after some time, it is necessary to understand if the cause is the decaying of the road surface acoustical characteristics or if the site conditions are changed (for example an increasing of traffic density or of heavy vehicles, *etc.*).

Therefore, it is necessary to have a measurement protocol which is able to evaluate the road surface effectiveness, its time evolution and without being influenced by traffic conditions, noise barriers *etc.* The experience gained within LEOPOLDO project leads to identify the CPX method, applied with the differential criterion, as the most suitable protocol to test a road surface, in space and in time. Choosing the most appropriate reference surface to apply the differential criterion is crucial, especially to the time stability investigation. For example, in the LEOPOLDO project even if the reference surfaces used are all of the same type commonly used in Tuscany, they provide different $L_{CPX}$ absolute values varying in about 4 dB(A). This does not allow the comparison between different road surface types by means of the $L_{CPX}$ differential values, and only the comparison between mitigation actions is allowed.

On the other side, CPX analysis allows to assess when SPB data are too influenced by surrounding conditions, e.g., see in site 1. Moreover, $L_{CPX}$ values as a function of the distance allows to evaluate the surface spatial regularity, with a measurement time significantly shorter than which requested by roadside methods.

The SPB method is not able to evaluate the road surface homogeneity. In addition, results are useful just to describe the stretch in front of the measurement position and they cannot be considered representative of the whole installation when highly spatial inhomogeneous, as in case of Site 1.

Being the laying homogeneous, the SPB method could be useful to compare different road surface types, but it suffers the sample variability, which corresponds to high associated uncertainties. Moreover, without a reference data set (provided from a reference surface or from the CPX method, as in case of the LEOPOLDO project), it suffers too much the surrounding conditions. The too high uncertainties and the possibility to chance upon an outlier due to surrounding conditions smooth over the SPB values, masking the aimed information. In particular this may be crucial when influencing results of *ante-operam* measurement (see site 4). Anyway, the SPB method has the advantage to use the actual car fleet as source, whilst the CPX method uses just one or few particular tires, and the gain in low tire/road noise assessed by means of the CPX method not necessarily is well representative of the roadside benefit due to the surface laid on an ordinary road.

## 6. Conclusions

In Tuscany, the LEOPOLDO project was planned in order to develop innovative noise mitigation techniques to be used in action plans for road infrastructures, based on new type of pavement layers: the project implementation is part of the required environmental policies to mitigate road noise pollution implemented by the Tuscany Region and other European Community member states, in accordance with the Directive of the European Parliament and Council 2002/49/EC [6].

The first task of the LEOPOLDO project was to study experimental low noise road surfaces, taking into account also the environmental compatibility, in order to define criteria on which choosing the most suitable surface to use when a noise mitigation action is needed. Thus, six experimental road surfaces have been proposed, laid in different contexts and monitored over four years. The performance of low-noise road surfaces has been monitored using three measurement methods, aiming to evaluate their suitability for a correct assessment of the effectiveness of a pavement laid as mitigation planning. Evaluation of roadside noise through the SPB method has shown unreliable results due to the influence of surrounding conditions, even using a modified procedure. Results have pointed out that the best evaluation of surfaces performance in the lowering of road/tire noise is achieved through CPX method and a modified measurement protocol that evaluate differential values in order to assure the comparability of performances along time.

The experience of the LEOPOLDO project improved the knowledge in the design and the characterization of low-noise road surfaces. And a further project—named LEOPOLDO II—is ongoing in order to develop further skills.

## Acknowledgments

Results come from measurement sessions performed within the Leopoldo Project funded by Transportation Ministry and Tuscany Region. This paper could not be written to its fullest without the support of different funds and people that supported the analysis within these 8 years. Special thanks to Luca Alfinito, Luca Nencini and Riccardo Zei for their precious work.

After the official end of the project further analysis become possible with the financial support of Ecopneus s.c.p.a., which funded a research project carried out by CNR-IPCF and its spin-off iPOOL s.r.l.

## Author Contributions

Experimental measurements, analysis and interpretation of the results as well as conclusions have been conducted by all the co-authors. The manuscript has been written by Luca Teti, artworks have been prepared by Mauro Cerchiai, the whole paper has been revised by Gaetano Licitra with approval by the others co-authors.

## Conflicts of Interest

The authors declare no conflict of interest.

## References

1. Sandberg, U.; Ejsmont, J.A. *Tyre/Road Noise Reference Book*; INFORMEX: Kisa, Sweden, 2002.
2. Sandberg, U.; Kalman, B.; Nilsson, R. *SILVIA Project Report—Design Guidelines for Construction and Maintenance of Poroelastic Road Surfaces*; Project Deliverable SILVIA-VTI-005-02-WP4-141005, SILVIA-CONTRACT N-GRD2-2000-31801-SI2.335701; Swedish National Road and Transport Research Institute (VTI): Linköping, Sweden, 2005.
3. *Harmonised, Accurate and Reliable Prediction Methods for the EU Directive on the Assessment and Management Of Environmental Noise*; IST-2000-28419; European Commission: Utreche, The Netherlands, 2000.
4. *Improved Methods for the Assessment of the Generic Impact of Noise in the Environment*; SSPI-CT-2003-503549-IMAGINE; Netherlands Organisation for Applied Scientific Research (TNO): Delft, The Netherlands, 31 May 2006.
5. *Predisposizione Delle Linee Guida Per La Progettazione Ed Il Controllo Delle Pavimentazioni Stradali Per La Viabilità Ordinaria*; University of Pisa: Pisa, Italy, 2006. (In Italian)
6. *The Environmental Noise Directive*; 2002/49/EC; European Commission: Brussels, Belgium, 2002.
7. Licitra, G.; Cerchiai, M.; Teti, L.; Alfinito, L. Road pavement description by psycho-acoustical parameters from CPX data. In Proceedings of the 38th International Congress and Exposition on Noise Control Engineering (INTER-NOISE 2009), Ottawa, Canada, 23–26 August 2009.
8. Licitra, G.; Cerchiai, M.; Alfinito, L. Acoustical performances of new generation road pavements developed in LEOPOLDO Project. In Proceedings of the 8th European Conference on Noise Control (Euronoise 2009), Edinburgh, UK, 26–28 October 2009.
9. Licitra, G.; Cerchiai, M.; Teti, L.; Ascari, E.; Fredianelli, L. Durability and variability of the acoustical performance of rubberized road surfaces. *Appl. Acoust.* **2015**, in press.
10. Ascari, E.; Licitra, G.; Cerchiai, M.; Teti, L. Low frequency noise impact from road traffic according to different noise prediction methods. *J. Sci. Total Environ.* **2015**, *505*, 658–669.
11. Losa, M.; Leandri, P.; Licitra, G. Mixture design optimization of low-noise pavements. *J. Transp. Res. Board* **2013**, *23–72*, 25–33.
12. Losa, M.; Leandri, P.; Cerchiai, M. Improvement of pavement sustainability by the use of crumb rubber modified asphalt concrete for wearing courses. *Int. J. Pavement Res. Technol.* **2012**, *5*, 395–404.

13. *DELIBERAZIONE di Giunta Regionale 11 marzo 2013, n. 157, BURT part 2—n.12 20-3-2013*; Regione Toscana: Florence, Italy, 2013. (In Italian)

14. *Delibera Giunta Regionale n.490 16-06-2014. Comitato regionale di coordinamento ex art. 15 bis, L.R. 89/98: linee guida regionali in materia di gestione degli esposti, di verifica di efficacia delle pavimentazioni stradali fonoassorbenti e/o a bassa emissività negli interventi di risanamento acustico e di gestione dei procedimenti di Valutazione di Impatto Acustico*; ARPAT (Agenzia regionale per la protezione ambientale della Toscana): Florence, Italy, 2014. (In Italian)

15. *ISO 11819-1 Acoustics—Measurement of the Influence of Road Surfaces on Traffic Noise—Part 1: Statistical Pass-By Method*; ISO: Geneva, Switzerland, 1997.

16. *ISO/DIS-11819-2 Method for Measuring the Influence of Road Surfaces on Traffic Noise—Part 2: Close-Proximity (CPX) Method*; ISO: Geneva, Switzerland, 2011.

17. *ISO 10534-1: 1996 Acoustics—Determination of Sound Absorption Coefficient and Impedance in Impedance Tubes—Part 1: Method Using Standing Wave Ratio*; ISO: Geneva, Switzerland, 1996.

18. *ISO 10534-1: 1998 Acoustics—Determination of Sound Absorption Coefficient and Impedance in Impedance Tubes—Part 2: Transfer-Function Method*; ISO: Geneva, Switzerland, 1998.

19. Jonasson, H. *Test Method for the Whole Vehicle*; Technical Report HAR11TR-020301-SP10; 2004

20. *ISO 1996-2:2010: Acoustics—Description, Measurement and Assessment of Environmental Noise—Part 2: Determination of Environmental Noise Levels*; ISO: Geneva, Switzerland, 2010.

21. Licitra, G.; Teti, L.; Cerchiai, M. A modified Close Proximity method to evaluate the time trends of road pavements acoustical performances. *Appl. Acoust.* **2014**, *76*, 169–179.

22. Bollard, K. *Report on the Status of Rubberized Asphalt Traffic Noise Reduction in Sacramento County*; County of Sacramento Public Works Agency—Transportation Division: Sacramento, CA, USA, 1999.

23. Bucka, M. *Asphalt Rubber Overlay Noise Study Update*; Technical Report AAAI Report 1272; County of Sacramento Public Works Agency: Sacramento, CA, USA, December 2002.

24. Treleaven, L.; Pulles, B.; Bilawchuk, S.; Donovan, H. Asphalt rubber—The quiet pavement? In Proceedings of the Annual Conference and Exhibition of the Transportation Association of Canada, Charlottetown, Canada, 17–20 September 2006.

# 10

# Morphology and Microstructure of NiCoCrAlYRe Coatings after Thermal Aging and Growth of an Al$_2$O$_3$-Rich Oxide Scale

**Giovanni Di Girolamo** [1,*], **Alida Brentari** [2] **and Emanuele Serra** [1]

[1]  ENEA, Materials Technology Unit, Casaccia Research Center, Rome 00123, Italy;
    E-Mail: emanuele.serra@enea.it
[2]  ENEA, Materials Technology Unit, Faenza Research Center, Faenza 48018, Italy;
    E-Mail: alida.brentari@enea.it

*  Author to whom correspondence should be addressed; E-Mail: giovanni.digirolamo@enea.it.

External Editor: Ugo Bardi

**Abstract:** The surface of metal parts operating at high temperature in energy production and aerospace industry is typically exposed to thermal stresses and oxidation phenomena. To this aim, plasma spraying was employed to deposit NiCoCrAlYRe coatings on metal substrates. The effects of early-stage oxidation, at ~1100 °C, on their microstructure were investigated. The partial infiltration of oxygen through some open pores and microcracks embedded in coating microstructure locally assisted the formation of a stable Al$_2$O$_3$ scale at the splat boundary, while the diffusion of Cr and Ni and the following growth of Cr$_2$O$_3$, Ni(Cr,Al)$_2$O$_4$ and NiO were restricted to Al depleted isolated areas. At the same time, a continuous, dense and well adherent Al$_2$O$_3$ layer grew on the top-surface, and was somewhere supported by a thin mixed oxide scale mainly composed of Cr$_2$O$_3$ and spinels. Based on these results, the addition of Re to the NiCoCrAlY alloy is able to enhance the oxidation resistance.

**Keywords:** high temperature; coatings; atmospheric plasma spraying

# 1. Introduction

High-temperature coatings are commonly employed to protect the surface of metal components operating in energy production and aerospace industry, with the purpose to improve their resistance to oxidation and hot-corrosion phenomena, as well as to prolong their lifetime [1].

High-temperature coatings are also potential tools for increasing the efficiency of thermal engines. The application of ceramic thermal barrier coatings (TBCs) is particularly well-suited for this challenge, by reducing the heat flux and the temperature at the surface of the underlying component [2]. In this context, metal coatings can be successfully adopted to cover turbine components in order to improve their environmental resistance, without compromising their structural stability and mechanical strength [3–5]. Among them, the well-known MCrAlY (Me = Ni, Co or Ni/Co) coatings can be used as overlay coatings or in conjunction with an upper ceramic TBC [6,7].

Plasma spraying is a cost-effective technology to manufacture MCrAlY coatings. In this process, a gas plasma is employed as heat source to process the powder-based raw material. The powder particles are injected in the plasma jet by a carrier gas, and melted and accelerated toward the substrate, where they impact at high speed and quench, thus producing the build-up of a coating with unique microstructure.

The thermal exposure at elevated temperature promotes the formation of a thin oxide layer on the surface of MCrAlY coating during service [8,9]. However, the extended formation of secondary mixed oxides in conjunction with $Al_2O_3$ has been reported to be detrimental for coating durability. Indeed, their fast growth induces thermal stresses which can develop cracking and delamination, leading to the spallation of the upper TBC [10–13].

It has been reported that the addition of Re to MCrAlY based alloy is able to increase the related oxidation resistance as well as the mechanical properties. The improvement has been somewhat explained in terms of better high-temperature diffusion behavior of Al, that decreasing the extent of β-NiAl phase depletion within the coating and thus the oxidation rate and the scale spallation [14–16]. A significant enhancement in thermal fatigue resistance has been also noticed, since the addition of Re seemed to have a beneficial effect on the mechanism of crack formation and propagation [17]. Otherwise, other investigators reported that the addition of Re to the metallic alloy involved lower thermal expansion coefficient and higher hardness, but lower oxidation resistance [15].

The aim of this study was to develop atmospheric plasma sprayed NiCoCrAlYRe coatings with enhanced oxidation resistance by reducing the in-flight oxidation of the starting metal particles during processing and thus assisting the formation of an approximately dense and continuous $Al_2O_3$ rich scale during early stage isothermal exposure, in order to partially prevent the fast formation of extended mixed oxides and to enhance the oxidation resistance of these systems at high temperature. The microstructural and morphological changes were investigated by scanning electron microscopy (SEM) and energy dispersive spectroscopy (EDS).

# 2. Experimental Section

NiCoCrAlYRe coatings with thickness of about 150 μm were deposited on Ni-superlloy disks (φ = 25 mm, thickness = 4 mm) starting from a commercial powder feedstock (Sicoat 2453, Siemens

AG, Mulheim, Germany). An atmospheric plasma spraying (APS) system equipped with F4-MB plasma torch (Sulzer Metco, Wolhen, Switzerland) with 6 mm internal diameter nozzle was used for powder processing. The substrates were sand blasted with alumina abrasive powder (Metcolite F, Sulzer Metco, Westbury, NY, USA) to increase their surface roughness and to improve the mechanical adhesion between coating and substrate. The grain size of the abrasive particles was in the range between 350 and 1100 μm, as reported by the supplier. The blasting parameters employed were: air pressure = 0.45 MPa, gun-substrate distance = 10 mm, angle = 45° and spray time = 10 s. The specimens were then cleaned in ethanol, placed on a rotating sample holder and coated by successive torch passes. The processing parameters are reported in Table 1.

**Table 1.** Plasma spraying parameters used in this work.

| Plasma spraying parameters | Value |
|---|---|
| Current [A] | 600 |
| Voltage [V] | 67 |
| Substrate tangential speed [mm/s] | 1041 |
| Gun velocity [mm/s] | 4 |
| Primary gas Ar flow rate [slpm *] | 55.5 |
| Secondary gas $H_2$ flow rate [slpm *] | 9.5 |
| Spraying distance [mm] | 130 |
| Carrier gas flow rate [slpm *] | 2 |
| Powder feed rate [g/min] | 52 |
| Distance torch-injector [mm] | 6 |
| Injector diameter [mm] | 1.8 |

* slpm—standard liters per minute.

The surface roughness of substrates and coatings was measured using a 3D optical surface profilometer (New View 5000 system, Zygo Corporation, Middlefield, CT, USA). This white-light interferometer system allowed for study of the topography of the coating surface, which was previously metallized by Al thin film vacuum deposition (thickness = 40 nm), in order to allow reflection. The measurements were performed on areas of $0.7 \times 0.5$ μm$^2$. The interferogram of the coating surface was then processed and transformed to three-dimensional high-resolution surface image by frequency domain analysis. The arithmetic mean roughness value ($R_a$) was calculated as the average deviation of the surface profile from the mean line, defined by Equation (1):

$$R_a = \frac{\int_{x=0}^{x=L} |y(x)| dx}{L} \tag{1}$$

where $y$ is the deviation of the surface profile from the centerline. The measured substrate roughness was found to be $6.9 \pm 1.1$ μm. In turn, the surface roughness $R_a$ of the as-sprayed coatings was found to be $13.5 \pm 0.4$ μm. After plasma spray processing, the surface roughness doubled, depending on the size of the powder particles as well as on their momentum and flattening upon impact on the substrate surface.

After deposition the coated samples were heated at 5 °C/min, held at the maximum temperature of 1110 °C for 48 h and then cooled inside the furnace at 5 °C/min.

As-sprayed and aged samples were then cut by low-speed diamond saw and their cross sections were mounted in vacuum in two-part polymer, polished with diamond suspensions and finished to 0.25 μm. The morphology and the microstructure of as-sprayed and oxidized coating cross sections were analyzed by scanning electron microscope (SEM-Field Emission Gun, Leo Gemini mod. 1530, Carl Zeiss, Oberkochen, Germany) equipped with energy dispersive spectroscopy (EDS). EDS analysis was used to measure the composition of powder and coatings as well as to map the distribution of the constituents.

## 3. Results and Discussion

As shown in the SEM picture reported in Figure 1, the powder feedstock is composed of spherical particle agglomerates. The inset in the left corner of the same micrograph shows the detailed surface morphology of an agglomerate, characterized by grains with size lower and higher than 1 μm and with spherical, quasi-spherical and elongated shape. Some dendritic structures can be also observed. The grey contrast could be associated to fluctuations in Al content. The particle size distribution measured by SEM observations was in the range between 45 and 65 μm, while the average particle size was $54 \pm 5$ μm.

**Figure 1.** SEM spherical morphology of NiCoCrAlYRe particle agglomerates. Grains with submicrometer size can be observed on the surface.

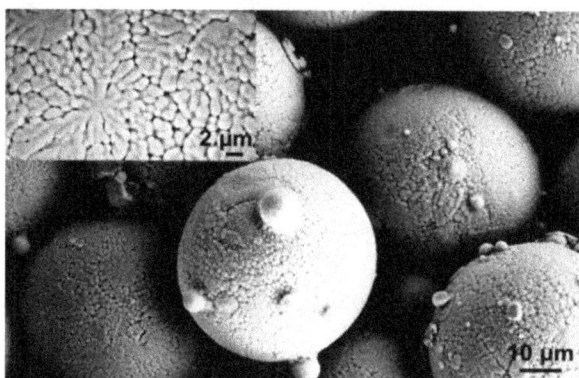

In turn Figure 2 shows the EDS map of powder particles, suggesting the presence of significant amount of Re and Y. The measured chemical composition (wt%) of the powder feedstock can be summarized as 52Ni 10Co 24Cr 10Al 1Y 3Re.

Figure 3a,b shows the cross sectional microstructure of plasma sprayed NiCoCrAlYRe coating at different magnification. The coating exhibits a lamellar microstructure composed of overlapped splats, originated by deformation and quenching of molten droplets at the substrate surface during deposition process. Some pores and splat boundaries can be observed. The former are related to filling defects produced by gas entrapped in the molten droplets during coating build-up, whereas the splat boundaries derived from weak bonding between the overlapped splats, depending on the temperature distribution along the same ones during cooling to room temperature.

A large and approximately uniform distribution of dark grey β-NiAl precipitates can be observed within the light-grey γ matrix, as detectable in Figure 3b. At splat boundary very restricted Al depletion and coalescence of β precipitates are also noticed. It is worth noting that these precipitates represent an Al reservoir for next selective oxidation.

**Figure 2.** Energy dispersive spectroscopy (EDS) map showing the composition of NiCoCrAlYRe powder particles.

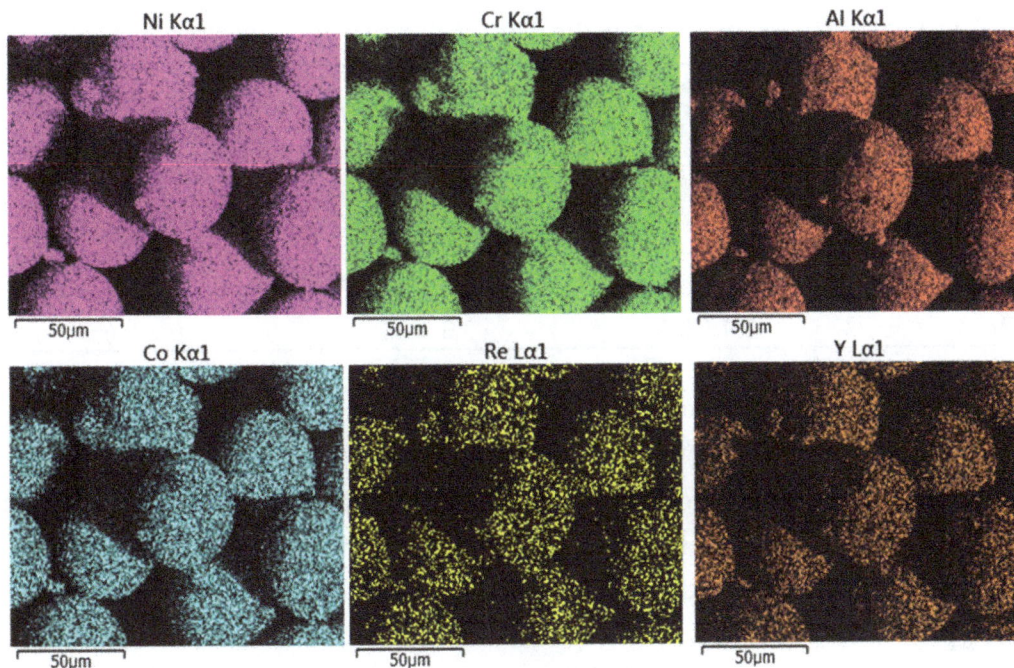

**Figure 3. (a)** Cross sectional SEM microstructure showing two-phase microstructure and typical defects of as-sprayed NiCoCrAlYRe coating; and **(b)** SEM micrograph showing coarser and finer β precipitates dispersed in the γ matrix.

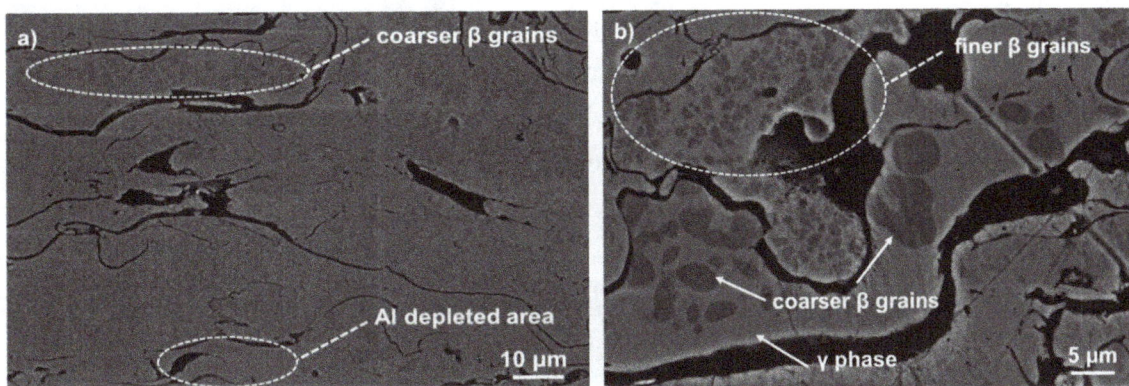

According to EDS analyses reported in Table 2, the areas rich with β precipitates contain about 18 wt% Al and are partially depleted in Cr and Re with respect to the γ phase, because of limited solubility of Re in the same β phase [18]. Indeed, the β phase contains 14 wt% Cr and 1 wt% Re, whereas the γ phase approximately contains 25 wt% Cr and 3 wt% Re.

It has been reported that high retention of β phase is usually obtained using HVOF (High Velocity Oxy Fuel) spraying, while APS coatings are mainly composed of γ phase because of higher oxidation rate experienced by the sprayed particles during processing [19]. However, in the present case the composition of the plasma gas mixture employed (total gas mixture flow = 65 slpm and Ar/H$_2$ ratio ~6) allowed for prevention of overheating and oxidation of the surface of the molten particles during spraying as well as the Al depletion. Indeed, the average content of Al did not decrease after spraying

(~10 wt%), as well as the average wt% Re, so that the composition of the as-sprayed coating was approximately equal to that of the starting powder particles.

**Table 2.** Chemical composition of as-sprayed NiCoCrAlYRe coating, matrix and β-rich areas.

| Chemical element | As-sprayed coating (wt%) | γ Matrix (wt%) | β Phase rich area (wt%) |
|---|---|---|---|
| Ni | $52.0 \pm 0.2$ | $51.2 \pm 0.5$ | $58.0 \pm 1.4$ |
| Co | $10.0 \pm 0.1$ | $10.0 \pm 0.1$ | $9.0 \pm 0.3$ |
| Cr | $23.9 \pm 0.1$ | $24.7 \pm 1.1$ | $14.0 \pm 1.6$ |
| Al | $10.5 \pm 0.1$ | $10.0 \pm 1.0$ | $18.0 \pm 1.0$ |
| Y | $0.7 \pm 0.3$ | $0.7 \pm 0.3$ | – |
| Re | $2.9 \pm 0.3$ | $3.3 \pm 0.2$ | $1.0 \pm 0.6$ |

Figure 4 shows a localized area in coating cross section which suffered internal oxidation during next high-temperature aging. The thermal exposure in air environment promoted the partial infiltration of oxygen through the open pores and microcracks located along coating thickness. Therefore, in some regions near the coating surface the penetration of oxygen promoted the Al diffusion from the β-NiAl phase and the following formation of a dark grey $Al_2O_3$ scale at splat boundary (the surface of metal particles), which typically surrounds the voids. In restricted areas the outward diffusion of Ni and Cr promoted further chemical reactions, leading to the growth of some mixed oxide scales on the $Al_2O_3$ surface. The growth of secondary oxide phases generally occurred in restricted areas where more pronounced Al depletion did not allow the formation of a stable and protective $Al_2O_3$ layer, whereas it was less significant in the areas characterized by growth of a continuous and well adherent $Al_2O_3$ layer.

**Figure 4.** A localized area characterized by internal oxidation within the cross sectional microstructure of NiCoCrAlYRe coating. After high-temperature aging dark grey alumina flakes grow at splat boundary and are somewhere supported by grey spinel-type porous oxides, whereas NiO appears in form of light grey precipitates embedded in the spinel phase.

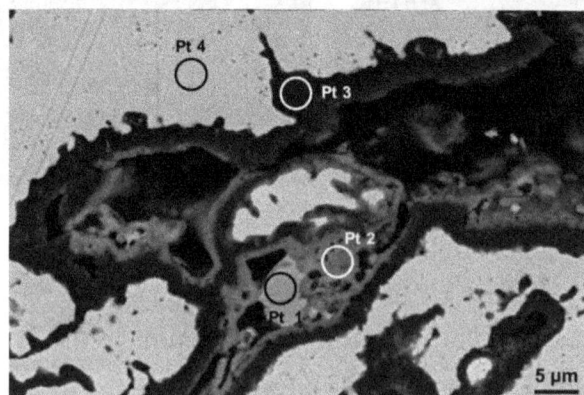

Figure 5 reports the EDS maps related to the area presented in Figure 4, demonstrating the formation of $Al_2O_3$ oxide scale at splat boundary as well as the formation of $Cr_2O_3$, $Ni(Cr,Al)_2O_4$ (medium grey phase) and NiO (light grey) at lesser extent in any localized areas with lower Al content [20]. It should be noted that the maps of Ni, Co, Cr and Re are well matched, thus suggesting the uniform distribution of these elements within the γ matrix. Figure 6 reports the EDS spectra related to different regions identified in coating microstructure, and denoted by circled areas in Figure 4. These spectra are compatible with the presence of NiO, $Cr_2O_3$ + spinels, $Al_2O_3$ and $\gamma$ matrix, respectively. In the Pt2 area a significant amount of Al (~8 wt%) was also found, this suggesting the presence of $Al_2O_3$ and $NiAl_2O_4$. In turn the presence of Cr suggested the formation of $Cr_2O_3$ and $NiCr_2O_4$. Otherwise, the wt% of Al in the depleted $\gamma$ phase was found to be close to 4%, this suggesting a depletion of Al to form $Al_2O_3$ oxide at splat boundary. A slight decrease of Cr content has been noticed in the areas near the oxide layer, due to Cr diffusion and formation of its oxides. As clearly displayed in the elemental maps reported in Figure 5 as well as in the EDS analyses shown in Figure 6, the rhenium is uniformly distributed in the depleted $\gamma$ matrix and its wt% was found to be 3.9% ± 0.5%, slightly higher than the related content observed in the matrix before high-temperature exposure (3.3 wt%), this suggesting Re diffusion from the $\beta$ phase. Otherwise the presence of Re could not be observed in the oxidized areas. It is reasonable to suppose that Re was not directly affected by formation of its oxides, even if the formation of volatile $Re_2O_7$ could not be excluded. However, the addition of Re to the metal alloy played a positive role on the oxidation resistance, promoting better diffusion of Al and lower diffusion of other elements such as Ni and Cr, so that $Al_2O_3$ was the major oxidation product.

**Figure 5.** EDS maps showing the distribution of the elements in the oxidized region shown in Figure 4. Dark grey alumina scale grows at splat boundary and is somewhere supported by irregular grey mixed oxides.

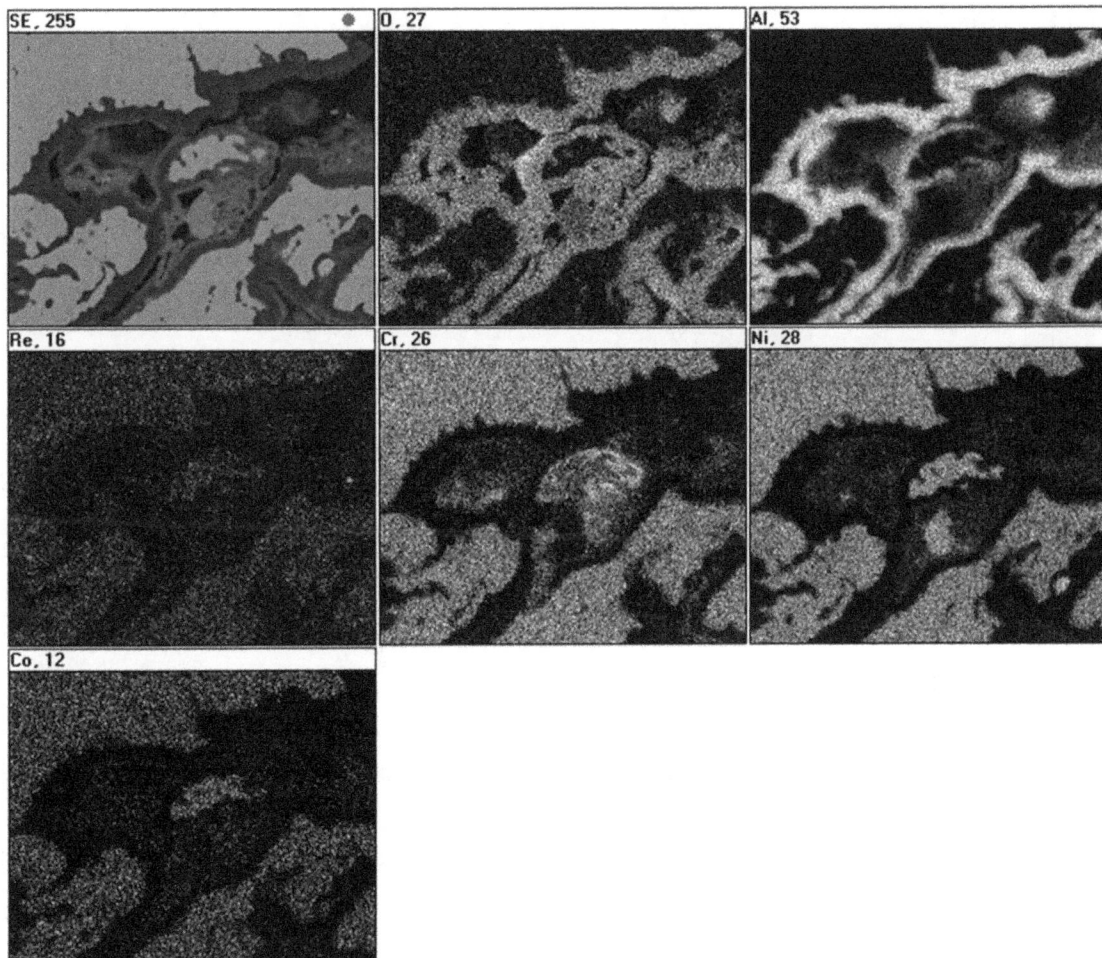

**Figure 6.** EDS spectra of different spot areas localized in coating microstructure, indicating the presence of (**a**) NiO; (**b**) $Cr_2O_3$ + spinels; (**c**) $Al_2O_3$; and (**d**) $\gamma$ matrix.

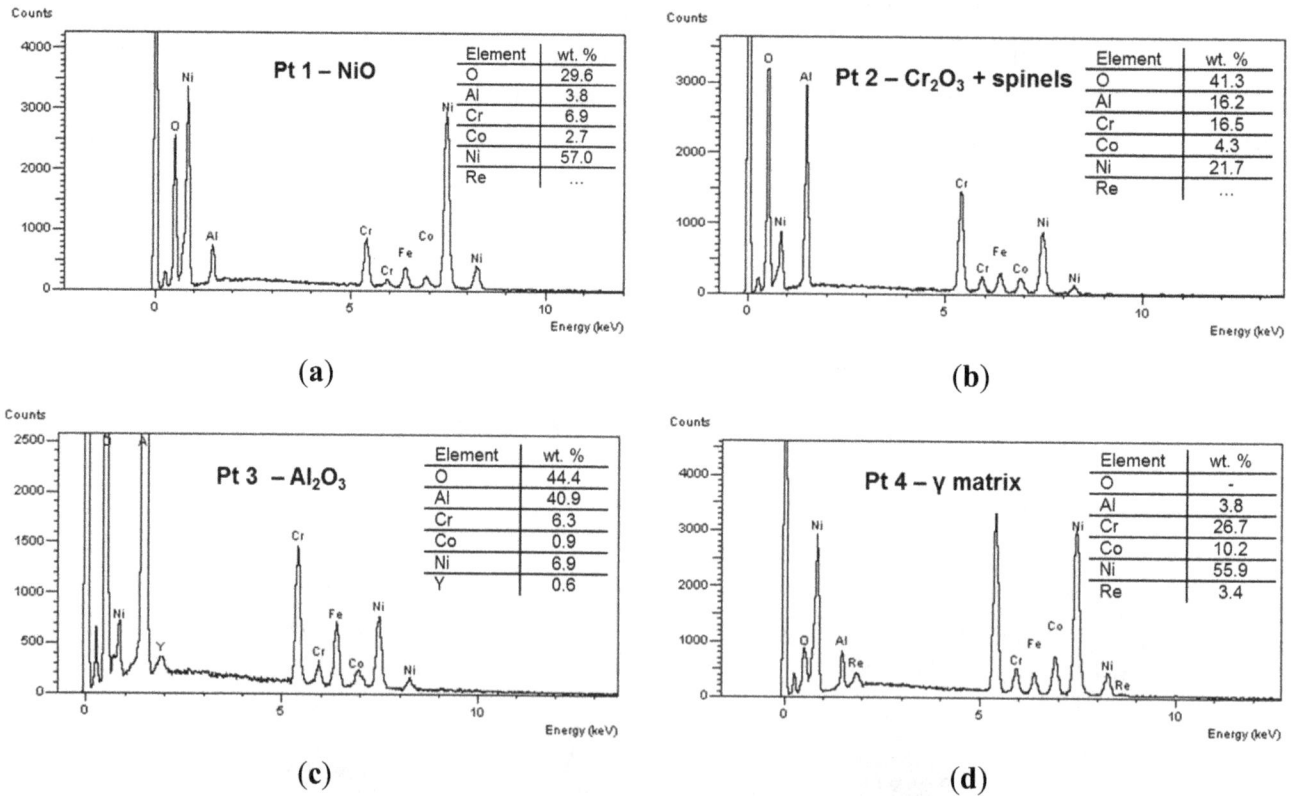

(a)

| Element | wt. % |
|---------|-------|
| O | 29.6 |
| Al | 3.8 |
| Cr | 6.9 |
| Co | 2.7 |
| Ni | 57.0 |
| Re | ... |

Pt 1 – NiO

(b)

| Element | wt. % |
|---------|-------|
| O | 41.3 |
| Al | 16.2 |
| Cr | 16.5 |
| Co | 4.3 |
| Ni | 21.7 |
| Re | ... |

Pt 2 – $Cr_2O_3$ + spinels

(c)

| Element | wt. % |
|---------|-------|
| O | 44.4 |
| Al | 40.9 |
| Cr | 6.3 |
| Co | 0.9 |
| Ni | 6.9 |
| Y | 0.6 |

Pt 3 – $Al_2O_3$

(d)

| Element | wt. % |
|---------|-------|
| O | - |
| Al | 3.8 |
| Cr | 26.7 |
| Co | 10.2 |
| Ni | 55.9 |
| Re | 3.4 |

Pt 4 – $\gamma$ matrix

The most significant chemical reactions occurring during high-temperature exposure can be summarized as follows:

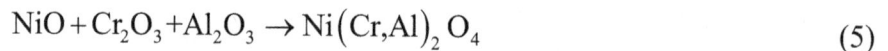

$$2Al + (3/2)O_3 \rightarrow Al_2O_3 \tag{2}$$

$$2Cr + (3/2)O_3 \rightarrow Cr_2O_3 \tag{3}$$

$$Ni + (1/2)O_2 \rightarrow NiO \tag{4}$$

$$NiO + Cr_2O_3 + Al_2O_3 \rightarrow Ni(Cr,Al)_2 O_4 \tag{5}$$

As also reported by other authors, the spinel phases were formed because of reaction between $Cr_2O_3$ (or $Al_2O_3$) and NiO [20]. Based on the XRD results and phase composition study reported in a previous work as well as on the EDS analysis herein discussed it could be supposed that the spinel phase was mainly composed of $NiCr_2O_4$ [21].

It should be noted that it was very arduous to provide the exact composition of the various oxide scales detected, because their composition locally changed and the small volume considered for EDS measurements affected the quantitative results.

Figure 7 shows the cross-sectional microstructure near coating top-surface after early stage oxidation and suggests the formation of a double oxide scale, composed of an inner nearly continuous and dense alumina layer, followed by an upper and more brittle thin mixed oxide layer. As the oxidation time increased the formation of the $Al_2O_3$ scale on the coating surface occurred according to Equation (2). $Al_2O_3$ was the most thermodynamically stable oxide, so that $Al_2O_3$ mainly tended to grow after a continuous TGO was developed. However, when Al was depleted to a certain extent near the coating top-surface Ni and Cr partially infiltrated through some cracks at nano scale across the growing Al2O3 layer, thus forming their oxides (NiO and Cr2O3) as well as spinels, via solid state reaction with pre-existing Al2O3 and C2O3 according to Equation (5). It is reasonable to suppose that the effect of the outward diffusion of Ni and Cr through the growing alumina layer was predominant with respect to that produced by inward oxygen infiltration, so that the mixed layer grew on the same Al2O3 layer. In addition, NiO and Cr2O3 could be formed at lesser extent in conjunction to Al2O3 during the first stage of thermal aging and then could react with the same Al2O3 to form spinels on the external surface.

**Figure 7.** Cross sectional SEM microstructure of oxidized NiCoCrAlYRe coating showing the formation of a double oxide scale on the top-surface, composed of an inner continuous Al2O3 scale supported by an upper mixed oxide layer.

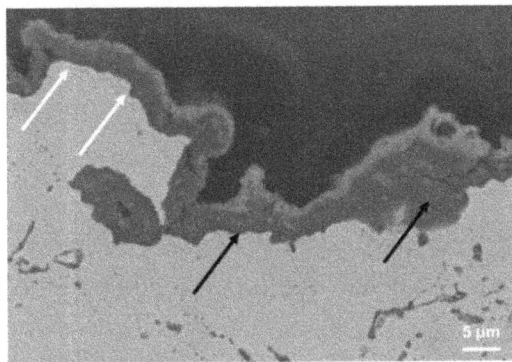

For this purpose the presence of the first stable alumina layer allowed for reduction of further oxygen infiltration, partially preventing the effect of internal oxidation, while the presence of the upper layer can represent a disadvantage, because some surface protrusions, derived from its fast and heterogeneous growth, and the thermal stresses herein originated can cause microcracking and delamination. However, if the coating had preserved a sufficient Al activity to allow the rehealing of the protective Al2O3 scale, the spallation of the outer mixed oxide layer does not represent a critical factor during the first hours of operation.

Otherwise, the same effect should be carefully taken in account when a ceramic TBC is applied on the NiCoCrAlYRe coating, because the thermal stresses generated at the interface and the horizontal microcraking can reduce the adhesion and thus promote unexpected TBC spallation during long-term service [22]. It is worth noting that the presence of a rough surface rich of asperities can partially impede the formation of a stable and continuous Al2O3 layer, allowing the fast formation of mixed oxide scale with embedded pores and microcracks. Indeed, as shown in Figure 7 the convex surface areas along the top-surface are characterized by the growth of a thin Al2O3 scale with an outer very thin spinel-type oxide scale (see the white arrows in the picture), while in the concave surface areas a thicker double oxide scale grew quickly, producing some surface protrusions, tensile stresses and microcracking (see the black arrows in the picture). High surface roughness increases the specific surface area for oxidation and also reduces the adhesion of the oxide scale during long-term service, as previously investigated [23]. The formation of mixed oxides has been also found in MCrAlY coatings deposited under vacuum atmosphere or by HVOF spraying and aged at 1000–1100 °C [24–26] and cannot be totally prevented, so that it is an important key issue to reduce their growth and to have a well adherent and dense oxide scale mostly composed of Al$_2$O$_3$ with low oxygen diffusivity, and able to reduce permeation of Ni and Cr as well as further oxygen infiltration. For this purpose the NiCoCrAlYRe particles have been herein processed to fabricate coatings with the above mentioned two phase structure, by reducing in-flight oxidation at their external surface and thus preventing Al depletion. The resulting high retention of β precipitates in the matrix allows for controlling at some extent the oxide scale formation during next high-temperature exposure. High process gas flow (Ar + H$_2$ = 65 slpm) and Ar/H$_2$ ratio (~6) were employed for melting and transport of melted particles in the plasma jet, in order to prevent metal surface oxidation during processing in air environment.

The simultaneous presence of different oxides can be also appreciated from the analysis of Figures 8 and 9. According to the EDS maps the presence of Al$_2$O$_3$ can be noticed as well as the formation of light grey structures particularly enriched in Cr. Figure 10a shows some metallic grains with size close to hundreds of nanometers and located below the TGO layer. The surface oxide layer was mechanically removed to observe their morphology and to calculate their composition, which was close to that of the depleted γ phase (see, for example, the graph 4 reported in Figure 6). Columnar and equiaxed grains of Al$_2$O$_3$ can be observed in the center of Figure 10b. In turn, Figure 10c shows some different structures larger than 1 μm and composed of a mixture of NiCr$_2$O$_4$ and NiO (see also the elemental map in Figure 8).

**Figure 8.** EDS maps of a localized region on the oxidized coating top-surface. The formation of mixed oxide scales composed of Ni, Al and Cr can be appreciated.

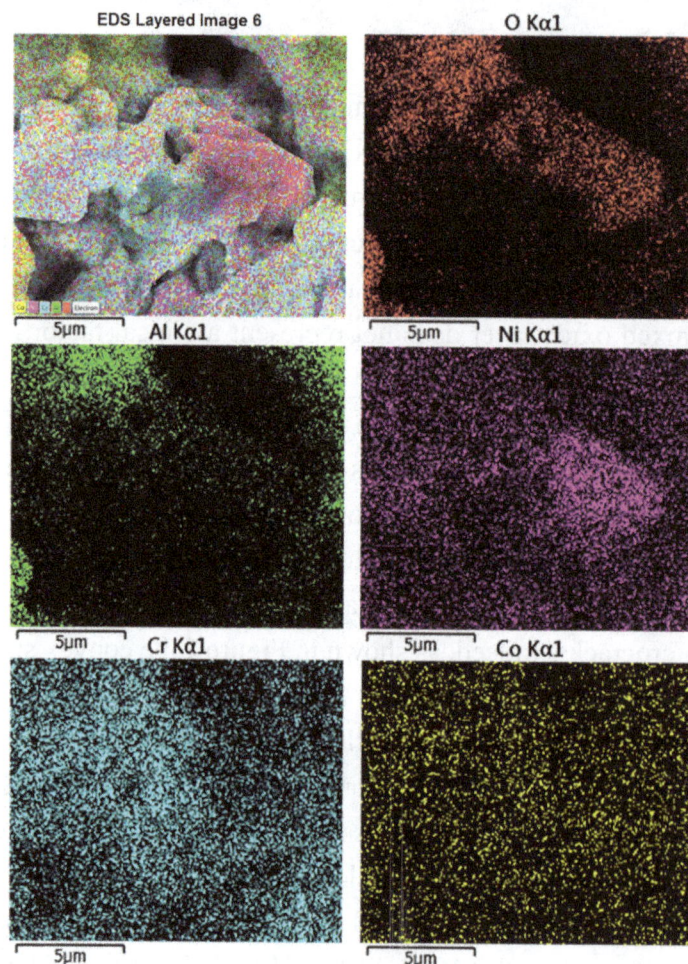

**Figure 9.** A localized area on oxidized coating top-surface characterized by the presence of mixed oxides grown on the surface of Al₂O₃ grains. NiO and spinels show polygonal and equiaxed shape, respectively.

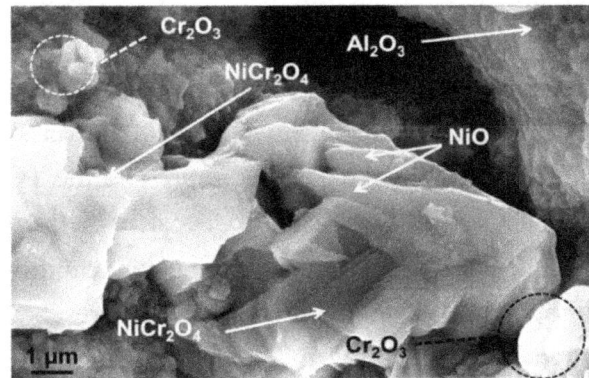

**Figure 10.** SEM morphology of (**a**) metal grains after removal of the oxide layer; (**b**) Al₂O₃ grains grown at the boundary of metal particles; and (**c**) mixed oxide scales with polygonal (NiO) and equiaxed shape (spinels).

Generally, the NiO scale exhibited granular or polygonal/elongated shape, while spinels and Cr₂O₃ were present in form of equiaxed grains. The presence of these mixed oxide structures was appreciated in restricted and small areas where Al exhibited a reduced activity and its concentration was not enough to promote the selective formation of the preferable Al₂O₃ scale [20]. Based on the results discussed herein, the addition of Re to the NiCoCrAlY alloy allows for prevention of Al depletion during processing, thus involving high retention of β phase in the as-sprayed coating. Therefore, it plays a beneficial role on the related oxidation resistance at high temperature by improving Al diffusion and reducing the outward diffusion of elements like Cr and Ni. In other words, the presence of Re in the microstructure is useful to promote the formation of a protective, well adherent, dense and quasi continuous Al₂O₃ layer at metal surface, so that the formation of mixed oxide scales is reduced at lower extent as well as the internal oxidation occurring at splat boundaries. Therefore, the β phase depletion and the oxidation rate of MCrAlYRe coatings are lower with respect to the conventional MCrAlY ones [6]. The addition of Re influences the morphology and the amount of the phases into the coating.

The studies concerning the composition and the microstructure of NiCoCrAlYRe overlay coatings can potentially affect their temperature capability and durability under different scenarios and be addressed to applications in power plants and aircraft engines. For this purpose, the systems used for propulsion typically experience multiple thermal cycles, whereas the power systems largely operate in isothermal mode with a few cycles. In addition the oxidation rate and the TGO growth become more critical when a ceramic TBC is applied on the metal coating, because the formation of the TGO at the interface between bond coat and top coat strongly affects the adhesion and the failure of the same TBC. Apart from the different operating conditions (temperature, cycling, presence of corrosive media and coating architecture) the oxidation mechanism of the metal coatings—including type, thickness, morphology and adhesive strength of the TGO—should be controlled at some extent to enhance coating performance.

## 4. Conclusions

The relative dense microstructure of NiCoCrAlYRe coatings, deposited by atmospheric plasma spraying, was characterized by high retention of well dispersed β-NiAl precipitates in the γ matrix. β grain coalescence and no Al depletion were observed after processing. Then high-temperature aging produced some effects of oxidation and oxide scale formation. In localized areas the presence of open pores and splat boundaries assisted partial oxygen infiltration through the coating thickness, promoting the formation of a continuous and dense $Al_2O_3$ scale at splat boundary. The diffusion of Cr and Ni assisted minor formation of $Cr_2O_3$, $Ni(Cr,Al)_2O_4$ and NiO on the surface of the same $Al_2O_3$ scale. At the same time, the oxidation of coating top-surface promoted the growth of an oxide scale, mainly composed of an inner continuous and well adherent $Al_2O_3$ layer, somewhere followed by an upper mixed oxide scale. The alumina layer was composed of grains with size of hundreds of nanometers, while the above mentioned mixed oxides tended to form complex micronsized/nanosized structures with polygonal and equiaxed shape. Based on these results, the addition of Re to the NiCoCrAlY alloy allowed for prevention of Al depletion during processing, thus involving high retention of β phase in the as-sprayed state. Moreover, it was able to improve Al diffusion and to partially prevent the outward diffusion of elements as Cr and Ni during high-temperature exposure. This was beneficial for promoting the growth of a protective, well adherent, dense and quasi continuous $Al_2O_3$ layer at the metal surface, so that the formation of mixed oxide scales was reduced at lower extent as well as the β phase depletion and the internal oxidation of the coating. For this purpose, further studies are needed to get a full understanding of the complex effect of Re addition on the oxidation resistance of NiCoCrAlYRe coatings.

## Acknowledgments

The authors wish to thank C. Blasi for her valuable contribution during plasma spraying.

## Authors Contribution

G.D. designed the experiments and analyzed the data; A.B. and E.S. contributed analysis tools and performed microstructural analyses.

## Conflicts of Interest

The authors declare no conflict of interest.

## References

1.  Davis, J.R. *Handbook of Thermal Spray Technology*; ASM International: Materials Park, OH, USA, 2004.

2.  Clarke, D.R.; Levi, C.G. Materials design for the next generation thermal barrier coatings. *Annu. Rev. Mater. Res.* **2003**, *33*, 383–417.

3.  Taylor, M.P. An oxidation study of an MCrAlY overlay coating. *Mater. High Temp.* **2005**, *22*, 433–436.

4.  Pollock, T.M.; Lipkin, D.M.; Hemker, K.J. Multifunctional coating interlayers for thermal-barrier systems. *MRS Bull.* **2012**, *7*, 923–931.

5.  Di Girolamo, G.; Pagnotta, L. Thermally sprayed coating for high-temperature applications. *Rec. Pat. Mater. Sci.* **2011**, *4*, 173–190.

6.  Di Girolamo, G.; Alfano, M.; Pagnotta, L.; Taurino, A.; Zekonyte, J.; Wood, R.J.K. On the early stage isothermal oxidation of APS CoNiCrAlY coatings. *J. Mater. Eng. Perf.* **2012**, *21*, 1989–1997.

7.  Haynes, J.A.; Ferber, M.K.; Porter, W.D. Thermal cycling behavior of plasma-sprayed thermal barrier coatings with various MCrAlX bond coats. *J. Therm. Spray Technol.* **2000**, *9*, 38–48.

8.  Daroonparvar, M.R.; Hussain, M.S.; Mat Yajid, M.A. The role of formation of continues thermally grown oxide layer on the nanostructured NiCrAlY bond coat during thermal exposure in air. *Appl. Surf. Sci.* **2012**, *261*, 287–297.

9.  Mercer, C.; Hovis, D.; Heuer, A.; Tomimatsu, T.; Kagawa, Y.; Evans, A.G. Influence of thermal cycle on surface evolution and oxide formation in a superalloy system with a NiCoCrAlY bond coat. *Surf. Coat. Technol.* **2008**, *202*, 4915–4921.

10. Ni, L.Y.; Liu, C.; Huang, H.; Zhou, C.G. Thermal cycling behavior of thermal barrier coatings with HVOF NiCrAlY bond coat. *J. Therm. Spray Technol.* **2011**, *20*, 1133–1138.

11. Chen, W.R.; Wu, X.; Marple, B.R.; Patnaik, P.C. The growth and influence of thermally grown oxide in a thermal barrier coating. *Surf. Coat. Technol.* **2006**, *201*, 1074–1079.

12. Evans, A.G.; Mumm, D.R.; Hutchinson, J.W.; Meier, G.H.; Petit, F.S. Mechanisms controlling the durability of thermal barrier coatings. *Prog. Mater. Sci.* **2001**, *46*, 505–553.

13. Rabiei, A.; Evans, A.G. Failure mechanisms associated with the thermally grown oxide in plasma-sprayed thermal barrier coatings. *Acta Mater.* **2000**, *48*, 3963–3976.

14. Czech, N.; Schmitz, F.; Stamm, W. Improvement of MCrAlY coatings by addition of rhenium. *Surf. Coat. Technol.* **1994**, *68–69*, 17–21.

15. Liang, J.J.; Wei, H.; Zhu, Y.L.; Sun, X.F.; Hu, Z.Q.; Dargush, M.S.; Yao, X.D. Influence of Re on the properties of a NiCoCrAlY coating alloy. *J. Mater. Sci. Technol.* **2011**, *27*, 408–414.

16. Beele, W.; Czech, N.; Quadakkers, W.J.; Stamm, W. Long term oxidation tests on a Re-containing MCrAlY coating. *Surf. Coat. Technol.* **1997**, *94–95*, 41–45.

17. Czech, N.; Schmitz, F.; Stamm, W. Microstructural analysis of the role of rhenium in advanced MCrAlY coatings. *Surf. Coat. Technol.* **1995**, *76–77*, 28–33.

18. Huang, W.; Chang, Y.A. A thermodynamic description of the Ni-Al-Cr-Re system. *Mater. Sci. Eng. A* **1999**, *259*, 110–119.

19. Di Ferdinando, M.; Fossati, A.; Lavacchi, A.; Bardi, U.; Borgioli, F.; Borri, C.; Giolli, C.; Scrivani, A. Isothermal oxidation resistance comparison between air plasma sprayed, vacuum plasma sprayed and high velocity oxygen fuel sprayed CoNiCrAlY bond coats. *Surf. Coat. Technol.* **2010**, *204*, 2499–2503.

20. Liang, G.Y.; Zhu, C.; Wu, X.Y.; Wu, Y. The formation model of Ni-Cr oxides on NiCoCrAlY-sprayed coating. *Appl. Surf. Sci.* **2001**, *257*, 6468–6473.

21. Di Girolamo, G.; Brentari, A.; Blasi, C.; Pilloni, L.; Serra, E. High-temperature oxidation and oxide scale formation in plasma sprayed CoNiCrAlYRe coatings. *Metall. Mater. Trans. A* **2014**, *45*, 5362–5370.

22. Wright, P.K.; Evans, A.G. Mechanisms governing the performance of thermal barrier coating. *Curr. Opin. Solid State Mater. Sci.* **1999**, *4*, 255–265.

23. Gil, A.; Shemet, V.; Vassen, R.; Subanovic, M.; Toscano, J.; Naumenko, D.; Singhiser, L.; Quadakkers, W.J. Effect of surface condition on the oxidation behavior of MCrAlY coatings. *Surf. Coat. Technol.* **2006**, *201*, 3824–3828.

24. Tang, F.; Ajdelsztajn, L.; Kim, G.E.; Provenzano, V.; Schoenung, J.M. Effects of surface oxidation during HVOF processing on the primary stage oxidation of a CoNiCrAlY coating. *Surf. Coat. Technol.* **2004**, *185*, 228–233.

25. Poza, P.; Grant, P.S. CoNiCrAlY coatings after heat treatment and isothermal oxidation. *Surf. Coat. Technol.* **2006**, *201*, 2887–2896.

26. Saeidi, S.; Voisey, K.T.; McCartney, D.G. The effect of heat treatment on the oxidation behavior of HVOF and VPS CoNiCrAlY coatings. *J. Therm. Spray Technol.* **2009**, *18*, 209–216.

# 11

# Recent Photocatalytic Applications for Air Purification in Belgium

**Elia Boonen\* and Anne Beeldens**

Belgian Road Research Center (BRRC), Woluwedal 42, 1200 Brussels, Belgium;
E-Mail: a.beeldens@brrc.be

\* Author to whom correspondence should be addressed; E-Mail: e.boonen@brrc.be

**Abstract:** Photocatalytic concrete constitutes a promising technique to reduce a number of air contaminants such as $NO_x$ and VOC's, especially at sites with a high level of pollution: highly trafficked canyon streets, road tunnels, the urban environment, *etc.* Ideally, the photocatalyst, titanium dioxide, is introduced in the top layer of the concrete pavement for best results. In addition, the combination of $TiO_2$ with cement-based products offers some synergistic advantages, as the reaction products can be adsorbed at the surface and subsequently be washed away by rain. A first application has been studied by the Belgian Road Research Center (BRRC) on the side roads of a main entrance axis in Antwerp with the installation of 10.000 m² of photocatalytic concrete paving blocks. For now however, the translation of laboratory testing towards results *in situ* remains critical of demonstrating the effectiveness in large scale applications. Moreover, the durability of the air cleaning characteristic with time remains challenging for application in concrete roads. From this perspective, several new trial applications have been initiated in Belgium in recent years to assess the "real life" behavior, including a field site set up in the Leopold II tunnel of Brussels and the construction of new photocatalytic pavements on industrial zones in the cities of Wijnegem and Lier (province of Antwerp). This paper first gives a short overview of the photocatalytic principle applied in concrete, to continue with some main results of the laboratory research recognizing the important parameters that come into play. In addition, some of the methods and results, obtained for the existing application in Antwerp (2005) and during the implementation of the new realizations in Wijnegem and Lier (2010–2012) and in Brussels (2011–2013), will be presented.

**Keywords:** TiO$_2$; photocatalysis; concrete pavements; air purification; nitrogen oxides surface treatment

---

## 1. Introduction

Emission from the transport sector has a particular impact on the overall air quality because of its rapid rate of growth: goods transport by road in Europe (EU-27) has increased by 31% (period 1995–2009), while passenger transport by road in the EU-27 has gone up by 21% and passenger transport in air by 51% in the same period [1]. The main emissions caused by motor traffic are nitrogen oxides (NO$_x$), hydrocarbons (HC) and carbon monoxide (CO), accounting for respectively 46%, 50% and 36% of all such emissions in Europe in 2008 [2].

These pollutants have an increasing impact on the urban air quality. In addition, photochemical reactions resulting from the action of sunlight on NO$_2$ and VOC's (volatile organic compounds) lead to the formation of "photochemical smog" and ozone, a secondary long-range pollutant, which impacts in rural areas often far from the original emission site. Acid rain is another long-range pollutant influenced by vehicle NO$_x$ emissions and resulting from the transport of NO$_x$, oxidation in the air into HNO$_3$ and finally, precipitation of (acid) NO$_3^-$ with harmful consequences for building materials (corrosion of the surface) and vegetation.

The European Directives [3] impose a limit to the NO$_2$ concentration in ambient air of maximum 40 µg/m$^3$ NO$_2$ (21 ppbV) averaged over 1 year and 200 µg/m$^3$ (106 ppbV) averaged over 1 h. These limit values gradually decreased from 50 and 250 in 2005 to the final limit in 2010.

Heterogeneous photocatalysis is a promising method for NO$_x$ abatement. In the presence of UV-light, the photocatalytically active form of TiO$_2$ present at the surface of the material is activated, enabling the abatement of pollutants in the air. The translation from laboratory results to real cases is starting. Different applications are implemented in Belgium in order to see the influence of the photocatalytic materials on real scale and to determine the durability of the air purifying capacity over time.

In the first part of this paper, the principle of photocatalytic concrete will be elaborated, followed by a description of the past laboratory research indicating important influencing factors for the purifying process. Next, an overview of the results regarding the first pilot project in Antwerp [2] is given, and finally, the different applications in Belgium that have recently been finished, will be discussed.

## 2. Photocatalytic Concrete: Purifying the Air through the Pavement

A solution for the air pollution by traffic can be found in the treatment of the pollutants as close to the source as possible. Therefore, photocatalytically active materials can be added to the surface of pavement and building materials [4]. Air purification through heterogeneous photocatalysis consists of different steps: under the influence of UV-light, the photoactive TiO$_2$ at the surface of the material is activated. Subsequently, the pollutants are oxidized due to the presence of the photocatalyst and precipitated on the surface of the material. Finally, they can be removed from the surface by the rain or cleaning/washing with water, see Figure 1.

**Figure 1.** Schematic of photocatalytic air purifying pavement.

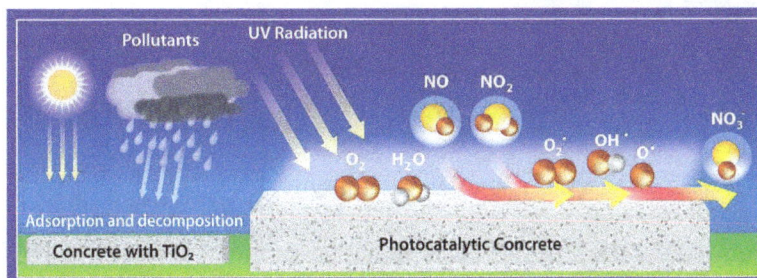

Heterogeneous photocatalysis with titanium dioxide ($TiO_2$) as catalyst is a rapidly developing field in environmental engineering, as it has a great potential to cope with the increasing pollution. Besides its self-cleaning properties, it is known since almost 100 years that titanium dioxide acts as a photo-catalyst that can decompose pollutants under UV radiation [5]. The impulse for the more widespread use of $TiO_2$ photocatalytic materials was further given in 1972 by Fujishima and Honda [6], who discovered the hydrolysis of water in the presence of light, by means of a $TiO_2$-anode in a photochemical cell. In the 1980s, organic pollution in water was also decomposed by adding $TiO_2$ and under influence of UV-light [7]. The application of $TiO_2$, in the photo-active crystalline phase anatase, as air purifying material originated in Japan in 1996 (e.g., [8]). Since then, a broad spectrum of products appeared on the market for indoor use as well as for outdoor applications. Regarding traffic emissions, it is important that the exhaust gases stay in contact with the active surface during a certain period. The street configuration, the speed of the traffic, the speed and direction of the wind, all influence the final reduction rate of pollutants *in situ*.

In the case of concrete pavement blocks [9,10], the anatase is added to the wearing layer of the pavers which is approximately 8 mm thick. In the case of cast-in-place concrete pavements, the $TiO_2$ is added in the top layer (40 mm thick). The fact that the $TiO_2$ is present over the whole thickness of this layer means that even if some surface wear takes place, for example by traffic or weathering, new $TiO_2$ will be present at the surface to maintain the photocatalytic activity (in contrast to the abrasion of a coating or dispersion layer for instance). The use of $TiO_2$ in combination with cement leads to a transformation of the $NO_x$ into $NO_3^-$, which is adsorbed at the surface due to the alkalinity of the concrete [11]. Thus, a synergetic effect is created in the presence of the cement matrix, which helps to effectively trap the reactant gases (NO and $NO_2$) together with the nitrate salt formed. Subsequently, the deposited nitrate can be washed away by rain or washing with water. In addition, these nitrates pose no real threat towards pollution of body waters because the resulting concentrations in the waste water are very low, even below the current limit values for surface and ground water [12].

Special attention is given here to the NO and $NO_2$ content in the air, since they are for almost 50% caused by the exhaust of traffic and are at the base of smog, secondary ozone and acid rain formation as indicated above. The photocatalytic oxidation of NO is usually assumed to be a surface reaction between NO and an oxidizing species formed upon the adsorption of a photon by the photocatalyst, e.g., a hydroxyl radical, both adsorbed at the surface of the photocatalyst [13]. It has been shown by various authors that the final product of the photocatalytic oxidation of NO in the presence of $TiO_2$ is nitric acid ($HNO_3$) while $HNO_2$ and $NO_2$ have been identified as intermediate products in the gas phase over the photocatalyst [2,4,11,13,14]. The resulting reaction pathway of the photocatalytic oxidation of NO has been discussed in several publications e.g., [2,4,13–16] most of which proposed the photocatalytic conversion of NO via $HNO_2$ to yield $NO_2$, which is subsequently oxidized by the attack of a hydroxyl radical to the final product $HNO_3$:

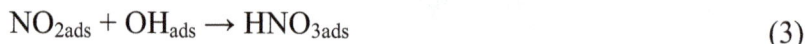

$$NO_{ads} + OH_{ads} \rightarrow HNO_{2ads} \tag{1}$$

$$HNO_{2ads} + OH_{ads} \rightarrow NO_{2ads} + H_2O_{ads} \tag{2}$$

$$NO_{2ads} + OH_{ads} \rightarrow HNO_{3ads} \tag{3}$$

Here, all nitrogen compounds adsorbed at the photocatalyst surface are assumed to be in equilibrium with the gas phase.

Until now, UV-light (in the UV-A spectrum) was necessary to activate the photocatalyst. However, recent research indicates a shift towards the visible light [17], for instance by doping the $TiO_2$ with transition metal ions or non-metallic anionic species, or forming reduced $TiO_x$. These techniques introduce impurities and defects in the band gap of $TiO_2$ thereby increasing the amount of visible light that can be absorbed and used in the photocatalytic process. This means that applications in tunnels and indoor environments become more realistic. Especially the application in tunnels is worth looking at due to the high concentration of air pollutants at these sites. One of the projects in Belgium is focusing on this subject [18].

## 3. Laboratory Results: Parameter Evaluation

Different test methods have been developed to determine the efficiency of photocatalytic materials towards air purification. An overview is given in [11]. A distinction can be made by the type of air flow; in the flow-through method according to ISO 22197-1 [19], the air, with a concentration of 1 ppmV of NO, passes once-only over the sample which is illuminated by a UV-lamp with light intensity equal to 10 $W/m^2$ in the range between 300 and 400 nm, as illustrated in Figure 2. Afterwards, the $NO_x$ (= sum of NO and $NO_2$) concentration is measured at the outlet and the reduction (in %) is calculated. It is also worth to note here that within Europe actions are underway to harmonize and develop new standards for photocatalyis [20]. In any case, the test procedure used for the current results is still based on the existing ISO standard.

**Figure 2. (a)** Schematic and **(b)** photo of measurement set-up based on ISO 22197-1:2007 [19] at Belgian Road Research Center (BRRC).

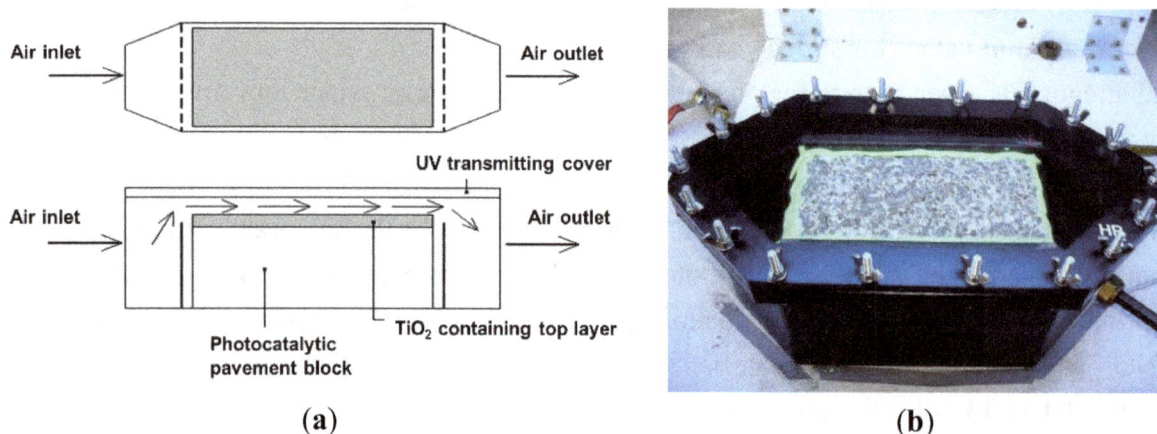

(a)                                                                                  (b)

The pre-treatment of the samples in the laboratory can be important to obtain reproducible results and mainly depends on the type of base material (e.g., concrete or paints). A typical test scheme according to the ISO standard is presented in Figure 3, where the following steps are applied to the sample: 0.5 h at 1 ppmV NO concentration, no light—5 h exposure to an air flow of 3 L/min with 1 ppmV NO and UV-illumination—0.5 h with UV-illumination and no exposure. A small increase with time of the $NO_x$ concentration is visible due to the deposit of the $NO_3^-$ at the surface.

**Figure 3.** Typical result obtained in the laboratory following the standard ISO test procedure.

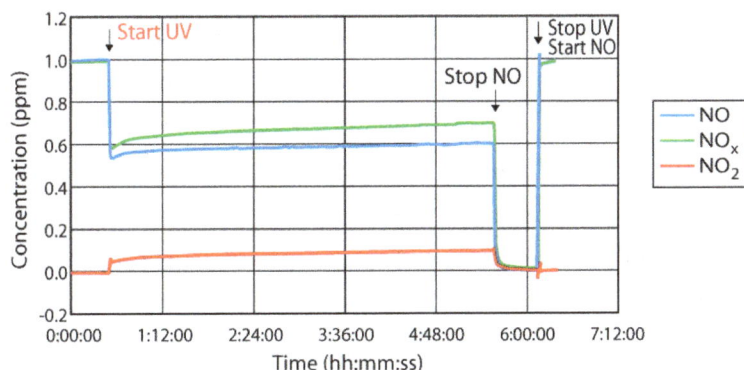

The influence of different important test parameters affecting the photocatalytic reaction has been investigated before [2] such as temperature, light intensity, relative humidity, contact time (controlled by surface area, flow velocity, height of air flow, *etc.*). For instance, the effect of relative humidity of the ingoing air is illustrated in Figure 4 for different materials including cementitious (concrete, mortar) and other (paint) substrates. Clearly, for cementitious materials the reduction of the $NO_x$ concentration in the outlet air decreases with increasing relative humidity (RH, %), an observation which was also found by other authors [21]. This probably has to do with the fact that the water in the atmosphere plays a role in the adhesion of the pollutants at the surface and with the competition effect that can arise between water molecules and $NO_x$ in the ambient air with increasing RH. For paints (acidic environment) though, it has been noticed that there is an optimum in RH where a maximal efficiency is obtained. Anyway, relative humidity proves to be an important limiting factor for photocatalytic applications in humid areas like Belgium. Temperature on the other hand, was found to have no significant influence on the $NO_x$ reduction within the ambient range (5–25 °C).

**Figure 4.** Effect of relative humidity on photocatalytic efficiency for different materials.

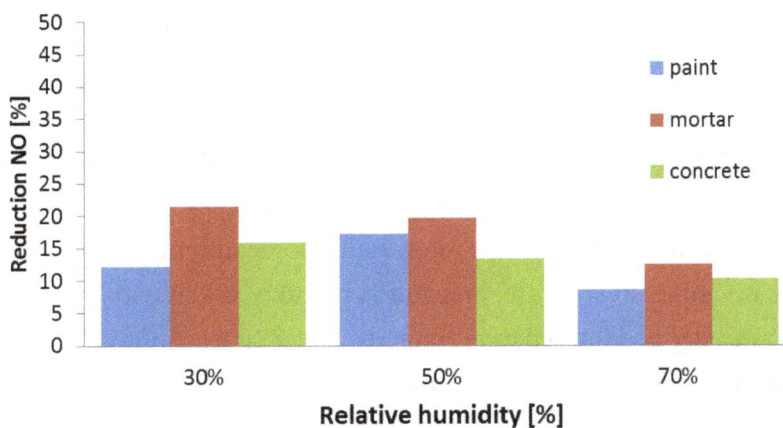

In general, it can be stated that the efficiency towards the reduction of $NO_x$ (in %) increases with a longer contact time (larger surface area, lower air velocity, smaller height of air flow, higher turbulence at the surface), a lower relative humidity (for cementitious materials) and a higher intensity of incident light. These are the conditions at which the risk of ozone formation in summer is the largest: higher sun light intensity, no wind and no rain. At these days, the photocatalytic reaction will be more pronounced.

## 4. Pilot Project in Antwerp

An important issue is the conversion of the results obtained in the laboratory to real applications. In order to see the influence of photocatalytic pavements in "real conditions", a first pilot section of 10.000 m² of photocatalytic pavement blocks was constructed in 2004–2005 on the parking lanes of a main axe in Antwerp [2]. Figure 5 depicts a view of the parking lane, where the photocatalytic concrete pavement blocks have been applied. Only the wearing layer (upper 5–6 mm) of the blocks contains anatase $TiO_2$ mixed in the mass of the concrete layer. The exact composition could not be given by the manufacturer (Marlux, Tessenderlo, Belgium) at that time in view of confidentiality. In spite of the fact that the surface applied on the Leien of Antwerp is quite important, one has to notice the relatively small width of the photocatalytic parking lanes in comparison with the total street: $2 \times 4.5$ m *versus* a total width of 60 m.

**Figure 5.** Separate parking lanes at the Leien of Antwerp with photocatalytic pavement blocks.

In order to check the durability of the photocatalytic efficiency, pavement blocks were taken from the lane after different periods of exposure and measured in the laboratory with and without washing of the surface. Some of the results are presented in Figure 6. They indicate a good durability of the efficiency towards $NO_x$ abatement. The deposition of pollutants on the surface leads to a decrease in efficiency which can be regained after washing. Recently repeated measurements in 2010 indicate that even after more than five years of service life, the photocatalytic efficiency of the pavers is still present [22].

Besides the tests in the lab, on site measurements were also carried out. Since no reference measurements without photocatalytic material (prior to the application) exist, the interpretation of these results is rather difficult. Especially the influence of traffic, wind speed, light intensity and relative humidity are playing an important role. Detailed results can be found in [2]. In brief, the field measurements suggested a decrease in $NO_x$ concentration at the sites with photocatalytic materials, where a levelling out of the pollution peaks is visible. In any case, precaution has to be taken with the interpretation of data since these results are momentary and limited over time. However, at least, they gave an indication of the efficiency of the photocatalytic pavement materials *in situ*, and a basis to work on for future applications.

**Figure 6.** $NO_x$ reduction measured on two pavement blocks, before (hatched) and after (colored) washing the surface, taken on different locations (house nr. 30, 35, 37, 42, 48, 53) and at different times at the Amerikalei in Antwerp.

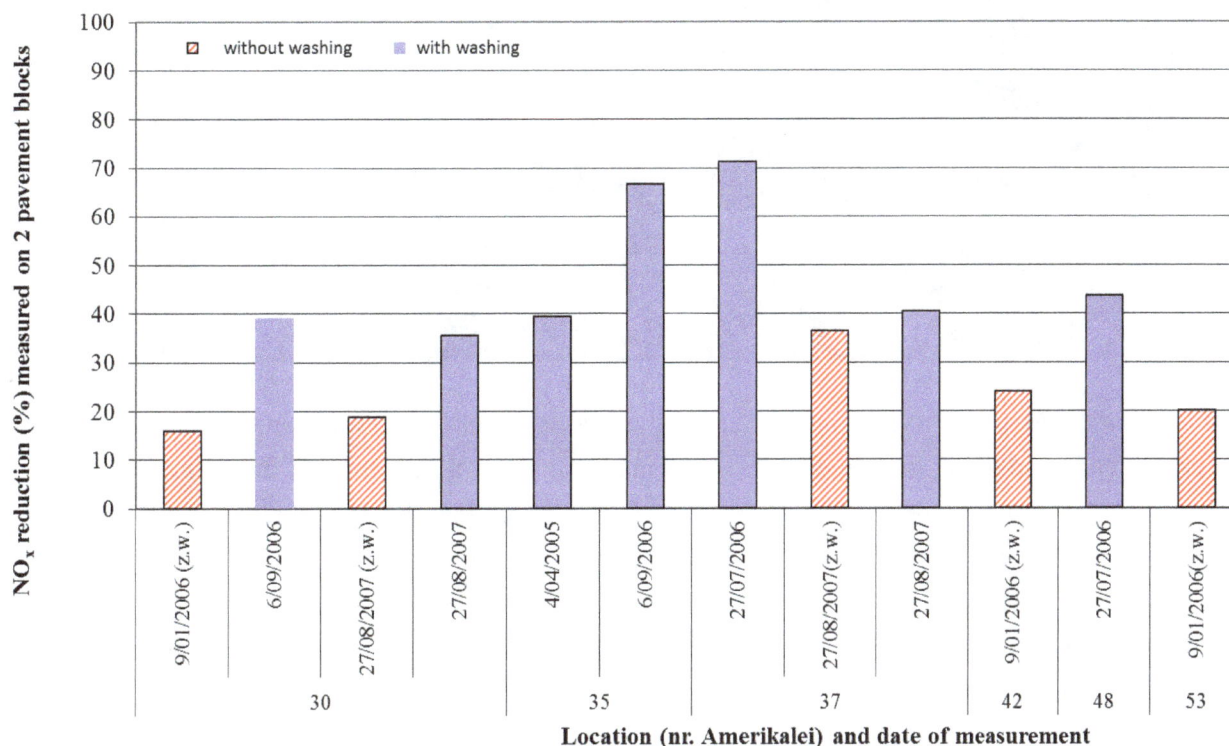

## 5. Recent Photocatalytic Applications in Belgium

Since the first application in Antwerp (2004–2005), much progress has been made within the photocatalytic research area. Newer, better and more efficient materials are constantly being developed, and action is more and more broadened also to visible light responsive materials [17]. This also led to new trial applications in which people have tried to establish the relation between the results in the laboratory and the real effect on site, see e.g., [23–25]. In this section an overview is given of two such recent projects in Belgium which were implemented in collaboration with the BRRC.

### 5.1. Life+ Project PhotoPAQ

The European Life+ funded project PhotoPAQ [18] was aimed at demonstrating the usefulness of photocatalytic construction materials for air purification purposes in an urban environment. Eight partners from five different European countries participated in the project.

In this framework, an extensive three-step field campaign was organized in the Leopold II tunnel in Brussels, from June 2011 till January 2013 [26,27]. A photocatalytic cementitious coating material (TX Active® white Skim Coat from CTG Italcementi Group) was applied on the side walls and roof (total area of about 2700 m²) of a tunnel section of about 160 m in length in one of the tunnel tubes directing to the city center. The air-purifying product was activated by a dedicated UV lighting system (including Supratec "HTC 241 R7s" light bulbs from Osram, see Figure 7). More details can be found in [26,27].

**Figure 7.** Application of the photocatalytic product and installation of the UV lamps in the Leopold II tunnel in Brussels, in the framework of PhotoPAQ.

Possible advantages of purifying the tunnel air may be, obviously, cleaner air to breathe, with a potentially reduced need of ventilation, but also (and maybe mainly) a reduction of the pollution impact of tunnel exhaust on the city air quality. During the field campaigns, the effect of the photocatalytic coating on the air pollution (including $NO_x$, VOC's, particulate matter, CO, *etc.*) inside the tunnel section was rigorously assessed.

The PhotoPAQ consortium mobilized a large panel of up-to-date instrumentations, installed in the tunnel for several weeks, aiming at characterizing the levels of pollution in this section of the Leopold II tunnel, with and without the air-purifying product (Figure 8).

**Figure 8.** Full characterization of the air quality inside the tunnel test section during the PhotoPAQ campaigns.

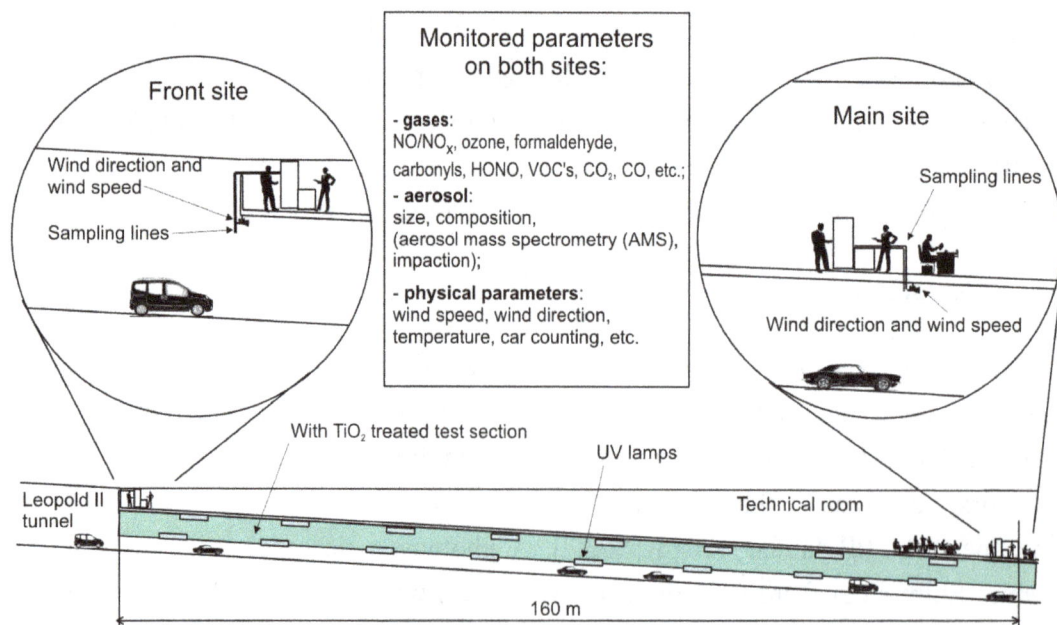

However, in contrast to first estimations based on laboratory studies, the results indicated no observable reduction of the pollution level, *i.e.*, the reduction of nitrogen oxides ($NO_x$, one of the major traffic related air pollutants) is below 2%, which is the experimental uncertainty of the measurements.

A severe de-activation of the photocatalytic material was observed inside the highly trafficked and strongly polluted Leopold II tunnel. In conjunction, final UV lighting intensity (only 2 W/m² UV-A) was below the targeted values (above 4 W/m²), which led to too low levels for proper activation inside the polluted tunnel environment. Another negative condition was the high wind speed (up to 3 m/s) inside the tunnel, limiting the contact time between pollutants and the active surface. Finally, January 2013 turned out to be an unusually wintry period causing cold and humid conditions inside the tunnel, with relative humidity ranging from 70% to 90%, which also reduces the activity of the photocatalytic material as shown before. Thus, all these issues together resulted in a reduction of the activity of the photocatalytic surfaces inside the harsh environment of the Leopold II tunnel, by a factor of 10 compared to the theoretical expectations. More details about the set-up and results of these extensive field campaigns inside the Leopold II tunnel are presented elsewhere [26,27].

Nevertheless, combining the knowledge gained during these campaigns and the laboratory based investigations performed by the PhotoPAQ consortium, numerical simulations (with the commercially available general purpose Computational Fluid Dynamics code ANSYS CFX®) were performed in order to estimate the possible best-case abatement of pollutants.

These calculations indicate that, under the best case scenario (proper level of UV light intensity higher than 4 W/m², relative humidity below 50%, and limited pollution to avoid passivation), the reduction of the $NO_x$ concentration may be expected to attain:

- ±3% for the 160 m long test section;
- ±12% for the entire Leopold II tunnel (*ca.* 3 km), if not affected by ventilation.

Despite the fact that the results were not as expected, the Leopold II field campaigns conducted by the PhotoPAQ team proved to be a unique real world and fully comprehensive assessment of the effect of photocatalytic air-purifying (road) construction materials on air pollution inside a tunnel environment. Based on the extensive experimental data set gathered and numerical model calculations, a valuable tool for extrapolation can be provided to estimate the expected pollution reduction in other urban tunnel sites, also for use by non-experts [18].

## 5.2. INTERREG Project ECO₂PROFIT

The broad environmental sustainability project ECO₂PROFIT dealt with the reduction of the emission of greenhouse gases and sustainable production of energy on industrial estates in the frontier area between Flanders and Holland. To reach these goals, several tangible demonstration projects were carried out on industrial sites in Belgium and the Netherlands. BRRC was involved in two such projects: "Den Hoek 3" in Wijnegem and "Duwijckpark" in Lier (both near Antwerp). Here, the regional development agency POM Antwerp was aiming to use a double layered concrete for the road construction, with recycled concrete aggregates in the bottom layer and photocatalytic materials ($TiO_2$) in the top layer, using photoactive cements and/or coatings. That way, air purifying and $CO_2$ reducing concrete roads could be built which are both innovating and energy efficient.

For these recently completed applications (2010–2011) BRRC was asked to set-up an elaborate testing program in the lab to help optimize the air purifying performance of the top layer, without interfering with other properties of the concrete (workability, strength, durability *etc.*). In the construction site of Wijnegem (Den Hoek 3), it was opted to use an exposed aggregates surface finish (with grain size between 0 and 6.3 mm) on the top layer for reasons of noise reduction and comfort of the road user. For the site in Lier (Duwijckpark) a brushed surface finishing was chosen to have more active cement at the surface. Indeed, the type of surface finishing and/or treatment of the pavement can have an effect on the photocatalytic efficiency, as shown in Figure 9 for three types of surface finishing: exposed aggregates, smooth (formwork side) and sawn surface. The results show that the exposed aggregates surface performs equally well as the smooth, formwork surface, but not as good as a sawn surface. This is the result of the combined action of less photoactive cement at the surface and a higher surface porosity (higher specific surface), two competing effects which in the end yield the final efficiency shown in Figure 9.

**Figure 9.** Effect of surface treatment on photocatalytic efficiency (only one type of "less" active product in mass).

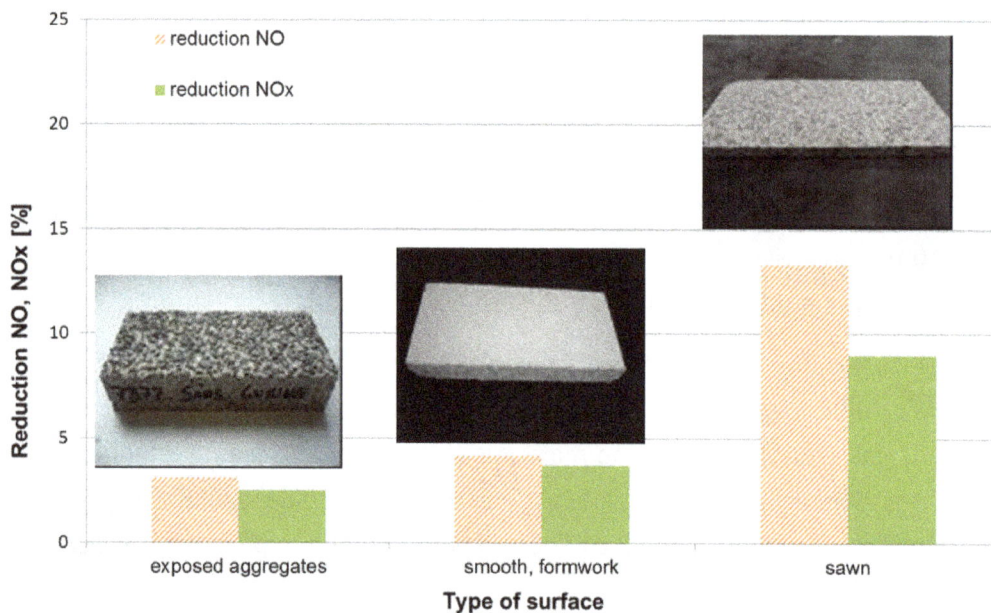

For the application of photocatalytic materials in a concrete road (and in general for any other type of application) a fundamental choice can be made between: mixing in the mass (e.g., $TiO_2$ in cement) and/or spraying on the surface (suspension of $TiO_2$). The former has the advantage of a more durable action since the $TiO_2$ will continuously be present, even after wearing of the top layer. On the other hand, the initial cost will be higher (higher $TiO_2$ content, necessity for double layered concrete) and only the $TiO_2$ at the surface will be active. In contrast, dispersing at the surface of a $TiO_2$ solution will provide a more direct action, and a lower initial cost (e.g., "ordinary" cement). In this case however, the longevity of the photocatalytic action could be questioned because of loss of adhesion to the surface in time. This fundamental choice was also investigated within the research program, together with the influence of several other parameters [28].

The effect of a curing compound for instance—generally applied to protect the young concrete against desiccation in Belgium and placed directly after concreting or after exposing the aggregates at the surface in case of denudation—is illustrated in Figure 10. From this, it appears the curing compound will initially inhibit the photocatalytic reaction, most likely because it is shielding off the "active" components from the pollutants in the air. Consequently, it is probable that the curing must disappear from the surface, *i.e.*, under influence of traffic or weathering, before the $TiO_2$ will reach its optimal air purifying performance. In case of a photocatalytic spray coating, this also means that it is best to apply the $TiO_2$ dispersion some months after the curing compound to have the most durable effect. Alternatively, the exposed aggregates concrete can be covered with a plastic sheet to prevent dehydration (in case the concrete surface is denuded).

**Figure 10.** (a) Application of curing compound on fresh concrete, and (b) Effect of curing compound on photocatalytic efficiency (different samples A–D, with photocatalytic $TiO_2$ in mass and/or applied as dispersion, with and without curing compound).

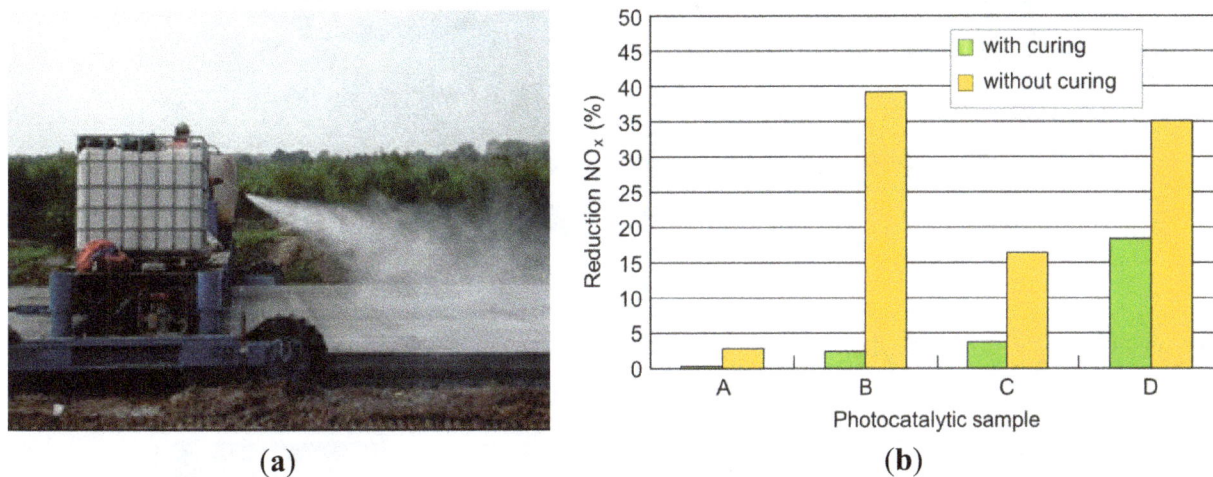

(a)　　　　　　　　　(b)

More detailed results of the laboratory research can be found in [28] and [29]. Based on the findings and the optimization of the concrete composition, a proper selection of photocatalytic materials and application techniques could be made, for the construction of double layered, photocatalytic concrete roads on the industrial zone "Den Hoek 3" in Wijnegem.

Double Layered Concrete at "Den Hoek 3" in Wijnegem

The concrete pavement of the industrial zone in Wijnegem has been constructed between the 15th and 18th of March 2011. The concrete was placed in two layers, wet-in-wet, with an interval time of approximately 1 hour. The bottom layer had a thickness of 180 mm, while the top layer was designed to be 50 mm. For the concrete of the bottom layer, 57% of the coarse aggregates were replaced by recycled concrete aggregates. For the top layer with $TiO_2$, commercially available white cement with 4% $TiO_2$ pre-mixed (by weight) was applied (CBR, Belgium, Heidelberg Cement Group). Two slip form pavers were used to place the concrete. As can be seen in Figure 11a, the color of the top layer is much lighter, due to the use of white cement and the presence of the $TiO_2$ (about 0.8 wt% of the top layer).

More information on the concrete composition, the execution and the results obtained in the lab as well as on site can be found in [28] and [29]. Besides the photocatalytic concrete roads, photocatalytic pavement blocks were also used for the bicycle lanes, parking spaces and foot paths.

Since this was a completely new industrial zone, it was not possible to have measurements on site before putting the photocatalytic concrete in place. An overview of the project is given in Figure 12. Immediately after concreting, a retarding agent was sprayed on the surface to be able to wash out the top surface after 24 h, to create an exposed aggregates surface finish (see Figure 11b). In order to prevent dehydration of the concrete during the first days, some parts of the road have been treated with curing compound; the other zones were covered with a plastic sheet. This way, the influence of the curing compound on the short and long term photocatalytic efficiency could be investigated.

**Figure 11.** (**a**) Construction of double layered concrete pavement at industrial zone "Den Hoek 3" in Wijnegem; (**b**) Detail of exposed aggregates surface finish of the top layer.

(**a**)                                                                            (**b**)

**Figure 12.** (**a**) Situation plan of the new industrial zone "Den Hoek 3" in Wijnegem, Belgium (Google Maps); (**b**) "On site" testing of photocatalytic efficiency.

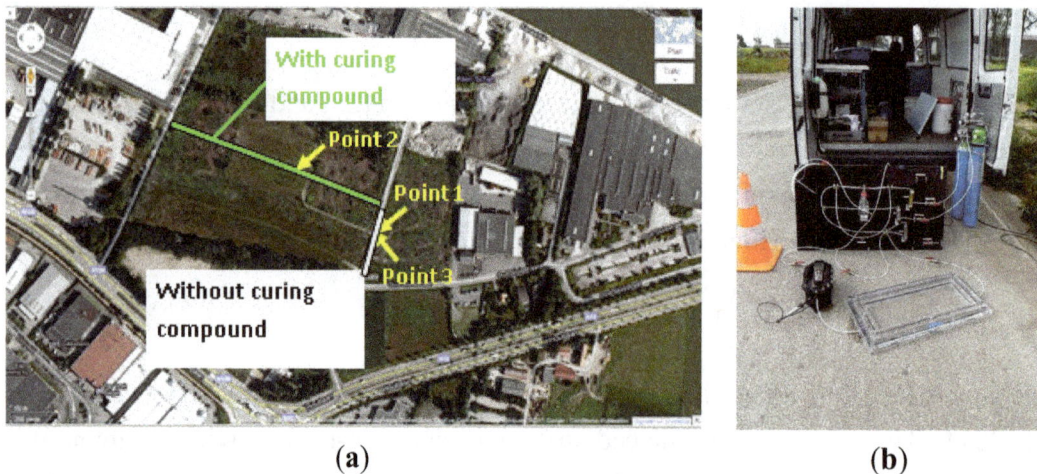

(**a**)                                                                            (**b**)

The photocatalytic efficiency of the top layer was measured in two ways: in the laboratory on cores taken from the surface at the places indicated in Figure 12a, and "on site" with a special measuring set-up, shown in Figure 12b. This "on site" test is developed to evaluate the photocatalytic properties of the concrete pavement over time and to compare the different sites (with and without curing, for example). It does not measure the overall purification of the air around the pavement but enables to measure the durability of the photocatalytic efficiency.

The set-up consists of a Plexiglas frame, screwed air-tight on the test surface (concrete pavement), and is covered with a UV-transparent glass lid. The input NO-concentration (1 ppmv), relative humidity (50% RH) and air flow (3 L/min) are taken similar to the laboratory set-up. However, the total area covered by the box is somewhat larger (700 × 300 mm²) to have a representative surface, and natural (varying!) sunlight is used in first instance to activate the surface. First results of the measurements on site are given in Figures 13 and 14, and were collected 5 months after the placement of the concrete (August 2011) at the places indicated in Figure 12a (points 1 and 2).

**Figure 13.** $NO_x$ concentration measured at the outlet for zone with curing compound, 5 months after concreting (point 2, August 2011).

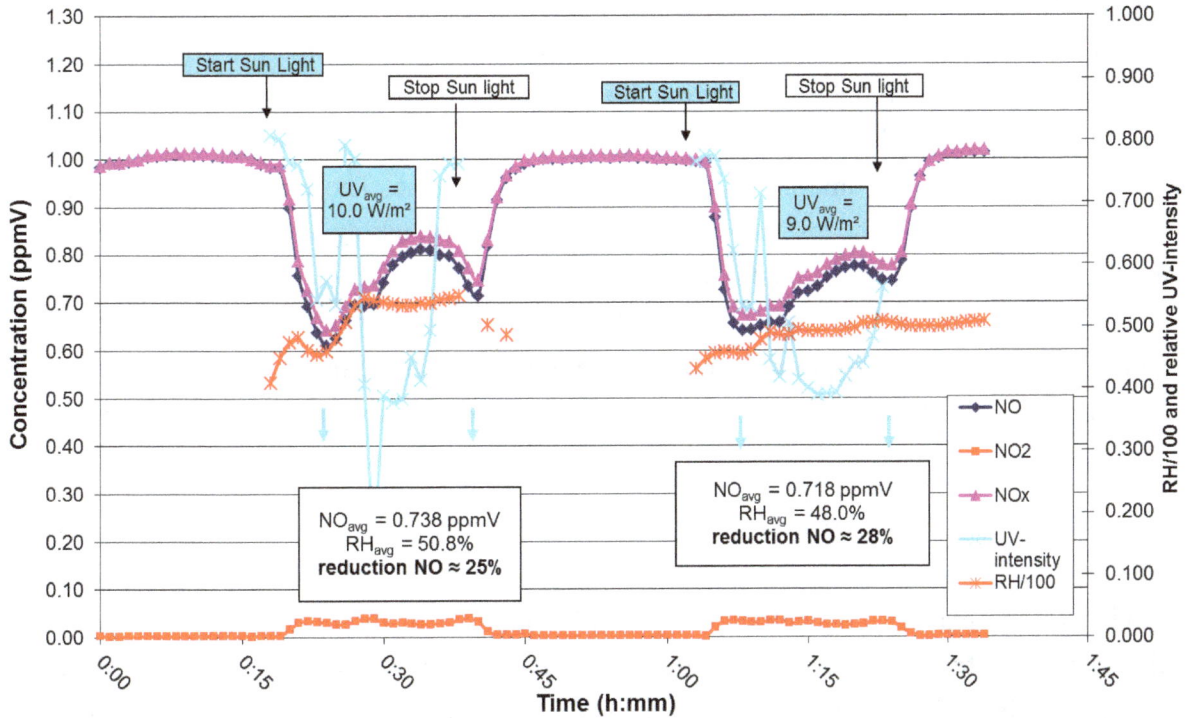

**Figure 14.** $NO_x$ concentration measured at the outlet for zone without curing compound, 5 months after concreting (point 1, August 2011).

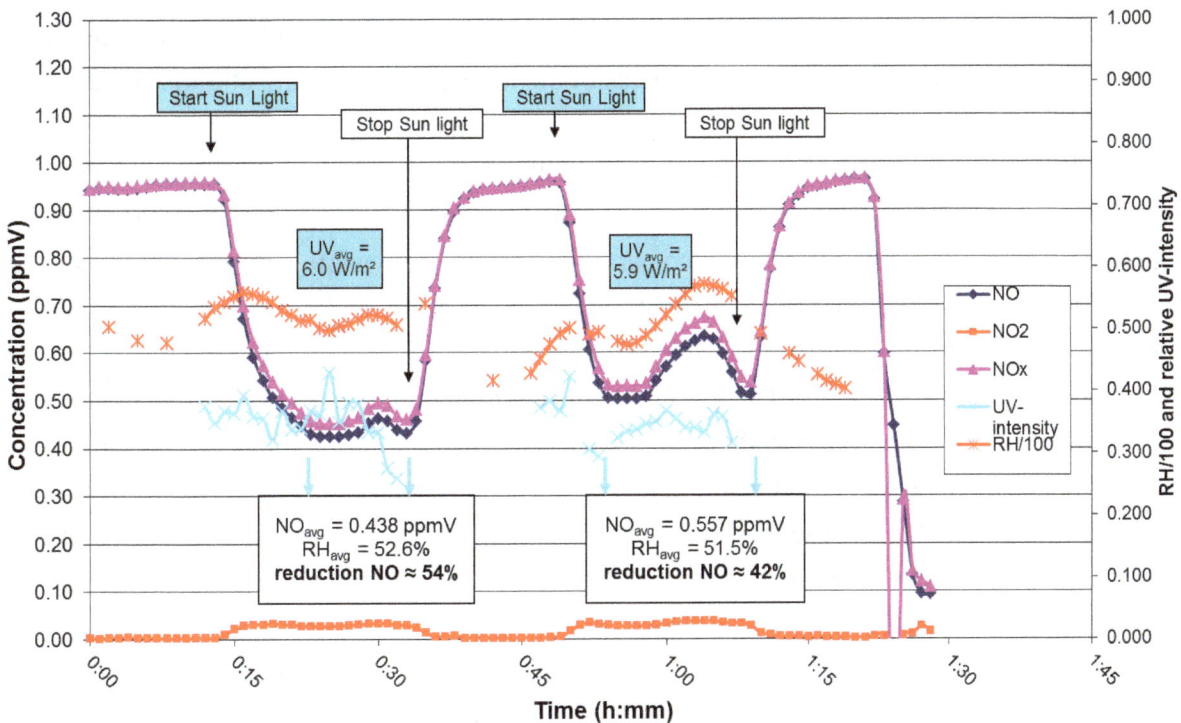

First of all, the results shown in Figures 13 and 14 indicate a large influence of the relative humidity (red curves). The $NO_x$ abatement is lower when the relative humidity increases and higher when RH decreases again. The influence of the sun light intensity (measured through the UV intensity, light blue lines) is also visible, but on a different scale: variations over a shorter period of time do not influence the $NO_x$ concentration immediately; it is the average sun light intensity over a longer period that is determining the attained $NO_x$ abatement for the photocatalytic process.

Furthermore, the reduction in NO concentration is significantly lower for the zone with curing compound, indicating it is still (slightly) inhibiting the reaction: average reduction of 27% (with curing) *versus* 48% (without curing). Nevertheless, the effect of applying a curing compound on the fresh concrete (to protect against dehydration) seems to diminish over time. These results obtained on site (year 2011) are also in line with the results obtained in the laboratory, taking into account the difference in surface, relative humidity and light intensity [28].

In order to correctly compare the results between the lab and the field, the photocatalytic activity for $NO_x$ (= sum of NO and $NO_2$) is expressed in terms of the photocatalytic deposition velocity in [m/h] under the assumption of a first order uptake kinetics and negligible transport limitations from the gas phase to the solid surface [30]:

$$k_R = \ln\left(\frac{c_0}{c_t}\right)\frac{F}{A} \tag{4}$$

where $c_0$ and $c_t$ are the reactant concentration at the inlet and exit of the photo-reactor, respectively. In fact, this parameter refers to a first order reaction rate coefficient independent of the applied flow rate $F$ and the active surface ($A$) to volume ratio of the used reactor (lab or on site). In the lab work [28], average values for the $NO_x$ deposition velocity $k_{R,NOx}$ of 0.25 and 0.70 m/h were obtained with and without curing compound respectively, which is in nice agreement with the results on site for 2011 (see further in Table 1).

The measurements on site are also repeated over time in order to see the influence of ageing and traffic on the photocatalytic efficiency. Recent measurements performed in the summer of 2012 for instance, are shown in Figure 15. Here, measurements were performed using an external UV-lamp (10 W/m²) as well as natural sunlight to activate the photocatalyst present in the pavement. It appears the activity under sun light is somewhat higher compared to the UV lamp only. This could be due to the fact that the applied $TiO_2$ (in the active cement) is also partially active under visible light and/or is excited by the shorter UV wavelengths (UV-B, UV-C) present in the sun spectrum.

On the other hand, the measurement of the UV-intensity comprised in the sun light could be erroneous because of the radiometer used here. This is only calibrated for specific UV-A lamps (between 300 and 400 nm) applied in the geometry of the lab set-up which differs substantially from these exterior tests. The activity observed under natural, varying sun light though, is still very interesting from the view point of practical application. The use of the external UV-lamp with constant light intensity in turn, allows making a more absolute comparison of the photocatalytic activity between different zones and for different times.

In any case, the results of Figure 15 reveal already that the efficiency of this kind of photocatalytic application ($TiO_2$ integrated in the cement) appears to decrease in time: on average 34% NO-reduction (after 17 months) *versus* 48% (after 5 months). Possible causes could reside in the covering of the $TiO_2$ at the surface by dirt, the detachment of the $TiO_2$ from the surface or the deposition of products from chemical reactions which can take place at the surface.

**Figure 15.** $NO_x$ concentration measured at the outlet for zone without curing compound, 17 months after concreting (point 3, August 2012).

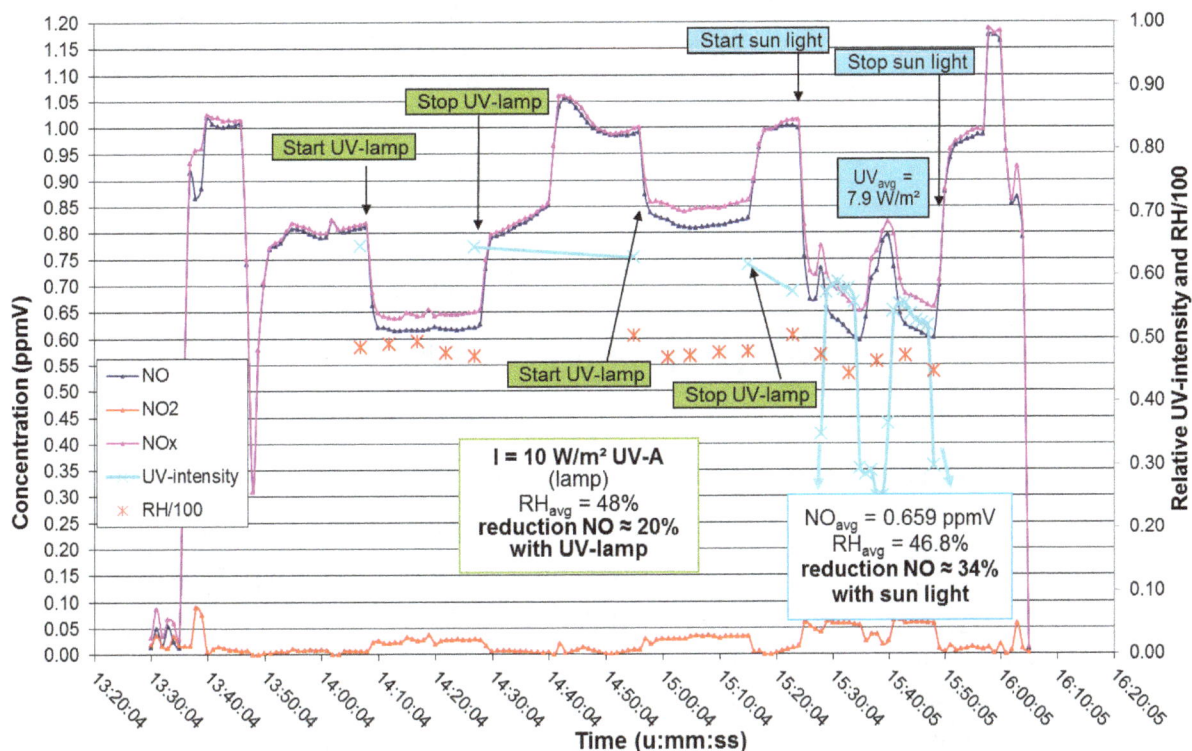

In this respect, in October 2012 an aqueous $TiO_2$ dispersion (Eoxolit® consisting of a mixture of two different types of $TiO_2$ particles with a total concentration of 40 g/L $TiO_2$) was also applied on the surface in some parts of the roads on the industrial zone in Wijnegem, as shown in Figure 16a, for the purpose of comparative measurements. In total four different zones were considered:

- Zone 1 = double layered concrete (0/6.3 mm on top) without $TiO_2$;
- Zone 2 = single layered concrete (0/20 mm) without $TiO_2$;
- Zone 3 = double layered concrete with $TiO_2$ (active cement) and without curing compound;
- Zone 4 = double layered concrete with $TiO_2$ (active cement) and with curing compound.

The photocatalytic dispersion was applied with a dose of approximately 1 L per 5 m² on a total of 800 m², followed by spraying of a hydrophobic product for optimal functioning of the coating (manufacturers' guidelines). Important to mention however, is the fact that at the time of application there was a severe pollution with soil and dirt at the surface of the pavement in some zones due to the presence of a grinding installation plant at the site. This most certainly had an impact on the efficiency of the $TiO_2$ suspension (see further). Subsequently, provisional controls of the photocatalytic efficiency have been carried out to check the separate action of the two types of photoactive materials (mass and dispersion), and to further assess the durability of the air purifying performance. Most recent measurements on the site in Wijnegem were performed in the summer of 2013, at the measurement points (1–9) indicated in Figure 16b. All results obtained up till now (2011–2013) are summarized in Table 1.

**Figure 16.** (**a**) Application of photocatalytic dispersion on part of the roads at industrial zone "Den Hoek 3" (October 2012); (**b**) Localization of measurement points for "on site" testing (Google Maps).

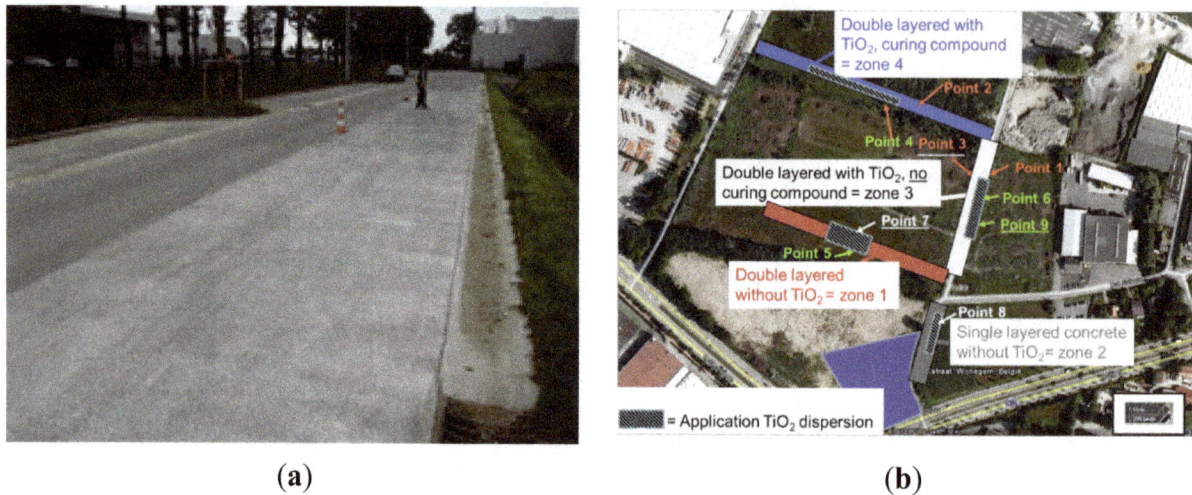

(**a**)                                                                                                  (**b**)

**Table 1.** Summary of results in time for photocatalytic activity measured on site in Wijnegem.

| Zone | $k_{R,NO}$ ($k_{R, NOx}$) [m/h] | | | | |
|------|------|------|------|------|------|
|      | Sun light | | | UV-lamp (10 W/m²) | |
|      | 2011 | 2012 | 2013 | 2012 | 2013 |
| 4: **with** curing compound, active cement (point 2, 4) | 0.30 (0.26) | 0.09 (0.07) | – | 0.06 (0.04) | – |
| 3: **without** curing compound, active cement (point 1 and 3) | 0.70 (0.66) | 0.39 (0.34) | 0.38 (0.28) | 0.21 (0.19) | 0.22 (0.18) |
| 4: **with** curing, active cement +TiO$_2$ dispersion (point 4) | – | – | 0.82 (0.62) | – | 0.28 (0.22) |
| 3: **without** curing, active cement +TiO$_2$ dispersion (points 6 and 9) | – | – | 0.27 (0.20) | – | 0.21 (0.15) |
| 1: double layered, **no** active cement + TiO$_2$ dispersion (point 7) | – | – | 0.32 (0.27) | – | 0.15 (0.13) |
| 2: single layered, **no** active cement + TiO$_2$ dispersion (point 8) | – | – | 0.14 (0.13) | – | 0.08 (0.07) |

First of all, when comparing the measurements on the surface of the pavement at points 1 and 3 (*cf.* Figure 16b) in 2013 with these of 2012, a very similar result can be noticed: a photocatalytic deposition velocity for NO$_x$ of *ca.* 0.2 m/h under UV light. This indicates that the decreasing trend in photocatalytic activity for the concrete with "active" cement (see evolution 2011–2012) seems to be stabilized in 2013.

Furthermore, the measured efficiency for points 1 and 3 (in 2013) appears to differ little or nothing with the one for points 6 and 9, with application of the photocatalytic coating (TiO$_2$ dispersion) on the pavement surface. Here, the TiO$_2$ dispersion did not produce a significant added value (yet) in terms of photocatalytic air purifying action. Only for point 4 (active cement with curing compound, after application of TiO$_2$ dispersion) one can notice a strong improvement of the photocatalytic efficiency (deposition velocity of *ca.* 0.8 m/h for NO under sun light and nearly 0.3 m/h under UV ). Possibly, the pollution of the surface at the time of application has played an important part causing the adhesion of the coating to be far from optimal.

For points 7 and 8 (concrete without active cement, but with $TiO_2$ dispersion on the surface), the activity is not significantly better either (or even less) compared to the "pure" concrete with active cement. In addition, point 8 (single layered concrete 0/20) reveals much smaller photocatalytic reactivity than point 7 (double layered concrete with top layer 0/6.3): deposition velocity of 0.08 m/h *versus* 0.15 m/h for NO reduction under UV. This probably has to do with the stronger adhesion of the coating on the surface of the finer (0/6.3) double layered concrete compared to the coarser (0/20) single layered concrete.

Finally, a measurement on site was also performed for the newly constructed pavements at the industrial zone in Lier, which have a different surface finishing as illustrated in Figure 17a. The results of this measurement, 20 months after construction, are shown in Figure 17b.

**Figure 17. (a)** Double layered photocatalytic concrete pavement with brushed surface finish at industrial zone "Duwijckpark" in Lier; **(b)** $NO_x$ concentration measured at the outlet for the site in Lier (active cement + curing compound), 20 months after concreting (August 2013).

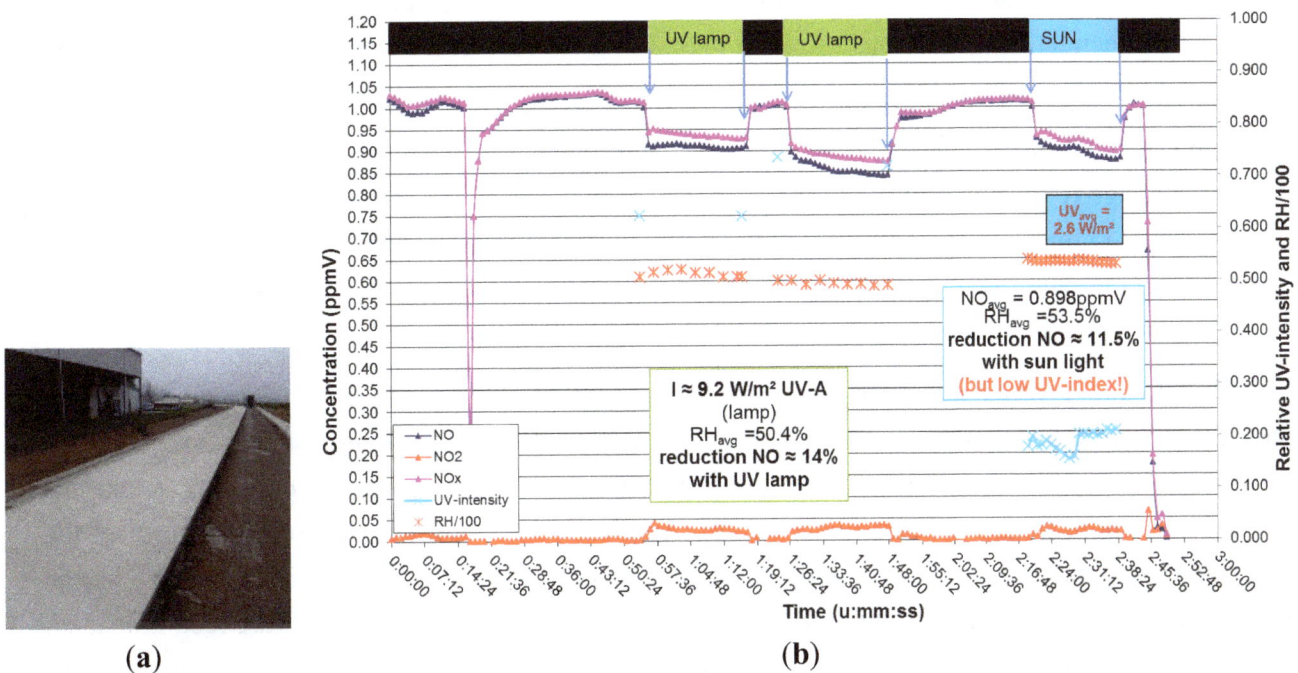

(a)                                                                 (b)

In comparison with the measurements of Wijnegem in 2013 (see Table 1), a slightly lower photocatalytic reaction is observed in Lier, which among others is due to the use of a curing compound (for the brushed surface) and the lower $TiO_2$ content (less cement used). However, if we make the comparison with the zone with curing compound in Wijnegem (*cf.* point 4 in zone 4) measured in 2012 (17 months after construction), a significantly better result under UV light is obtained in Lier: deposition velocity for NO of 0.14 m/h in Lier *versus* 0.06 m/h in Wijnegem. This higher activity probably has to do with the brushed surface finish instead of exposing the aggregates (*cf.* Figure 9). In any case, these measurements confirm the photocatalytic action 20 months after construction of the concrete pavement.

## 6. Conclusions and Perspectives

Photocatalytic (TiO$_2$ containing) paving materials with the potential of reducing air pollution by traffic are being used more frequently on site in horizontal as well as in vertical applications, also in Belgium. Laboratory results indicate a good efficiency towards the abatement of NO$_x$ in the air by using these innovative materials. The durability of the photocatalytic action also remains mostly intact, though regular cleaning (by rain) of the surface is necessary. The relative humidity (RH) is an important parameter, which may reduce the efficiency on site. If the RH is too high, the water will be adsorbed at the surface and prevent the reaction with the pollutants.

The translation from the laboratory results to the "on-site efficiency" is still a difficult and critical factor, because of the great number of parameters involved. Hence, there is still a need for large scale applications to demonstrate the effectiveness of photocatalytic materials in "real life" and evaluate the durability of the air purifying action, such as the European Life+ project PhotoPAQ and the industrial zones "Den Hoek 3" in Wijnegem en "Duwijckpark" in Lier. These recent applications in Belgium show already some interesting results.

It seems the use of photocatalytic cement-based coatings inside road tunnels is not mature for application on a large scale yet. From the experience gained during the Leopold II tunnel campaigns in Brussels, recommendations for the proper use of these innovative materials can be made though, such as:

-   Optimized coating application for low surface roughness and minimizing dust adsorption;
-   High UV light intensity levels in the order of magnitude of 10 W/m$^2$;
-   Low average relative humidity of tunnel air ($\leq$ 60%);
-   High enough photocatalytic activity, with threshold values defined from lab studies;
-   Low average wind speed (< 2 m/s) in the tunnel for increased reaction time of pollutants;
-   High surface to volume ratio (smaller sized tunnel tubes).

For the double layered photocatalytic concrete pavements using active cement, an efficiency comparable to the one measured in the laboratory is obtained initially; though it seems to decrease somewhat in time due to dirt build-up and other deposits on the surface, the air purifying action has stabilized after more than two years (2011–2013). Application of a curing compound—to protect the fresh concrete against desiccation—initially strongly reduces the photocatalytic activity and also has an impact on the long term. Use of a plastic sheet to protect the young concrete is therefore recommended. Furthermore, the exposed aggregates technique is not ideal for the photocatalytic efficiency since in this case a lot of aggregates are present at the surface and the TiO$_2$ is only present in the paste. The application of a brushed surface finish could lead to a better result.

Use of a photocatalytic coating (TiO$_2$ dispersion) on the surface of the concrete pavement does not produce an added value for the air purifying action compared to mixing in the mass, despite the good results in the laboratory. This probably has to do with the loss of adhesion in time and the filthiness of the surface at the time of application. Possibly, the coating is partially washed away with the dirt. In addition, better results are obtained on the finer, double layered concrete (0/6.3) than for the coarser, single layered concrete (0/20) which could be due to the better adhesion of the coating on the surface.

Durability of the photocatalytic action in time (for products mixed in the mass and/or applied on the surface) and optimization of the adhesion of photo-active coatings on the concrete surface, are topics that need to be investigated further.

Finally, the best results will be achieved by modeling the environment, validating the models by measurements on site, followed by an implementation of the different influencing parameters to assess the real life effect. One must bear in mind that photocatalytic applications are not effective everywhere; "good" contact between the airborne pollutants and the active surface is crucial and factors such as wind speed and direction, street configuration and pollution sources all play a very important role.

## Acknowledgements

The authors wish to thank IWT Flanders (Institute for the Promotion of Innovation by Science and Technology in Flanders), FPS Economy (Federal Public Service), Life+ and EFRO (European Union), INTERREG and the Ministry of the Brussels-Capital Region—Brussels Mobility for the (financial) support of the different projects.

## Author Contributions

E.L.B and A.B. both coordinated and supervised the different research projects involving photocatalytic applications; E.L.B. prepared the manuscript. All authors read and approved the manuscript.

## Conflicts of Interest

The authors declare no conflict of interest.

## References

1.  European Commission. *EU Energy and Transport in Figures, Statistical Pocketbook*; Publications Office of the European Union: Brussels, Belgium, 2011.
2.  Beeldens, A. Air purification by pavement blocks: Final results of the research at the BRRC. In Proceedings of Transport Research Arena—TRA 2008, Ljubljana, Slovenia, 21–24 April 2008.
3.  Directive 2008/50/EC of the European Parliament and of the Council on ambient air quality and cleaner air for Europe. *Off. J. Eur. Union* **2008**, L152:1–L152:44.
4.  Chen, J.; Poon, C. Photocatalytic construction and building materials: From fundamentals to applications. *Build. Environ.* **2009**, *44*, 1899–1906.
5.  Renz, C. Lichtreaktionen der Oxyde des Titans, Cers und der Erdsäuren. *Helv. Chim. Acta* **1921**, *4*, 961–968.
6.  Fujishima, A.; Honda K. Electrochemical photolysis of water at a semiconductor electrode. *Nature* **1972**, *238*, 37–38.
7.  Fujishima, A.; Rao, T.N.; Tryk, D.A. Titanium dioxide photocatalysis. *J. Photochem. Photobiol. C* **2000**, *1*, 1–21.
8.  Sopyan, I.; Watanabe, M.; Murasawa, S.; Hashimoto, K.; Fujishima, A. An efficient $TiO_2$ thin-film photocatalyst: Photocatalytic properties in gas-phase acetaldehyde degradation. *J. Photochem. Photobiol. A* **1996**, *98*, 79–86.

9. Cassar, L.; Pepe, C. Paving Tile Comprising an Hydraulic Binder and Photocatalyst Particles. EP-Patent 1600430 A1, 1997.

10. Murata, Y.; Tawara, H.; Obata, H.; Murata, K. $NO_x$-Cleaning Paving Block. EP-Patent 0786283 A1, 1996.

11. Ohama, Y.; Van Gemert, D. *Application of Titanium Dioxide Photocatalysis to Construction Materials*; Springer: Dordrecht, The Netherlands, 2011.

12. Saubere Luft Durch Pflastersteine Clean Air by Airclean®. Available online: http://www.ime.fraunhofer.de/content/dam/ime/de/documents/AOe/2009_2010_Saubere%20Luft%20durch%20Pflastersteine_s.pdf (accessed on 25 July 2014).

13. Dillert, R.; Stötzner, J.; Engel, A.; Bahnemann, D.W. Influence of inlet concentration and light intensity on the photocatalytic oxidation of nitrogen(II) oxide at the surface of Aeroxide® $TiO_2$ P25. *J. Hazard. Mater.* **2012**, *211–212*, 240–246.

14. Laufs, S.; Burgeth, G.; Duttlinger, W.; Kurtenbach, R.; Maban, M.; Thomas, C.; Wiesen, P.; Kleffmann, J. Conversion of nitrogen oxides on commercial photocatalytic dispersion paints. *Atmos. Environ.* **2010**, *44*, 2341–2349.

15. Devahasdin, S.; Fan, C.; Li, J.K.; Chen, D.H. $TiO_2$ photocatalytic oxidation of nitric oxide: Transient behavior and reaction kinetics. *J. Photochem. Photobiol. A* **2003**, *156*, 161–170.

16. Ballari, M.M.; Yu, Q.L.; Brouwers, H.J.H. Experimental study of the NO and $NO_2$ degradation by photocatalytically active concrete. *Catal. Today* **2011**, *161*, 175–180.

17. Fujishima, A.; Zhang, X. Titanium dioxide photocatalysis: Present situation and future approaches. *Comptes Rendus Chim.* **2006**, *9*, 750–760.

18. PhotoPAQ (2010–2014) Life+ Project. Available online: http://photopaq.ircelyon.univ-lyon1.fr/ (accessed on 25 July 2014).

19. *ISO 22197-1:2007 Fine Ceramics (Advanced Ceramics, Advanced Technical Ceramics)—Test Method for Air-Purification Performance of Semi Conducting Photocatalytic Materials—Part 1: Removal of Nitric Oxide*; International Standards Organization (ISO): Geneva, Switzerland 2007.

20. CEN Technical Committee 386 "Photocatalysis" Business Plan—(internet) Draft BUSINESS PLAN CEN/TC386 PHOTOCATALYSIS. Available online: http://standards.cen.eu/BP/653744.pdf (accessed on 28 July 2014).

21. Hüsken, G.; Hunger, M.; Brouwers, H.J.H. Experimental study of photocatalytic concrete products for air purification. *Build. Environ.* **2009**, *44*, 2463–2474.

22. Beeldens, A.; Boonen, E. Photocatalytic applications in Belgium, purifying the air through the pavement. In Proceedings of the XXIVth World Road Conference, Mexico City, Mexico, 26–30 September 2011.

23. Maggos, Th.; Plassais, A.; Bartzis, J.G.; Vasilakos, Ch.; Moussiopoulos, N.; Bonafous, L. Photocatalytic degradation of $NO_x$ in a pilot street canyon configuration using $TiO_2$-mortar panels. *Environ. Monit. Assess.* **2008**, *136*, 35–44.

24. Gignoux, L.; Christory, J.P.; Petit, J.F. Concrete roadways and air quality—Assessment of trials in Vanves in the heart of the Paris region. In Proceedings of the 12th International Symposium on Concrete Roads, Sevilla, Spain, 13–15 October 2010.

25. Guerrini, G.L. Photocatalytic performances in a city tunnel in Rome: $NO_x$ monitoring results. *Constr. Build. Mater.* **2012**, *27*, 165–175.

26. Boonen, E.; Akylas, V.; Barmpas, F.; Boréave, A.; Bottalico, L.; Cazaunau, M.; Chen, H.; Daële, V.; De Marco, T.; Doussin, J.F.; *et al.* Photocatalytic de-pollution in the Leopold II tunnel in Brussels, Part I: Construction of the field site. *Constr. Build. Mater.* **2014**, Submitted.

27. Gallus, M.; Akylas, V.; Barmpas, F.; Beeldens, A.; Boonen, E.; Boréave, A.; Bottalico, L.; Cazaunau, M.; Chen, H.; Daële, V.; *et al.* Photocatalytic de-pollution in the Leopold II tunnel in Brussels, Part II: NO$_x$ abatement results. *Constr. Build. Mater.* **2014**, Submitted.

28. Boonen, E.; Beeldens, A. Photocatalytic roads: From lab testing to real scale applications. *Eur. Transp. Res. Rev.* **2013**, *5*, 79–89.

29. Beeldens, A.; Boonen, E. A double layered photocatalytic concrete pavement: A durable application with air-purifying properties. In Proceedings of 10th International Conference on Concrete Pavements (ICCP), Quebec, Canada, 8–12 July 2012.

30. Ifang, S.; Gallus, M.; Liedtke, S.; Kurtenbach, R.; Wiesen, P.; Kleffmann, J. Standardization methods for testing photo-catalytic air remediation materials: Problems and solution. *Atmos. Environ.* **2014**, *91*, 154–161.

# 12

# Deposition of High Conductivity Low Silver Content Materials by Screen Printing

**Eifion Jewell \*, Simon Hamblyn, Tim Claypole and David Gethin**

College of Engineering, Swansea University, Swansea SA2 8PP, UK;
E-Mails: s.m.hamblyn@swasnea.ac.uk (S.H.); t.c.claypole@swansea.ac.uk (T.C.);
d.t.gethin@swansea.ac.uk (D.G.)

\* Author to whom correspondence should be addressed; E-Mail: e.jewell@swansea.ac.uk

**Abstract:** A comprehensive experimental investigation has been carried out into the role of film thickness variation and silver material formulation on printing capability in the screen printing process. A full factorial experiment was carried out where two formulations of silver materials were printed through a range of screens to a polyester substrate under a set of standard conditions. The materials represented a novel low silver content (45%–49%) polymer material and traditional high silver content (65%–69%) paste. The resultant prints were characterised topologically and electrically. The study shows that more cost effective use of the silver in the ink was obtained with the low silver polymer materials, but that the electrical performance was more strongly affected by the mesh being used (and hence film thickness). Thus, while optimum silver use could be obtained using materials with a lower silver content, this came with the consequence of reduced process robustness.

**Keywords:** printed silver; screen printing; mesh ruling; ink formulation

## 1. Introduction

In total the silver ink market is estimated to be worth around $760 Million in 2012 [1]. Silver inks fall into three main categories, silver solutions, micro particle inks (particle size > 1 μm) and nano particle inks (<1 μm). The nano particle inks are of increasing interest as they offer lower sintering temperatures (for improved conductivity, which are appropriate to polymer substrates [2]. Nano silver materials have seen widespread research and development in inkjet [2], gravure [3] and flexographic printing [4].

Although the use of nano silver inks in inkjet and other printing processes is rising, thick film screen printing remains the dominant application technology in the market [5]. The patterning method of choice for large particle silver inks is screen printing which offers an ability to accurately deposit thick films with a large rheological window with minimum pressure over a wide area at economical production rates.

During the material formulation process the balance between silver quantity, particle size (and structure) must be considered in order to provide the required conductivity while maintaining film integrity, adhesion, flexibility and process compatibility. This requires a balance of conductive element, binder and solvent. The increase in the cost of silver over recent years, has led to development formulations which permit a reduction in silver quantity, without sacrificing conductivity or printability. Although the use of silver as a conductive material is widely reported, few published studies have examined the interactions between macro material properties, mesh characteristics and printed feature quality (predominantly topology and conductivity).

Screen printing of large particle silver materials is now the process of choice for metalizing silicon Photovoltaic (PV) cells, flexible circuits including membrane switch manufacture, Radio Frequency Identification (RFID) aerials and is also commonly proposed as a means of manufacturing many products within the nascent printed electronics arena. The increase in the cost of silver over recent years has led to the silver being the dominant factor in the bill of materials of many products and a prime driver in developing formulations with a reduced silver quantity, without sacrificing conductivity or printability. Although the use of silver as a conductive material for device manufacture (PV, sensors and RFID) by screen printing is widely reported, few published studies have examined the interactions between macro material properties, mesh characteristics and printed feature quality, particularly for systems which are cured at low temperature (<130 °C).

For lower temperature cured materials which are screen printed, the literature available on material or process parameters is sparse. As directly applicable literature is limited, it is pertinent to review associated relevant areas which have examined models for screen printing, solder paste printing through stencils and material/screen printing parametric studies whose objectives are within the scope of the present study. Theoretical studies of the process have made little advancement in the development of reliable screen printing process models. Early models showed some predicted trends which were in agreement with experimental observation but even the most recent developments require significant oversimplification (e.g., material viscoelasticity is ignored) which limit practical use [6,7]. Establishing process/material characteristics and their impact on printed performance therefore is reliant on experimental studies.

The importance of material flow properties and its interaction with the image carrier has been identified as the primary factor in determining feature rendition in solder paste printing [8] (which utilises a non-woven stencil and high temperature sintering). Much of the literature relates to silver materials which are used for metallization in the silicon PV industry where materials are fired at high temperature. The role of silver particle size and particle packing on silicon PV performance has been addressed [9] but this did not directly measure the impact of size on the material flow properties or printed feature conductivity. The flow characteristics of silver materials have been carried out but the subsequent relationship with the printed line quality has not been studied [10].

The role of the screen has been examined and this showed some general trends between the printed line quality and the mesh characteristics with line resistance reducing and line aspect ratio increasing as the screen open area proportion increased [11], although the scope of the study and analysis carried out was limited. The influence of squeegee pressure found some correlation between the applied pressure and the silver deposit characteristics [12] but this was limited to one mesh type and hence a limited film thickness. The net effect of formulation on the reliability of circuits has been examined [13] but this again was carried out at one mesh/film thickness. For carbon materials, the interactions between material properties mesh, structure and their impact on printed deposit topology showed non linear relationships between mesh ruling and sheet resistance, line resistance and film topology [14].

The review of the literature has demonstrated that there is a void in the understanding of the relationship between mesh type used, the properties of the silver materials (which are cured at low temperature) and the properties of the final printed film. At present, there is limited opportunity for developing models based on the exact physical mechanisms that take place during the printing process and linking this to material characteristics. In order to develop further understanding an investigation was therefore carried out which examined the role of mesh, physical silver ink characteristics on the topology and conductivity of fine lines. The investigation had three primary aims. Firstly, it aimed identify the role of silver content and binder properties on the microstructure of the cured films and their subsequent impact on electrical performance. Secondly, it aimed to establish the capabilities of lower silver content materials and their sensitivity to changes in film thickness. Finally, it aimed to create a reference dataset which could be used as a design tool which would identify material requirements and process settings given design specifications or likely print results given materials and process settings. These outputs would be significant benefit to the material formulator, device designer or process engineer.

## 2. Experimental Section

In order to investigate the relationship between formulation and film thickness a full factorial experimental design was employed where six silver materials were printed through 10 screens. The silver materials represent two formulation types, Table 1. One represents a traditional paste material used for rigid and flexible circuit printing while other is a novel polymer ink which has been developed for applications which are more cost sensitive.

The polymer family of materials possesses a gel like consistency utilising a low silver content (between 45 wt.% and 49 wt.%) material in gel like binder/solvent. Usually such a low silver content would not produce such a conductive printed film as conductive percolation through the cured silver film could not be guaranteed. The gel binder used allows a greater solvent content without sacrificing rheological limits which subsequently leads to a more compact cured film than that which is achieved with conventional materials. Solids contents were measured using a Perkin Elmer TGA while volumetric calculations are based on the primary organic materials shown in manufacturer's data [15,16] and standard densities for each component. From the densities and TGA, volumetric proportions for the wet and dry films can be calculated. Particle sizes were measured by SEM sections and plan views of the cured film. Analysis of the images was carried out using Image J software using visually assessed contrast enhancement (for edge detection) and blob analysis.

**Table 1.** Silver material properties.

| Notation | Polymer | Paste |
|---|---|---|
| Material as provided | – | – |
| Ink silver content (wt.%) | 45, 47 and 49 | 65, 67 and 69 |
| Total solids (wt.%) | 51–55 | 85–89 |
| Dry film for each 1 μm wet (mid range material) | 0.175 | 0.66 |
| Dry film ratio for same wet film thickness | 1 | 3.8 |
| SEM of dry film | | |
| Silver: Binder mass ratio | 7.8 : 1 | 3.4 : 1 |
| Silver: Binder volume ratio | 0.7: 1 | 0.3 : 1 |
| Mean particle size (μm) | 4.8 | 4.5 |

As dried and printed (for the 47% and 67% materials respectively)

The exact formulation of the materials are not disclosed in the material datasheets and are subject to commercial confidentiality. For each formulation type the varying solids content was established by dispersing the silver within a binder/solvent mixture at maximum solids content, *i.e.*, the maximum quantity of silver which could be added before the silver was no longer held in a stable homogenous suspension which was considered to within the viscosity range for the process. Solvent (butyl digol) was then added to individual batches of each base material at maximum silver concentration in a step wise manner in order to achieve the necessary silver content. Thus, within each formulation type each material differs only by the proportion of solvent in the formulation.

The film thickness in screen printing is predominantly dictated by the mesh characteristics of the screen [14]. Thus in order to examine the role of film thickness the materials were printed with 10 screens (5 polyester and 5 stainless steel). These screens represent the range of mesh rulings, and subsequent ink deposit, commonly used for screen printing conductive materials, Table 2. A nominal stencil thickness of 12 μm was used on all of the meshes and all screens were mounted with the mesh warp direction at 22.5° to the print direction.

A DEK 248 digitally controlled screen printing machine was used to print the experiment. This allows accurate and repeatable PC control settings to be applied to each experimental setup. Standard conditions were established for the printing which allowed deposition at each experimental condition. Six prints were produced at each experimental condition with any residual material being discarded. The screen and squeegee were cleaned within the press to ensure consistent set up. The materials were printed to 300 mm × 300 mm white stabilized PET substrate. The substrate was 250 μm thick with a $R_a$ of 0.15 μm +/−0.02 μm.

**Table 2**. Mesh specifications.

| Material | Mesh Ruling (threads/inch) | Mesh ruling (threads/cm) | Mesh opening (μm) | Thread diameter (μm) | Theoretical ink volume cm$^3$/m$^2$ |
|---|---|---|---|---|---|
| Polyester | 123 | 48 | 133 | 70 | 45 |
| Polyester | 156 | 61 | 90 | 64 | 30 |
| Polyester | 195 | 77 | 77 | 48 | 28 |
| Polyester | 255 | 100 | 57 | 40 | 21 |
| Polyester | 305 | 120 | 45 | 34 | 16 |
| Stainless Steel | 145 | 57 | 118 | 56 | 55 |
| Stainless Steel | 200 | 77 | 90 | 40 | 43 |
| Stainless Steel | 250 | 97 | 63 | 36 | 32 |
| Stainless Steel | 300 | 114 | 56 | 32 | 28 |
| Stainless Steel | 325 | 125 | 50 | 30 | 24 |

The prints were dried immediately after printing using a SC Technical hot air dryer at 120 °C with a residence time of 6 min. Of the six prints produced at each condition, samples 4, 5 and 6 were measured in order to reduce uncertainty. The printed design consisted of 30 mm long lines between 50 μm and 1000 μm in the print and transverse print direction. The results focus on the large solid area and cured 200 μm line which was reproduced consistently throughout the experiments and whose volume could be measured accurately. The sheet resistance was measured in a 50 mm × 50 mm solid layer section and were carried out using a Keithley 2400 source-meter operating in 4 point mode with a probe spacing of 5 mm with subsequent conversion to sheet resistance [17]. Each line resistance measurement represents an average of three measurements, while the sheet resistance represents an average of nine measurements.

A Veeco NT 2000 White light interferometer was used to measure the printed line topology, width height and line volume. All measurements were carried out at ×5 optical resolution with a total vertical scan length of 30 μm. This yields a measurement area of 1.2 mm × 0.93 mm, sampling every 1.67 μm in the X–Y direction. The smooth nature of the substrate allowed clear reference surfaces on either side of the line, allowing the geometric characteristics and cross section of the line to be readily established. In order to take into account topological variations along the line length, mean characteristics were taken from each of the 480 pixels.

Material rheology was measured using a Bohlin Gemini HR nano rheometer. In order to examine the viscous and elastic properties of the ink at low shear a dynamic strain sweep with an angular frequency of 1 Hz was carried out. This technique was adopted as viscous and elastic behaviour at low shear has been postulated as a key indicator of the material transfer process [18] and the method adopted had shown important trends [14]. Measurements were carried out using a parallel plate geometry with a diameter of 20 mm and sample gap of 70 μm at 25 °C and were repeated three times for each material.

## 3. Results and Discussion

### 3.1. Material Rheology

Whilst the silver content and solvent/binder proportions are factors which determine the electrical and physical characteristics of the final cured printed structure, the physical action of the material transfer to the substrate is dictated by the rheology [10,14,18] Both materials possess highly pseudoplastic characteristics with a reduction in viscosity as the shear stress is increased, Figure 1a. The polymer material shows a gradual reduction in both elastic and loss modulus (and hence viscosity) with increasing applied stress with the liquid behaviour becoming dominant over solid behaviour at a shear stress between 10 and 100 Pa. The paste silver posseses a higher moduli at low shear stress followed by a rapid reduction in elastic moduli in the 1–10 Pa shear stress range resulting in the material exhibiting predominately liquid behaviour at a shear stress of 10 Pa. The most significant difference between the material sets is evident in the low stress behaviour of the traditional paste which possesses complex, storage and loss moduli which are at least an order of magnitude higher than the polymer material. The transition from solid to liquid dominated behaviour (as depicted by tan δ) is far more sudden for the paste inks and occurs at a shear stress which is an order of magnitude lower than the polymer material. For both materials, there is a step wise increase in elastic and viscous modulus as the silver content is increased. The impact of this formulation and rheology on electrical and physical characteristics of the print are presented later in the results.

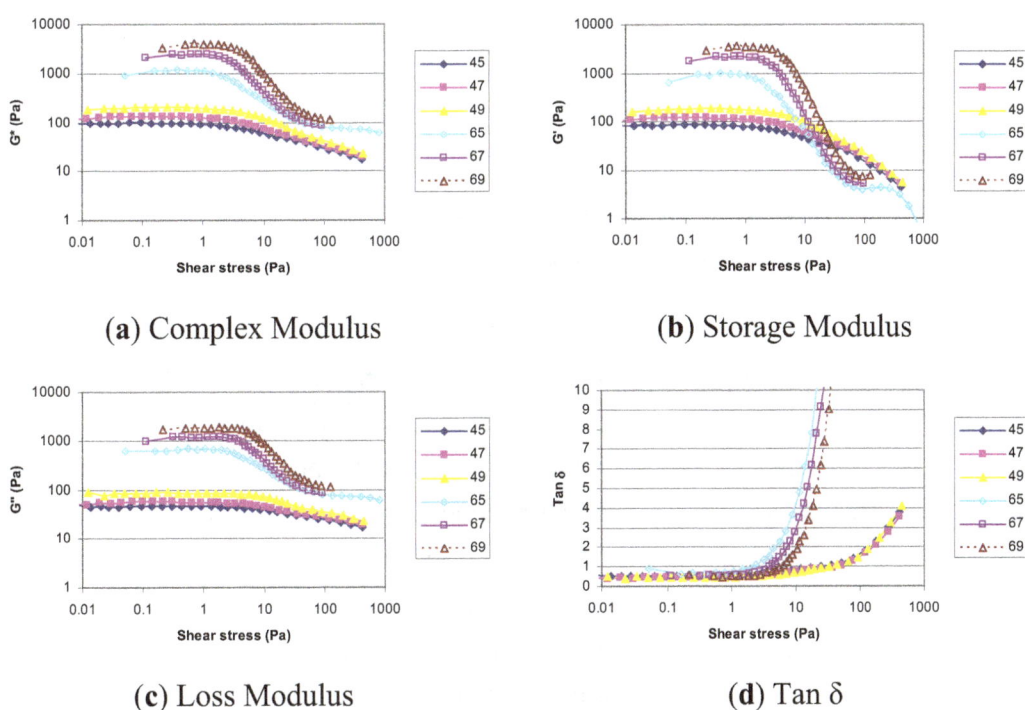

**(a)** Complex Modulus

**(b)** Storage Modulus

**(c)** Loss Modulus

**(d)** Tan δ

**Figure 1.** Rheological characteristics for each silver material **(a)** viscosity over medium shear range, elastic **(b)** and viscous **(c)** modulii and **(d)** Tan δ.

## 3.2. Printed Material Performance

As the materials have distinct performance in terms of rheology and printing performance, the presentation of the results will examine each material in turn before drawing comparisons between the material sets.

### 3.2.1. Polymer Inks

Increasing the mesh ruling resulted in a reduction sheet resistance as a result of the reduction in film thickness for the polymer inks, Figure 2. The sheet resistance is lower (typically less than 0.1 Ω/sq) for the stainless steel meshes at comparable mesh rulings as a result of the larger internal volume available within the mesh to contain the material. For both mesh types, the sheet resistance reduces as the silver content increases. For the stainless steel ink, there is a gradual increase as the mesh ruling increases and then shows a sudden increase with the 325 threads/cm mesh which has been attributed to the appearance of mesh patterning/marking in the printed solid film with this material. This topological surface phenomena is a result of a number of complex interacting parameters and is of importance to many applications [19].

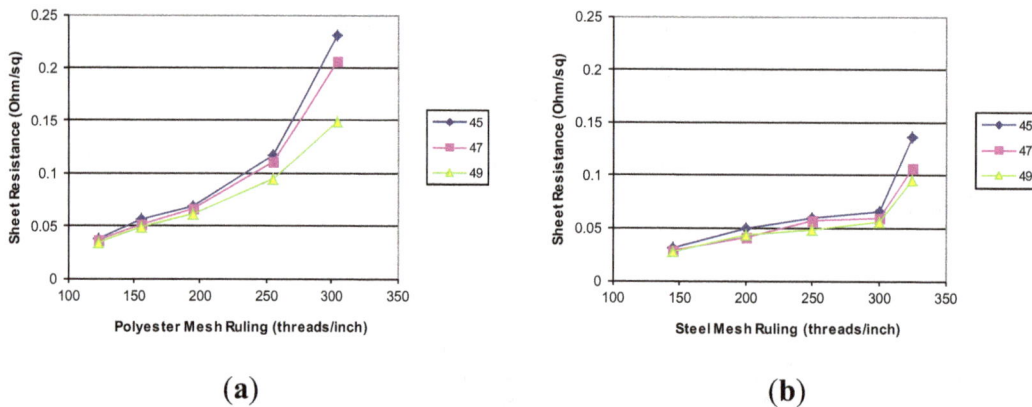

(a)                                                                                         (b)

**Figure 2**. The effect of mesh ruling on the polymer sheet resistance printed through the (**a**) polyester and (**b**) stainless steel meshes for each silver proportion.

The cured cross sectional area effectively represents the total quantity of material transferred to the substrate once it is cured. This represents a more valid measurement than film thickness as it takes account non rectangular nature and topologically complex nature of the printed lines. For the polymer ink set, increasing the mesh ruling reduced the cross sectional area of the nominally 200 μm line for the polyester, Figure 3a, and steel meshes, Figure 3b. This reduction in ink in cross sectional area is in line with the reduction in available free volume within the mesh for material transfer. The volume reduction is around 50%–60% over the mesh ruling range investigated. For 49% silver this represents a reduction from 1100 μm$^2$ to 560 μm$^2$ while the 45% silver material reduces from 1000 μm$^2$ to 425 μm$^2$. The line cross sectional area with the lower silver content materials in line with the reduced solids content.

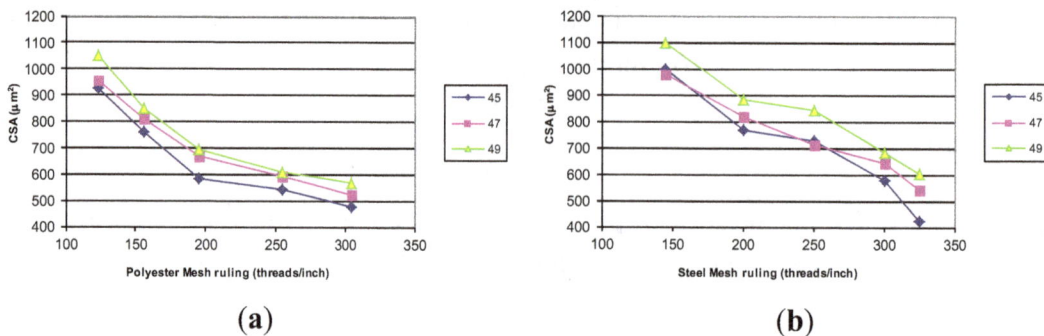

(a)                                                                                         (b)

**Figure 3**. The effect of mesh ruling on the cured cross sectional area of the polymers inks for a nominally 200 μm line (**a**) polyester and (**b**) stainless steel meshes for each silver content.

The line width is an important metric as it dictates feature density. The nominally 200 µm wide line is rendered wider than its nominal value for almost all mesh types, Figure 4. The lines are between 10 and 130 µm larger than nominal for the polyester mesh while the effect of the mesh ruling is less noticeable for the stainless steel screens. The increase in line width is most evident with the lowest solids material and reflects greater slumping of the lower viscosity material. There is a sudden reduction in printed line width at the highest mesh ruling, for the finest stainless steel screen which is attributed to a reduction in the overall film thickness (as shown by the cross sectional area in Figure 3b). This limits the quantity of wet material which slumps during the printing process. This tends to suggest that there is some minimum critical volume/height at which slumping does not occur.

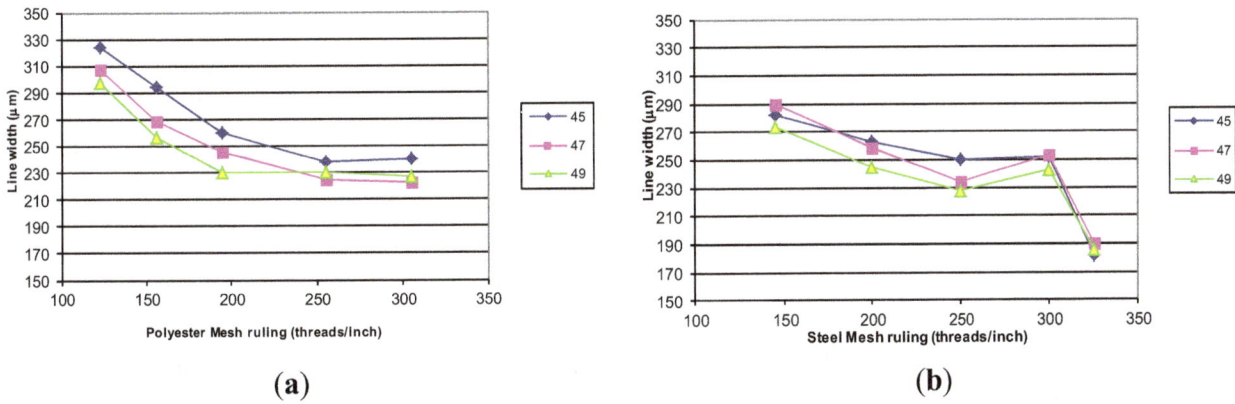

**(a)**                                                    **(b)**

**Figure 4.** The effect of mesh ruling on the cured printed line width of the polymers inks for a nominally 200 µm line (**a**) polyester and (**b**) stainless steel meshes for each silver content.

## 3.2.2. Paste Inks

The paste inks show similar trends to the polymer ink in respect that the sheet resistance increases with mesh ruling, Figure 5. There is also a clear trend with a reduction in the sheet resistance as the silver content of the material is increased, Figure 5 Sheet resistances are approximately the same as those obtained with the polymer materials at coarse mesh rulings (around 0.05 Ω/sq).

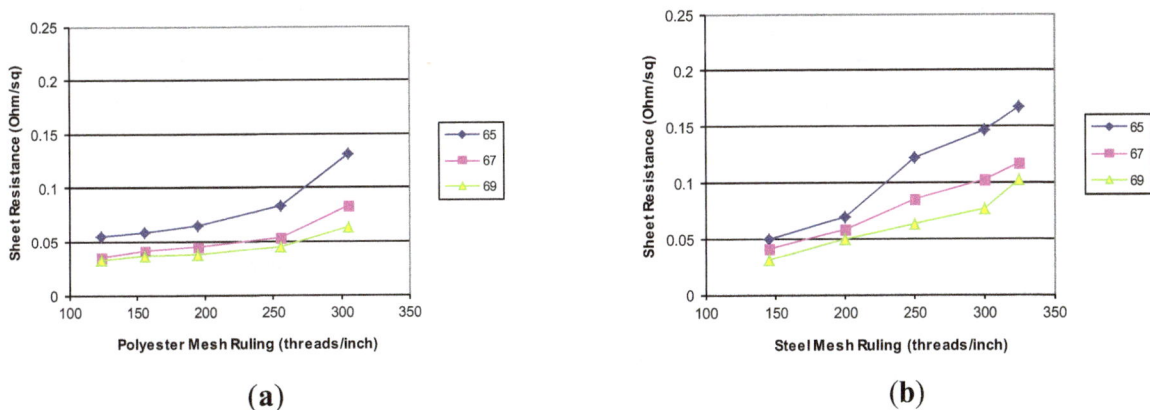

**(a)**                                                    **(b)**

**Figure 5.** The effect of mesh ruling on the paste sheet resistance printed through the (**a**) polyester and (**b**) stainless steel meshes for each silver proportion.

The printed cross sectional area of the paste materials is significantly higher for the paste inks compared to the polymer inks with cross sectional areas ranging from 2800 $\mu m^2$ to 1900 $\mu m^2$ for the polyester mesh and 2500 $\mu m^2$ to 1800 $\mu m^2$ for the stainless steel mesh, Figure 6. For the paste inks, the cross sectional areas are higher for the stainless steel mesh than the comparable polyester mesh. The impact of the mesh ruling is less dramatic (typically reductions of 30% to 40%) than that observed with the polymer inks (Figure 3) where more proportional dramatic reductions in CSA were observed. There is a general trend that the cross section area is reduced for the lower silver content inks in line with reductions in solids content.

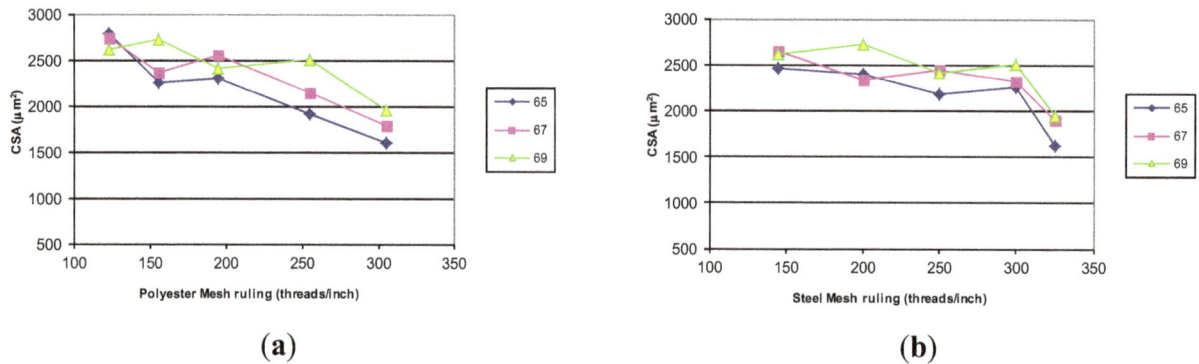

(a)                                                                 (b)

**Figure 6**. The effect of mesh ruling on the cured cross sectional area of the paste inks for a nominally 200 $\mu m$ line (**a**) polyester and (**b**) stainless steel meshes.

The impact of mesh ruling and solids content on the printed line width is less significant for the paste materials compared to the polymer materials, Figure 7. The line width increase is in the range of 10 and 50 $\mu m$ larger than nominal and shows a gradual reduction as the mesh ruling increases although no clear trends are observed between the printed line width and the material solids content.

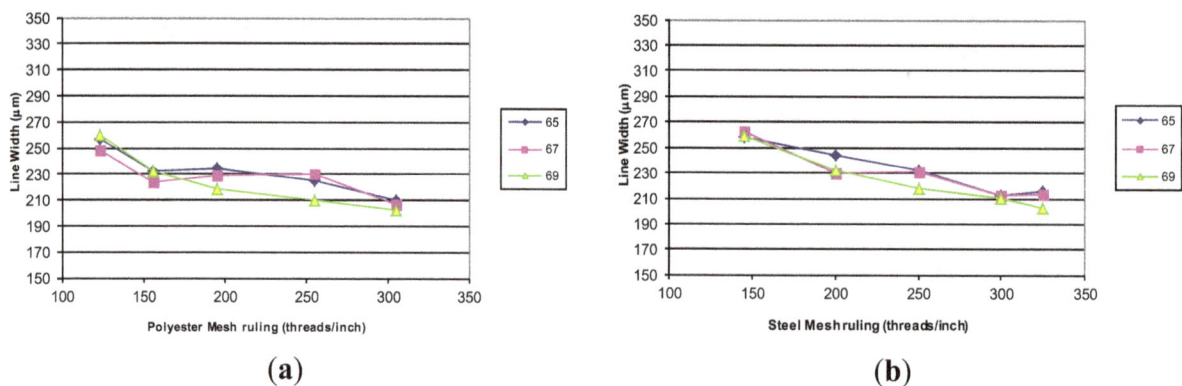

(a)                                                                 (b)

**Figure 7**. The effect of mesh ruling on the cured line width of the paste inks for a nominally 200 $\mu m$ line (**a**) polyester and (**b**) stainless steel meshes.

### 3.3. Comparison of Each Set of Materials

A clear difference between the between the two families is the cross sectional area (and hence film thickness) of the final cured film, Figure 8, which can be attributed to higher solids content in the paste materials. The polymer materials undergo a greater reduction in volume as the solvent content evaporates during the curing process. The polymer materials also exhibit a greater degree of line broadening with the lines consistently wider than that observed with the paste inks. This is attributed to the greater slumping of the lower viscosity material as it is released from the screen.

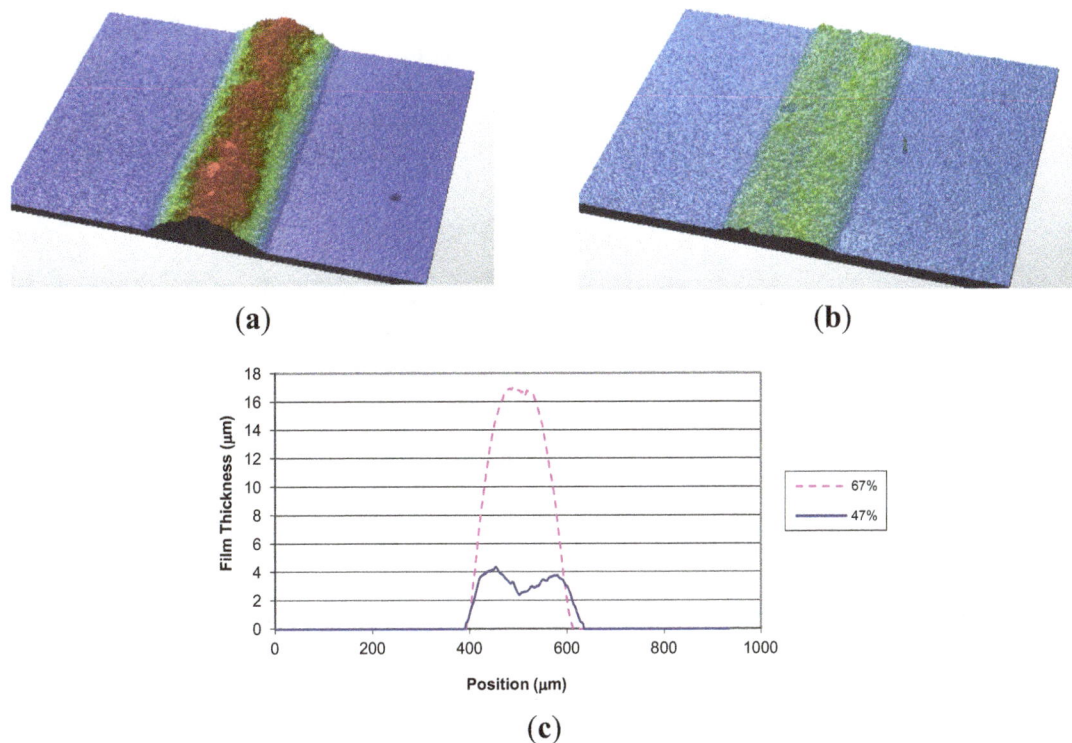

**Figure 8**. A comparison of the cross section of the printed lines through the polyester 195 threads/cm screen. (**a**) 3D view of 67% material; (**b**) 3D view of 47%; and (**c**) Mean cross section through line.

From the measured cross sectional area and resistance of the 200 µm line, it is possible to calculate the printed material resistivity and hence conductivity. The thinner film produced with the polymer ink delivers a printed feature where the silver particles are more densely packed. The net result of the denser packing of the silver in the film is to increase the conductivity of the cured material, Figure 9. Within each material set there is a gradual increase in the conductance of the printed film as the silver proportion is increased. For the polymer materials, this varies between $2.2 \times 10^6$ to $6.2 \times 10^6$ S/m (between 15 and 37 times bulk resistivity) while the paste materials show a variation between $5 \times 10^5$ and $1.6 \times 10^6$ S/m (between 35 and 148 times bulk resistivity).

For each material family, increasing the film thickness increases the conductivity but there is some data spread. This spread is associated with a number of features of the printed lines. Irregularities along the line length, lead to islands of materials which contribute to the calculated cross sectional area but contribute nothing to the line conductivity. These topological variations are particularly visible with the polymer materials where films are thinner and where the effect of non continuous topographic anomalies are greater, Figure 10.

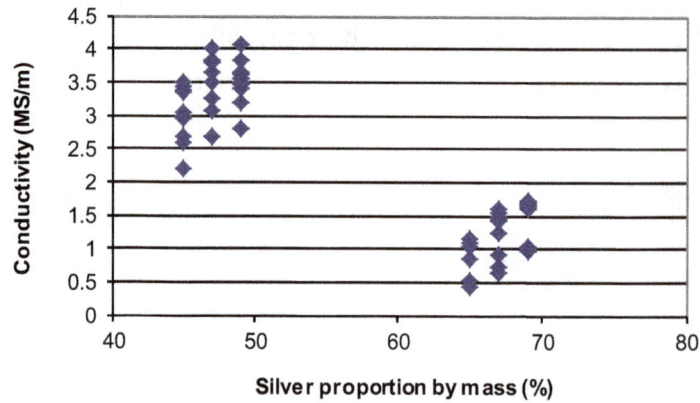

**Figure 9**. Spread of printed material conductivities obtained with each material/film thickness combination.

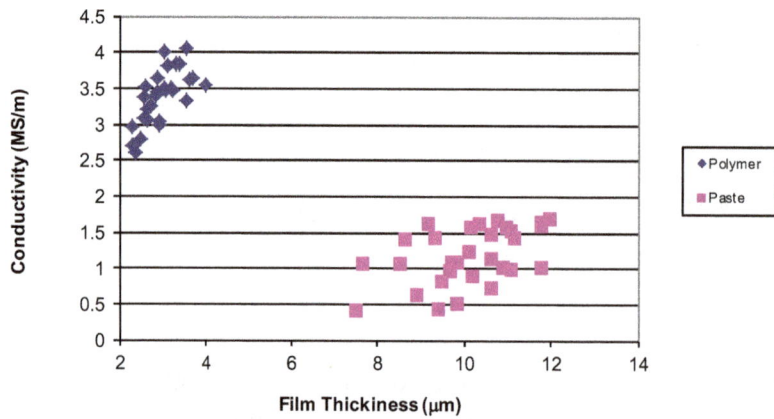

**Figure 10**. Spread of material conductivities obtained with each material/film thickness combination.

The increased conductivity of the polymer material demonstrates that it is possible to improve electrical performance by optimization of the binder system while maintaining a material which is capable of being processed in a conventional manner, *i.e.* the correct rheological and curing properties.

From commercial perspective, the relative merit of the materials and their universal applicability can be examined. A bill of materials (BOM) cost analysis of materials using data provided by the material supplier [15,16], shows that the dominant element of cost in the bill of materials is the silver, Table 3. The paste material (due to its higher silver content) has a 40% cost premium when compared to the polymer material. In practice, this premium is likely to lower as the manufacturing costs and operational overheads of the material supplier need also be considered.

**Table 3.** Cost analysis for bill of materials for the 47% and 67% silver content for each material family.

| Component | £/kg | Gel (@47% Ag) | Paste (@67% Ag) |
|---|---|---|---|
| Silver metals costs | 232 | 109 | 154 |
| Polymer | 12 | 0.72 | 2.4 |
| Solvent costs | 0.5 | 0.24 | 0.06 |
| Total | | 110 | 158 |

An useful means of examining the commercial is to examine the conductivity (S/m) obtained per unit BOM currency which effectively describes the economic efficiency of the material, When this is normalised such that it is relevant to all currencies, the maximum conductivity/unit currency is given by the lowest silver content material (45%), Figure 11. The 45% material also however produces the largest range of values, highlighting that it produces lines where topological anomalies are increased. Thus, the lower silver materials provide a more cost effective conductive structure, they are more highly sensitive to the mesh being used as the printed films are thinner. This is also reflected in the range of line widths which are produced with the polymer materials. For the polymer materials, the range of the conductivity produced is narrower as the silver proportion is increased highlighting that there is a trade off between improved process robustness and material cost.

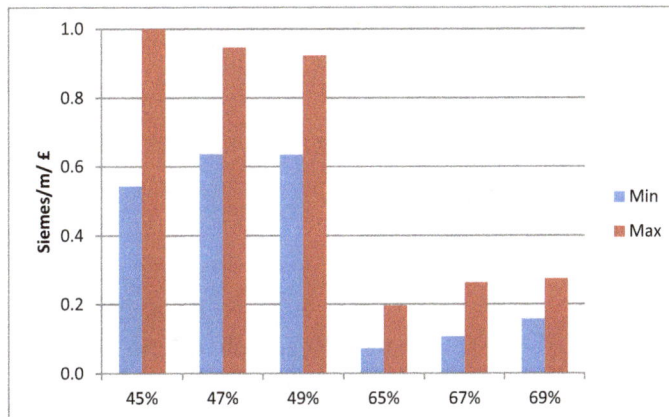

**Figure 11.** The range of conductivity/unit cost obtained with each material.

The creation of this dataset linking feature silver material properties and screen characteristics has developed a design tool which can be used to determine the optimum material properties and process settings in order to achieve the desired electrical performance (and *vice versa*). Work on the inclusion of additional parameters such as squeegee settings, mesh tension, screen–substrate distance is on going such that these secondary parameters can be incorporated into the design tool. The dataset also serves to form the basis of cost models for material use given defined electrical requirements. A further factor which determines the screen which is to be used is the resolution of the image required with finer features requiring finer meshes to be used. The drive for finer features in order to reduce device size is thus dictating the use of finer meshes. An investigation into the relationship between mesh ruling, material properties and fine line rendering capabilities is on-going.

## 4. Conclusions

An experimental investigation has been carried out on the effect of screen printing silver material formulation and mesh type on the characteristics of printed conductive structures. The study has shown that it is possible to formulate large particle silver materials which can provide higher conductivities by increasing metal density in the cured film while maintaining materials within the operational window of the screen printing process. The thinner film nature of these new materials does however make them more prone to variations in printed resistance which will likely have an impact on product robustness. The impact of finer meshes is to reduce the printed film thickness which is in line with expectations but the degree to which the film thickness and sheet resistance is affected is dependent on the material being printed. Further studies will examine the role of other process parameters and printed features.

## Acknowledgments

The authors would like to thank the Welsh Government and Gwent Electronic Materials for their support of this project.

## Author Contributions

E.J. performed some of the experiments, analysis and primary authored the paper, S.H. performed the remainder of the experiments and analysis. T.C and D.G. drove the research programme, aided in experimental design and co-authored the work.

## Conflicts of Interest

The authors declare no conflict of interest.

## References

1.  Silver Inks and Pastes Markets: 2014–2021; N-tech Research: Glen Allen, VA, USA, 23 June 2014.

2.  Smith, P.J.; Shin, D.-Y.; Stringer, J.E.; Reis, N.; Derby, B. Direct inkjet printing and low temperature conversion of conductive silver patterns. *J. Mater. Sci.* **2006**, *41*, 4153–4158.

3.  Sung, D.; de la Fuente Vornbrock, A.; Subramanian, V. Scaling and optimization of gravure-printed silver nanoparticle lines for printed electronics. *IEEE Trans. Compon. Packag. Technol.* **2010**, *33*, 105–114.

4.  Deganello, D.; Cherry, J.A.; Gethin, D.T.; Claypole, T.C. Patterning of micro-scale conductive networks using reel-to-reel flexographic printing. *Thin Solid Films* **2010**, *518*, 6113–6116.

5.  The Silver Inks and Pastes Market 2013; N-tech Research: Glen Allen, VA, USA, 7 January 2013.

6.  Owczarek, J.A.; Howland, F.L. A study of the off-contact screen printing process. I. Model of the printing process and some results derived from experiments. *IEEE Trans. Compon. Hybrids Manuf. Technol.* **1990**, *13*, 358–367.

7.  Taroni, M.; Breward, C.J.W.; Howell, P.D.; Oliver, J.M.; Young, R.J.S. The screen printing of a power-law fluid. *J. Eng. Math* **2012**, *73*, 93–119.

8.  Hoornstra, J.; Weeber, A.W.; de Moor, H.H.C.; Sinke, W.C. The Importance of Paste Rheology in Improving Fine Line, Thick Film Screen Printing of Front Side Metallization. In Proceedings of the 14th European Photovoltaic Solar Energy Conference (EPSEC), Barcelona, Spain, 1997; pp. 404–407.

9.  Hilali, M.M.; Nakayashiki, K.; Chandra, K.; Reedy, R.C.; Rohatgi, A.; Shaikh, A.; Kim, S.; Sridharan, S. Effect of Ag particle size in thick-film Ag paste on the electrical and physical properties of screen printed contacts and silicon solar cell. *J. Electrochem. Soc.* **2006**, *153*, A5–A11.

10. Lin, H.W.; Chang, C.P.; Hwu, W.H.; Ger, M.D. The rheological behaviours of screen-printing pastes. *J. Mater. Process. Technol.* **2008**, *197*, 284–291.

11. Buzby, D.; Dobie, A. Fine Line Screen Printing of Thick Film Pastes on Silicon Solar Cells. In Proceedings of the 41st International Symposium on Microelectronics (IMAPS 2008), Rhode Island, USA, 2–6 November 2008.

12. Faddoul, R.; Reverdy-Bruas, N.; Blayo, A. Printing force effect on conductive silver tracks: Geometrical, surface, and electrical properties. *J. Mater. Eng. Perform.* **2012**, *22*, 640–649.

13. Merilampi, S.; Laine-Ma, T.; Ruuskanen, P. The characterization of electrically conductive silver ink patterns on flexible substrates. *Microelectron. Reliab.* **2009**, *49*, 782–790.

14. Jewell, E.H.; Hamblyn, S.M.; Claypole, T.C.; Gethin, D.T. The impact of carbon content and mesh on the characteristics of screen printed conductive structures. *Circuit World* **2013**, *39*, 3–21.

15. GEM–Conductive Silver Polymer Paste; Product information sheet provided by Gwent Electronic Materials. Available online: http://www.gwent.org/gem_data_sheets/polymer_systems_products/flexible_conductor_and_membrane_touch_switch/45_per_cent_ag_c2080415p2.pdf (accessed on 4 May 2015).

16. GEM–Conductive Silver Ink; Product information sheet provided by Gwent Electronic Materials. Available online: http://www.gwent.org/gem_data_sheets/polymer_systems_products/flexible_conductor_and_membrane_touch_switch/65_per_cent_ag_c2110817d5.pdf (accessed on 4 May 2015).

17. Smits, F.M. Measurement of sheet resistivites with 4 point probes. *Bell System Tech. J.* **1958**, *37*, 711–718.

18. Abbott, S.J.; Gaskell, P.H.; Kapur, N. A new model for the screen printing process—From theory to practical insight. In Proceedings of 2nd International Symposium on Printing & Coating Technology, Swansea, UK, September 2000.

19. Petersen, I.; Hübner, G.; Claypole, T.C.; Jewell, E. Influence and interaction phenomena of screen printing machine settings on surface roughness. In Proceedings of International Association of Research Institutes for the Graphic Arts Industries (IARIGAI), Budapest & Debrecen, Hungary, 11–14 September 2011.

# 13

# Micromechanical Simulation of Thermal Cyclic Behavior of ZrO$_2$/Ti Functionally Graded Thermal Barrier Coatings

**Hideaki Tsukamoto**

Graduate School of Engineering, Nagoya Institute of Technology, Gokiso-cho, Showa-ku, Nagoya, Aichi, 466-8555, Japan; E-Mail: tsukamoto.hideaki@nitech.ac.jp

Academic Editor: Ugo Bardi

**Abstract:** This study numerically investigates cyclic thermal shock behavior of ZrO$_2$/Ti functionally graded thermal barrier coatings (FG TBCs) based on a nonlinear mean-field micromechanical approach, which takes into account the time-independent and dependent inelastic deformation, such as plasticity of metals, creep of metals and ceramics, and diffusional mass flow at the ceramic/metal interface. The fabrication processes for the FG TBCs have been also considered in the simulation. The effect of creep and compositional gradation patterns on micro-stress states in the FG TBCs during thermal cycling has been examined in terms of the amplitudes, ratios, maximum and mean values of thermal stresses. The compositional gradation patterns highly affect thermal stress states in case of high creep rates of ZrO$_2$. In comparison with experimental data, maximum thermal stresses, amplitudes and ratios of thermal stresses can be effective parameters for design of such FG TBCs subject to cyclic thermal shock loadings.

**Keywords:** functionally graded thermal barrier coatings (FG TBCs); thermal cycle; micromechanics; creep; compositional gradation

## 1. Introduction

Ceramic-metal functionally graded thermal barrier coatings (FG TBCs) have been attracting a great deal of attention for structures working under super high temperatures and temperature gradients. A unique blend of material properties make them good candidates for the use in various structural components demanding high strength, high resistance to heat, and low weight, which is of particular interest in the modern transportation and aerospace industries [1,2].

FG TBCs are advanced multiphase composites that are engineered to have a smooth spatial variation of material constituents. This variation results in an inhomogeneous structure with smoothly varying thermal and mechanical properties. The advantages of FG TBCs as an alternative to two dissimilar materials (ceramics and metal) joined directly together include smoothing of thermal stress distributions across the layers, minimization or elimination of stress concentrations and singularities at the interface corners and increase in bonding strength [3].

FG TBCs are macroscopically and microscopically heterogeneous. Macroscopic heterogeneity means the gradation of microstructures and material properties through the thicknesses of the FG TBCs, and microscopic heterogeneity is due to the fact that the composite materials are composed of several constituents, usually metals and ceramics. In order to make the best use of such heterogeneous materials, the formulation of the constitutive relation from the standpoints of these two different scales is required [4,5].

So far a number of analytical and computational methods to predict thermal stress states in functionally graded materials (FGMs) and design optimal FGMs have been proposed [6,7]. Some of them considered macroscopic heterogeneity, and used the simple rules of mixture such as Voghit and Reuss rules to derive effective properties of the composites. Some studies take into more account the microscopic heterogeneity. These studies are classified into two categories. One adopts finite element methods, which assume periodic microstructure of compositions in FGMs [8,9]. The other is an analysis based on micromechanics, one of which is the Eshelby's equivalent inclusion method [10–12] being effective and applicable to analyze high-temperature behavior of ceramic-metal composites. An earlier work by Wakashima and Tsukamoto [13,14] applied the mean-field micromechanical concepts to estimating the thermal stresses in a FGM plate. Up to date, many studies have been done based on such micromechanical concepts [15,16].

From the viewpoints of what kinds of inelastic deformations for each phase to be taken into account, some studies [17] considered plastic deformation of metal phase, which is the time-independent deformation, and some studies [18] took into consideration time-dependent deformation such as creep in FG TBC plates, for which the constitutive relations derived from experimental results were used. Recently, micromechanical approach considering plastic deformation of metal phase and creep of metal and ceramic phase as well as diffusional mass flow at the interface between metal and ceramic phases is proposed by the author [5].

In the past decade, ZrO2/Ti FG TBCs have been of potential for high-temperature applications in automobile and aerospace industries. ZrO2 has superior thermal and mechanical properties, which are effective for TBCs [19]. One of superior mechanical properties of ZrO2 is high fracture toughness due to stress-induced transformation (from tetragonal to monoclinic crystal structures under some stress conditions, which can be used in enhancement of fracture toughness of various ceramics and ceramic matrix composites [20–22].

The purpose of the current study is to numerically investigate thermal cyclic behavior of ZrO2/Ti functionally graded thermal barrier coatings (FG TBCs) based on a nonlinear mean-field micromechanical approach [5], which takes into account the time-independent and dependent inelastic deformation, such as plasticity of metals, creep of metals and ceramics, and diffusional mass flow at the ceramic/metal interface. The fabrication processes for the FG TBCs have been also considered. The effect of compositional gradation patterns and time-dependent inelastic deformation on micro-stress states in the

FG TBCs during thermal cycling has been examined in terms of the amplitudes, ratios and mean values of thermal stresses.

## 2. Mean-Field Micromechanics-Based Analysis

### 2.1. Basic Model

The model used in the study is based on the work by Tsukamoto [5]. A brief description of the model is made here. Macroscopically homogeneous composites with spherical particles are considered to be building blocks of the functionally graded thermal barrier coating (FG TBC) plate as shown in Figure 1. The building blocks are assumed to be subject to balanced bi-axial plane stresses. Here, the metal phase is assumed to be matrix and ceramic phase is particles, which are indicated by subscript 0 and 1, respectively, while the inversion of the relation of matrix and particles can be easily derived in the similar way. The inelastic deformation of constituents of the composites include creep with the strain, $\varepsilon^c$, plastic deformation with the strain, $\varepsilon^p$, and diffusional mass transport along the metal-ceramic interface with the eigen strain of the particle, $\varepsilon^d$. The in-plane and out-of-plane micro-stresses in each phase can be written as follows [5],

$$\sigma_0^{in} = 2(\beta_0 + 1/3\gamma_0)\sigma + 3f_1\beta^*(\alpha_1 - \alpha_0)\theta + 2f_1\gamma^* \left\{ (\varepsilon_1^c - \varepsilon_0^c) + \varepsilon^d - \varepsilon^p \right\} \tag{1}$$

$$\sigma_0^{out} = 2(\beta_0 - 2/3\gamma_0)\sigma + 3f_1\beta^*(\alpha_1 - \alpha_0)\theta - 4f_1\gamma^* \left\{ (\varepsilon_1^c - \varepsilon_0^c) + \varepsilon^d - \varepsilon^p \right\} \tag{2}$$

for the metal matrix (indicated by subscript 0), and

$$\sigma_1^{in} = 2(\beta_1 + 1/3\gamma_1)\sigma - 3f_0\beta^*(\alpha_1 - \alpha_0)\theta - 2f_0\gamma^* \left\{ (\varepsilon_1^c - \varepsilon_0^c) + \varepsilon^d - \varepsilon^p \right\} \tag{3}$$

$$\sigma_1^{out} = 2(\beta_1 - 2/3\gamma_1)\sigma - 3f_0\beta^*(\alpha_1 - \alpha_0)\theta + 4f_0\gamma^* \left\{ (\varepsilon_1^c - \varepsilon_0^c) + \varepsilon^d - \varepsilon^p \right\} \tag{4}$$

for the ceramic particle (indicated by subscript 1).

Here, $f_0$ and $f_1$ are the volume fraction, $\alpha_0$ and $\alpha_1$ are the coefficient of thermal expansion, and $\sigma$ is a macro-stress due to balanced bi-axial loadings. $\beta_0$, $\beta_1$, $\gamma_0$, $\gamma_1$, $\beta^*$ and $\gamma^*$ are micromechanical constants depending on the elastic constants and volume fraction of each phase, which were given in the work by Tsukamoto [5]. The equivalent micro-stresses in each phase are expressed by,

$$\sigma_0^{eq} = \left| \sigma_0^{in} - \sigma_0^{out} \right| = 2 \left| \gamma_0\sigma + 3f_1\gamma^* \left\{ \left( \varepsilon_1^c - \varepsilon_0^c \right) + \varepsilon^d - \varepsilon_0^p \right\} \right| \tag{5}$$

$$\sigma_1^{eq} = \left| \sigma_1^{in} - \sigma_1^{out} \right| = 2 \left| \gamma_1\sigma - 3f_0\gamma^* \left\{ \left( \varepsilon_1^c - \varepsilon_0^c \right) + \varepsilon^d - \varepsilon_0^p \right\} \right| \tag{6}$$

In this analysis, plastic and creep deformations are supposed to obey the associated flow rule in which both deformation potentials are taken equal to the von Mises-type yield function. Plastic deformation of metal phase is assumed to be expressed by the Swift's equation:

$$\sigma_0^{eq} = a(c + \varepsilon^{p^{eq}})^{n_p} \tag{7}$$

where $a$, $c$ and $n_p$ are constants. $\sigma_0^{eq}$ is the (equivalent) flow stress of metals. When the creep deformation of each phase is assumed to be controlled by grain-boundary diffusion (Coble creep), the constitutive equation is expressed as follows,

$$\dot{\varepsilon}^{c}_{coble}{}^{eq} = C\frac{\omega_{gb}D_{gb}\Omega}{kTd^3}\sigma^{eq} \tag{8}$$

$C$ is the geometric constant (~16), $D_{gb}$ the grain boundary diffusivity, $\omega_{gb}$ the grain boundary width, $\Omega$ the volume of a diffusing atom, $d$ the grain size and $k$ the Boltsman's constant. The inelastic strain $\dot{\varepsilon}^{d\,eq}$ by mass transport along the interface between metal and ceramic phases is expressed as follows [5],

$$\dot{\varepsilon}^{d\,eq} = C^{int}\frac{\omega_{int}D_{int}\Omega}{kTd_p^3}\sigma_1^{eq} \tag{9}$$

$C^{int}$ is the material constant derived from micromechanical considerations, $\omega_{int}$ the interface width for diffusion, $D_{int}$ the interfacial diffusivity, $\Omega$ the volume of diffusing atom and $d_p$ the diameter of the particle. Therefore, when considering the composites under plane-stress conditions, the constitutive equations can be described by

$$\dot{\sigma}(z,t) = \left\{ S^e(z) + S^p(z,t) \right\} \left\{ \dot{\varepsilon}(z,t) - \alpha(z)\dot{\theta}(z,t) - \varepsilon^{p(cd)}(z,t) - \varepsilon^{c-d}(z,t) \right\} \tag{10}$$

$\dot{\sigma}(z,t)$ is the plane stress rate, $S^e(z)$ the overall plane-stress elastic compliance, $S^p(z,t)$ the overall plane-stress plastic compliance, $\alpha(z)$ the overall in-plane thermal expansion coefficient, $\dot{\varepsilon}^{p(cd)}(z,t)$ the overall plastic strain rate due to the difference between creep abilities of each phase and interfacial diffusion, and $\dot{\varepsilon}^{c-d}(z,t)$ the overall creep strain rate. The details for mathematical expressions of these functions are given in the work by Tsukamoto [5]. $\dot{\sigma}(z,t)$ given in Equation (10) can be incorporated with the lamination theory.

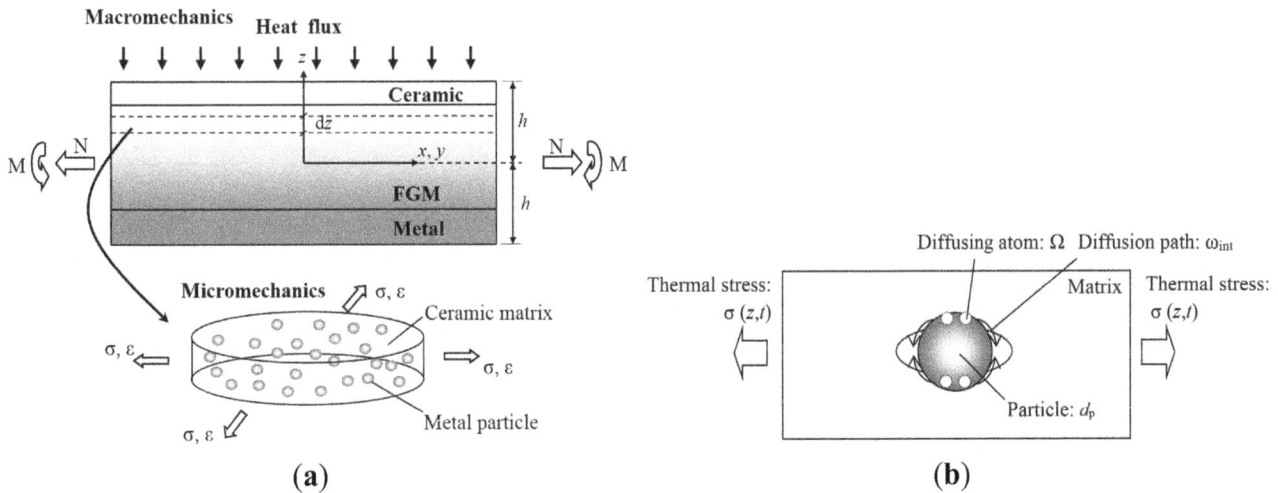

**Figure 1.** Schematic illustration of **(a)** a functionally graded thermal barrier coatings (FG TBCs) subjected to thermo-mechanical loadings and its building block of the thin layer composed of spherical particle-dispersed composites, and **(b)** mass transfer at the particle/matrix interface.

## 2.2. Simulation Input Data

Now let us consider cyclic thermal behavior of ZrO2/Ti FG TBCs to be analyzed. The FG TBCs have a thickness of 9 mm, in which the FGM part has a thickness of 3 mm and Ti substrate part has that of 6 mm. The FGM part (3 mm) consists of six composite layers with different compositions according to predetermined compositional gradation patterns of the FGMs. Figure 2 shows the compositional gradation patterns used in this study. Step-wised compositional gradation patterns are parametrically described using the following expression [4,5].

$$f_m(i)=1-f_c(i)=(i-1)^n/(P-1)^n \tag{11}$$

where, $f_m(i)$ and $f_c(i)$ are the volume fractions of metal (Ti) and ceramics (ZrO$_2$) phases in the $i$-th sub-layer, respectively. $P$ is the total number of sub-layers, which have the thickness of 0.5 mm for each in the FGM part. The exponent, $n$, is a parameter characterizing the compositional gradation pattern. For the gradation parameter $n$, $n = 1$ means the linear compositional gradation, $n > 1$ means the ceramic-rich gradation and $n < 1$ means the metal-rich gradation. In this study, FGM samples with three different compositional gradation patterns of $n = 0.5$, 1 and 2 were simulated to investigate the effects of compositional gradation patterns on the thermo-mechanical behavior.

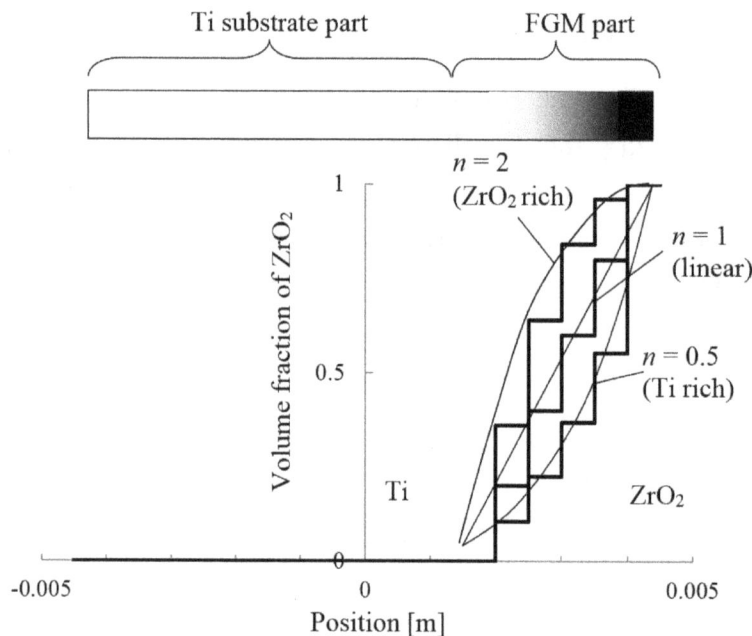

**Figure 2.** Compositional gradation patterns of FG TBCs.

The material property data of Ti and ZrO$_2$ used in the calculations are shown in Table 1. The total thickness of the samples is set at 9 mm (including 3 mm thickness for the FGM part and 6 mm thickness for the substrate part). The thermo-mechanical boundary conditions considered here are illustrated in Figure 3, and described as follows. The powders to be sintered is stuffed in dice (fully constraint in both in-plane and out-of-plane deformations) at 1400 °C. Under this condition, there is no stress in the samples. Then temperature of samples goes down to room temperature (R.T.) at the cooling rate of 100 °C/min under the mechanical boundary conditions of fully constraint deformation. At R.T., all the mechanical constraints are relieved (free mechanical constraint). After that, cyclic thermal shock tests start. The ceramic surface is assumed to be heated to 1200 °C for 1 min. After the temperature of the ceramic surface reaches 1200 °C, the temperature is hold for 60 s. Then the ceramic surface is exposed to the air with the heat transfer coefficient of 300 W/(m$^2$ K) and temperature of R.T. during cooling processes. The air-cooling processes take 60 s, and then the ceramic surface is suddenly cooled down to

R.T. During cyclic thermal shock tests Ti substrate sides are exposed to air with the heat transfer coefficient of 500 W/(m² K) all the time. This process corresponds to one cycle, which is repeated 4 times in the simulation like the current experiments. Based on the micromechanical model, the program codes for analysis of thermo-mechanical behavior of FGMs were developed using Fortran 95 programming languages. The simulation was conducted on laptop computer with normal specifications.

**Table 1.** Material property data of Ti and ZrO2 used in the calculation. Diameter of a particle is $40 \times 10^{-6}$ m, $D_{int} \times \omega_{int}$ is assumed to be the same value as $D_{gb} \times \omega_{gb}$ for Ti. Flow stress parameters in Swift equation for Ti are $a = 600$ MPa, $c = 0.3$ and $n = 1$.

| Property data | Ti | ZrO₂ |
|---|---|---|
| Young's moduluns [GPa] | 116 | 200 |
| Poisson's ration | 0.32 | 0.3 |
| CTE $10^{-6}$ [$K^{-1}$] | 8.6 | 10.0 |
| Thermal conductivity [$Wm^{-1} K^{-1}$] | 21.9 | 3.0 |
| Specific heat [$J kg^{-1} K^{-1}$] | 520 | 3000 |
| Density[$kg m^{-3}$] | 4506 | 5990 |
| *Coble creep parameters* | | |
| $D_{gb0}$(pre-exp. Term) $\times \omega_{gb}$ [$m^{-3} s^{-1}$] | $1.9 \times 10^{-7}$ | $0.29 \times 10^{-6}$ |
| Activation energy [$J mol^{-1}$] | $1.53 \times 10^{5}$ | $5.7 \times 10^{5}$ |
| Atomic volume [$m^{3}$] | $1.15 \times 10^{-29}$ | $4.66 \times 10^{-29}$ |
| Grain size[m] | $10.0 \times 10^{-6}$ | 0.1, 1 and $10.0 \times 10^{-6}$ |

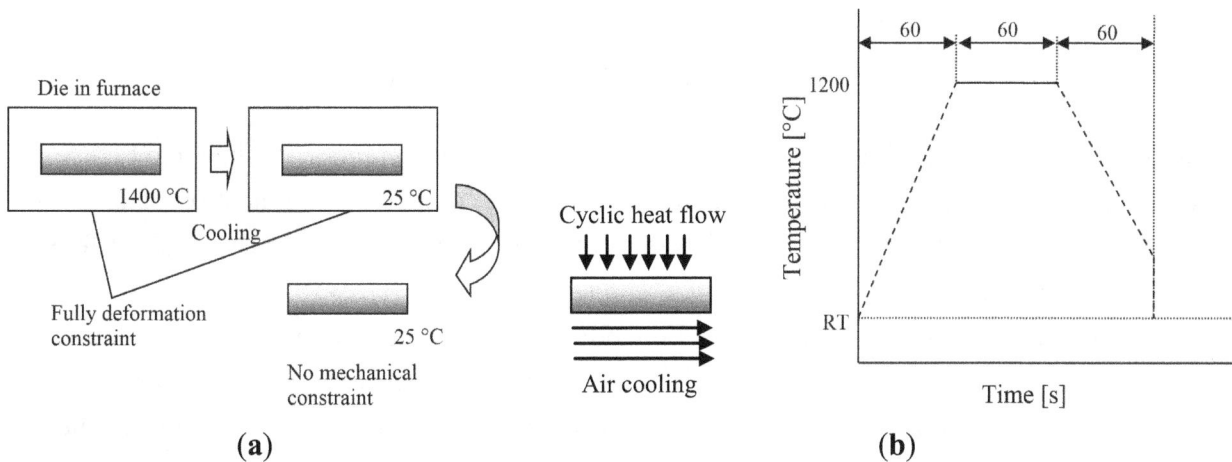

**Figure 3.** Schematic illustration of thermo-mechanical boundary conditions for fabrication and thermal cycling processes: (**a**) fabrication process, and (**b**) thermal cycling process, including a schematic illustration of heat flow and temperature profile for one cycle.

## 2.3. Simulation Results

Figure 4 shows calculation results of temperature transient of the ZrO2 surface and maximum stress transient in the ZrO2 surface layer during fabrication and cyclic thermal shock test processes. It is seen in Figure 4a,b that maximum stresses in the ZrO2 surface layer vary with temperature transients. The highest stresses can be reached just after the fabrication (at point A in Figure 4a,b). During cyclic thermal shock tests, the maximum stresses in the ZrO2 surface layer change with temperature transients, in which high maximum stress peaks are reached at R.T. Now let us define three parameters such as the mean stress, $\sigma_{mean}$, range, $\Delta\sigma$, and ratio, $R$, of the maximum thermal stresses. To begin with, it is set $\sigma_{max}$ the high peak value of maximum thermal stresses in the $ZrO_2$ surface layer at R.T. in a cycle, and $\sigma_{min}$ the low peak value of the maximum thermal stresses in the $ZrO_2$ surface layer at 1200 °C in a cycle. The mean stress, $\sigma_{mean}$, is defined by $(\sigma_{max} + \sigma_{min})/2$. The range of stress, $\Delta\sigma$, is defined by $\sigma_{max} - \sigma_{min}$. The ratio of stress, $R$, is defined by $\sigma_{min}/\sigma_{max}$.

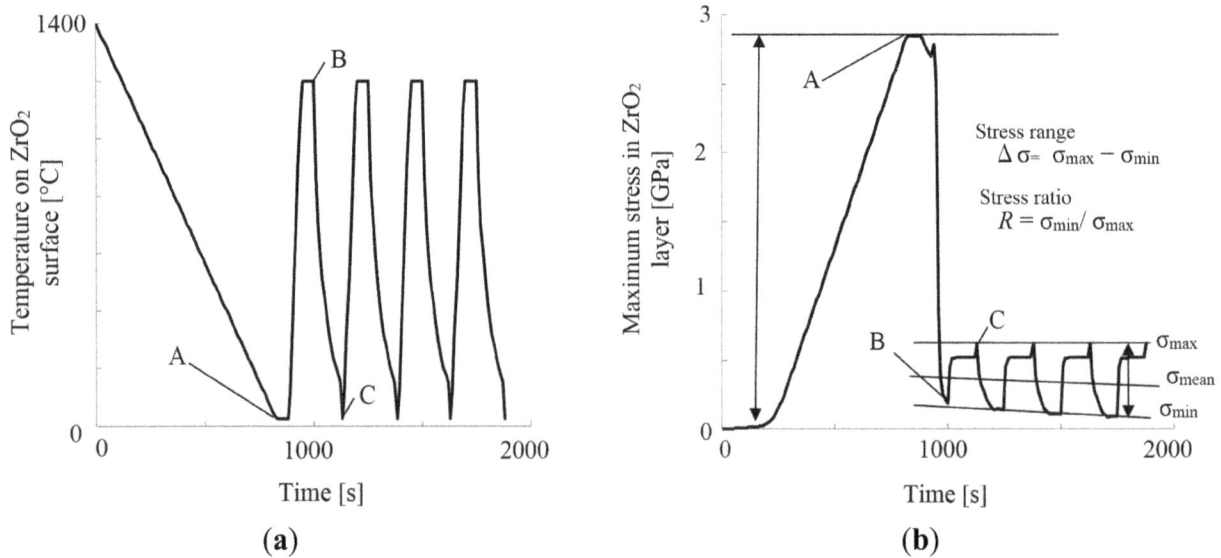

**Figure 4.** Temperature transient of $ZrO_2$ surface (**a**) and maximum thermal stress transient in $ZrO_2$ surface layer (**b**) during fabrication and cyclic thermal shock processes. Points A, B and C shown in the figures indicates the corresponding points, at which the time is the same.

Residual stress states in the $ZrO_2$ surface layer just after the fabrication process are shown in Figure 5, which includes the effect of creep ability (grain size) of $ZrO_2$ and compositional gradation patterns on the residual stress states. It is seen that there is almost no difference among the FG TBCs with different compositional gradation patterns, while creep rates represented by grain sizes (0.01, 1 and 10 μm) of $ZrO_2$ (corresponding to creep ability expressed by Equation 8) highly affect the residual stress states. In any cases, maximum residual stresses in $ZrO_2$ layers are tensile. In case of the grain size of $ZrO_2$ of 0.01 μm, corresponding to high creep rates maximum residual stresses are low compared to others, which is attributed to high creep rates of $ZrO_2$ leading to large creep deformation during the cooling process in the fabrication.

Figure 6 shows the maximum thermal stresses, $\sigma_{max}$, and mean thermal stresses, $\sigma_{mean}$, in $ZrO_2$ surface layer during cyclic thermal shock loading tests in case of $d_{ZrO_2} = 0.01$ μm and $d_{ZrO_2} = 10$ μm. The effect of creep of $ZrO_2$ can be apparently seen in the figures. For $\sigma_{max}$, the order for composition gradation patterns is different between $d_{ZrO_2} = 0.01$ μm and $d_{ZrO_2} = 10$ μm. In case that the creep rates of $ZrO_2$ is high ($d_{ZrO_2} = 0.01$ μm), FG TBCs with $n = 2$ shows higher $\sigma_{max}$ than others, while FG TBCs with $n = 0.5$ shows higher $\sigma_{mean}$ than others. Meanwhile, in case that creep rates of $ZrO_2$ is low ($d_{ZrO_2} = 10$ μm), FG TBCs with $n = 0.5$ shows higher $\sigma_{max}$ than others, and FG TBCs with $n = 0.5$ also shows higher $\sigma_{mean}$ than others.

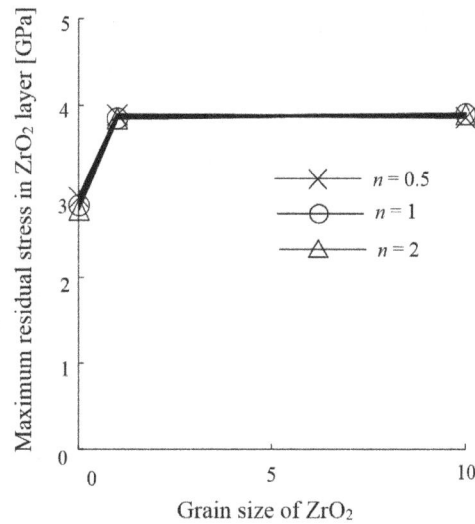

**Figure 5.** Residual stress states in the ZrO2 surface layer just after fabrication processes for functionally graded materials (FGMs) with $n = 0.5$, 1 and 2.

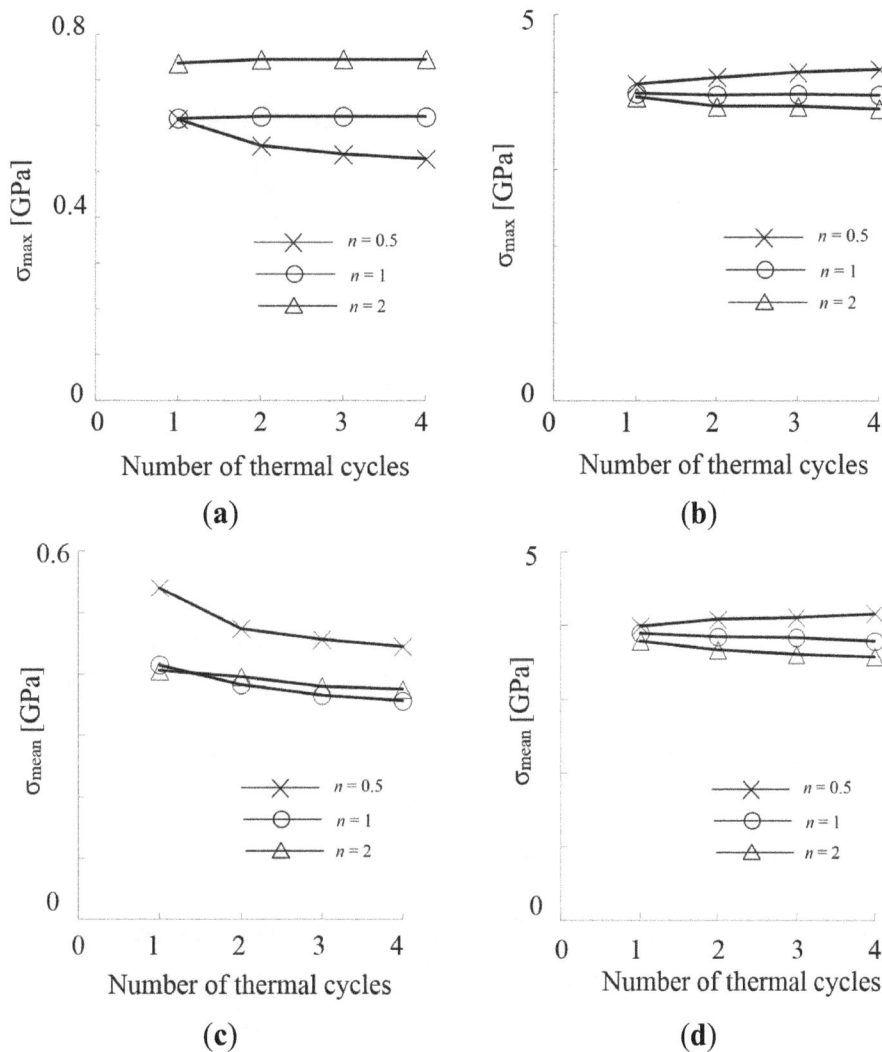

**Figure 6.** Maximum thermal stresses, $\sigma_{max}$, and mean thermal stresses, $\sigma_{mean}$, in ZrO2 surface layer during thermal shock cycling. (**a**) $\sigma_{max}$ for $d_{ZrO_2} = 0.01$ μm, (**b**) $\sigma_{max}$ for $d_{ZrO_2} = 10$ μm, (**c**) $\sigma_{mean}$ for $d_{ZrO_2} = 0.01$ μm, and (**d**) $\sigma_{mean}$ for $d_{ZrO_2} = 10$ μm.

Figure 7 shows the amplitudes (ranges), $\Delta\sigma$, and ratios, $R$, of thermal stresses in the $ZrO_2$ surface layer during cyclic thermal shock loading tests for $d_{ZrO_2} = 0.01$ $\mu$m and for $d_{ZrO_2} = 10$ $\mu$m. Creep of $ZrO_2$ largely affects $\Delta\sigma$ and R. For $\Delta\sigma$, in case of both low and high creep rates of $ZrO_2$, FG TBCs with $n = 2$ shows higher $\Delta\sigma$ than others. It is expected that when the thermal fatigue behavior is dominant, FG TBCs with $n = 2$ exhibit the lowest resistance to such cyclic thermal shock loadings. For R, in case of low creep rates of $ZrO_2$ ($d_{ZrO_2} = 10$ $\mu$m), FG TBCs with any compositional gradation patterns show almost the same and constant values, which are not also affected by number of thermal cycles.

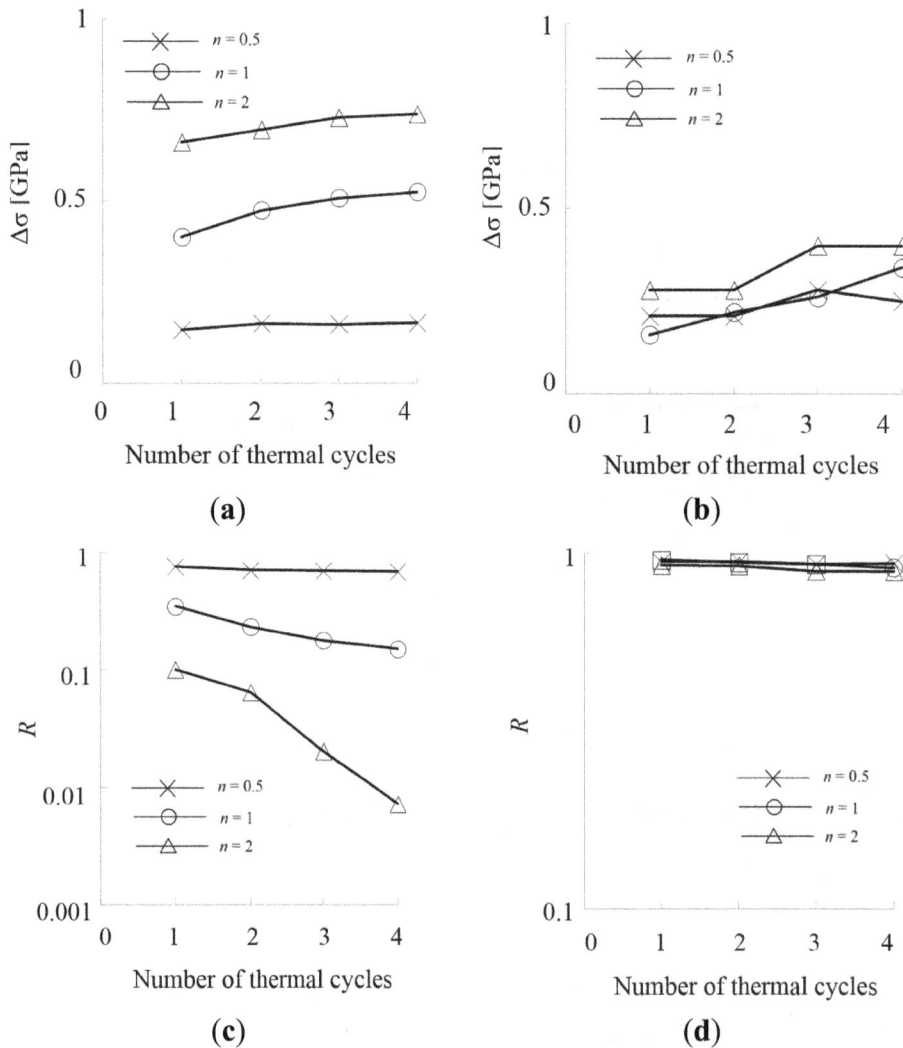

**Figure 7.** Amplitudes, $\Delta\sigma$, and ratios, $R$, of thermal stresses in $ZrO_2$ surface layer during thermal shock cycling. (**a**) $\Delta\sigma$ for $d_{ZrO_2} = 0.01$ $\mu$m, (**b**) $\Delta\sigma$ for $d_{ZrO_2} = 10$ $\mu$m, (**c**) $R$ for $d_{ZrO_2} = 0.01$ $\mu$m, and (**d**) $R$ for $d_{ZrO_2} = 10$ $\mu$m.

Maximum equivalent thermal stresses, $\sigma_{max}^{eq}$, in $ZrO_2$ surface layer during cyclic thermal shock loading tests for $d_{ZrO_2} = 0.01$ $\mu$m and for $d_{ZrO_2} = 10$ $\mu$m are shown in Figure 8. It is seen that in case of high creep rates of $ZrO_2$ ($d_{ZrO2} = 0.01$ $\mu$m), FG TBCs with $n = 2$ shows higher $\sigma_{max}^{eq}$ than others, while in case of low creep rates of $ZrO_2$ ($d_{ZrO_2} = 10$ $\mu$m), FG TBCs with $n = 0.5$ shows higher $\sigma_{max}^{eq}$ than others. This tendency is the same as that of $\sigma_{max}$.

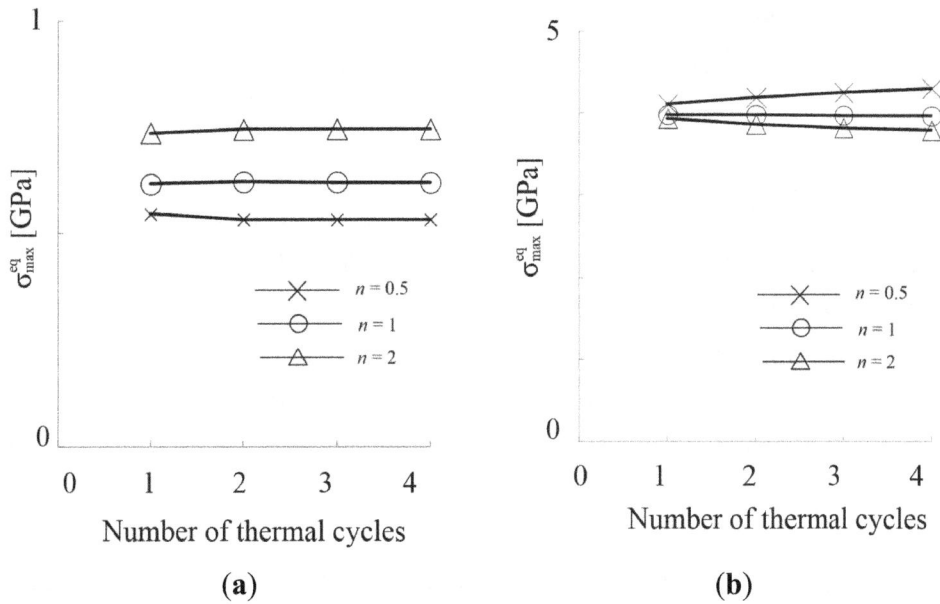

**Figure 8.** Maximum equivalent thermal stresses, $\sigma_{max}^{eq}$, in $ZrO_2$ surface layer during thermal shock cycling. **(a)** $\sigma_{max}^{eq}$ for $d_{ZrO_2} = 0.01$ μm; **(b)** $\sigma_{max}^{eq}$ for $d_{ZrO_2} = 10$ μm.

Amplitudes (ranges) of thermal stresses, $\Delta\sigma$, plotted against mean stresses, $\sigma_{mean}$ are shown in Figure 9. The data can be classified into two regimes. It can be seen that in case of high creep rates of $ZrO_2$ ($d_{ZrO_2} = 0.01$ μm), the values of $\sigma_{mean}$ are very low compared to values for low creep rates of $ZrO_2$ ($d_{ZrO_2} = 1$ and $10$ μm).

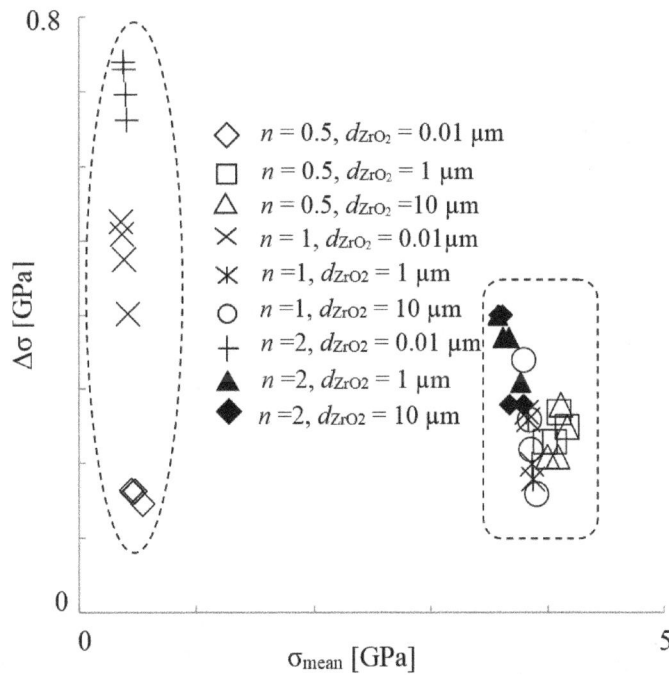

**Figure 9.** Amplitudes of thermal stresses, $\Delta\sigma$ plotted against mean stress, $\sigma_{mean}$.

$\Delta\sigma$ scatters over the wide ranges highly depending on compositional gradation patterns, n, in case of high creep rates of $ZrO_2$, in which higher value of n leads to higher $\Delta\sigma$. $\Delta\sigma - \sigma_{mean}$ graph is in general used for examining the fatigue limits. $\sigma_{mean}$ value is high compared to $\Delta\sigma$ value in case of low creep rates of $ZrO_2$, while $\sigma_{mean}$ and $\Delta\sigma$ values are similar and small in case of high creep rates of $ZrO_2$. These results are considered to be due to creep of $ZrO_2$ in cooling processes under cyclic thermal shock conditions.

## 3. Experimental Data

Now let us examine the simulation results including extracted micromechanical parameters shown above based on experimental data. Here, consider the FGM samples, which were fabricated using spark plasma sintering (SPS) methods [23]. The ingredient powders were $ZrO_2$ partially-stabilized by 3 mol% $Y_2O_3$. The average diameter of $ZrO_2$ powder is 26 nm, and that of Ti powder is less than 45 μm. The SPS was conducted in vacuum at 1400 °C under the uniaxial pressure of 30 MPa with the time duration of 20 min. The sintered FGM samples have a diameter of 20 mm and the same total and layer thicknesses as input data used in the simulation. The compositional gradation patterns of the FGMs are also the same as used in the calculations, that is, $n = 0.5$, 1 and 2. For thermal cycling, the profile of temperature of the $ZrO_2$ surface used in the experiments is the same as used in the simulation and shown in Figure 3, except for the cooling process. In the experiment, cooling was performed for the time duration of 1800 s, in which the temperature of $ZrO_2$ surface turned to R.T. (around 25 °C).

Figure 10 shows the relation between the total length of cracks in $ZrO_2$ surface layer of the FG TBC samples and number of thermal shock cycles. It can be seen in Figure 10 that the total length of cracks increases with increasing number of thermal shock cycles. The FGMs with higher $ZrO_2$ content, which corresponds to higher value of $n$, show higher total length of cracks generated on $ZrO_2$ surfaces and are easier to fracture in the FGMs under such cyclic thermal shock loading conditions. Among the tested samples, FGMs with $n = 0.5$ show the highest resistance to cyclic thermal shock loadings.

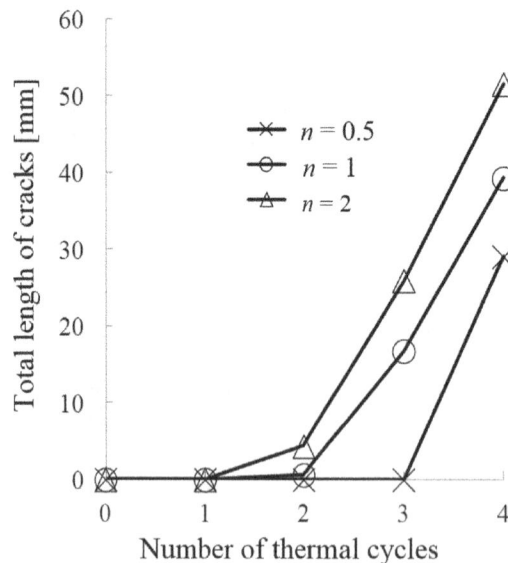

**Figure 10.** Cyclic thermal shock fracture behavior of FG TBCs. Relation between total length of cracks on $ZrO_2$ surface and number of thermal cycles is shown.

In the experiment, the average diameter of $ZrO_2$ powder is 26 nm, so it is reasonable to refer to simulation results for $d_{ZrO_2} = 0.01$ μm (10 nm). As to thermal fatigue behavior, it is considered with simulation results shown in Figure 6c that mean stresses, $\sigma_{mean}$, do not so much affect the cyclic thermal shock fracture behavior of the FG TBCs because even if $\sigma_{mean}$ is high, resistance to cyclic thermal shock loadings is high for $n = 0.5$. Meanwhile, it is seen in Figure 6a that higher n corresponds to higher $\sigma_{max}$, which is reasonable to explain experimental results on cyclic thermal shock fracture behavior of the FG TBCs. It is also seen in Figure 7a that higher $n$ corresponds to higher $\Delta\sigma$, which is reasonable to explain experimental results, in which higher n leads lower resistance to thermal shock loadings. For the ratio of maximum stress, $R$, it is seen in Figure 7c that higher n corresponds to lower $R$, which is also reasonable in connection with experimental data. With increasing number of thermal cycles, $R$ decreases in cases of $n = 1$ and 2, while $R$ is almost constant in case of $n = 0.5$. It is seen in Figure 8a that $\sigma_{max}^{eq}$ is also possibly a reasonable indicator for such fracture. In case of high creep rates of ZrO2 ($d_{ZrO_2} = 0.01$ μm), FG TBCs with $n = 2$ shows the highest values and FG TBCs with $n = 0.5$ show the lowest values in both $\sigma_{max}$ and $\sigma_{max}^{eq}$. This simulation result is reasonable to understand the current experimental data.

## 4. Discussion

Creep of ZrO2 is considered to largely affect the thermal shock fracture behavior of the FGMs. In comparison of simulation results with experimental data, micromechanical parameters, $\sigma_{max}$, $\sigma_{max}^{eq}$, $\Delta\sigma$ and $R$, are considered to be reasonable parameters for assessing cyclic thermal shock fracture behavior of ZrO2/Ti FGMs with high creep rate of ZrO2. According to Figure 9 showing the relation between $\Delta\sigma$ and $\sigma_{mean}$, in case of high creep rates of ZrO2, $\sigma_{mean}$ is very low compare to that in case of low creep rates of ZrO2, which is attributed to creep deformation of ZrO2 at the beginning of cooling processes under thermal shock loading conditions. As seen in Figures 6 and 8, absolute values of $\sigma_{max}$, $\sigma_{max}^{eq}$ and $\sigma_{mean}$ are largely different between high creep rates and low creep rates of ZrO2.

Because $\sigma_{max}$ and $\sigma_{max}^{eq}$ are considered to affect the cyclic thermal shock fracture behavior of the FGMs, such a difference need to be considered to engineer the FGMs with low creep rate of ZrO2 subject to cyclic thermal shock loadings. In general, it is considered that in case that $\sigma_{mean}$ is high, even small value of $\Delta\sigma$ largely affect the fatigue fracture behavior of the FGMs. Consequently the current simulation results are useful and effective to engineer such kinds of FG TBCs subject to cyclic thermal shock loadings. Parameters extracted from the simulation results are effective to use for predicting the cyclic thermal fracture criterion and design of the FG TBCs.

## 5. Conclusions

This study numerically investigates cyclic thermal shock behavior of ZrO2/Ti FG TBCs based on a nonlinear mean-field micromechanical approach, and some micromechanical parameters such as amplitudes, ratios, mean and maximum thermal stresses, and equivalent thermal stresses were examined to be used for engineering such FG TBCs. In the simulation, the fabrication processes for the FG TBCs were taken into account. The effect of creep and compositional gradation patterns on these micromechanical parameters, related to micro-stress states in the FG TBCs, were examined. The findings from the study can be summarized as follows.

1. In case of high creep rates of $ZrO_2$ ($d_{ZrO_2} = 0.01$ μm), $ZrO_2$-rich FG TBCs ($n = 2$) shows high maximum stresses, $\sigma_{max}$, in $ZrO_2$ layers, while Ti-rich FG TBCs ($n = 0.5$) shows high mean stresses, $\sigma_{mean}$. In case of low creep rates of $ZrO_2$ ($d_{ZrO_2} = 10$ μm), Ti-rich FG TBCs shows high $\sigma_{max}$ and Ti-rich FG TBCs with $n = 0.5$ also shows high $\sigma_{mean}$.

2. In case of both low and high creep rates of $ZrO_2$, $ZrO_2$-rich FG TBCs shows high amplitude of thermal stresses, $\Delta\sigma$. For thermal stress ratios, $R$, in case of low creep rates of $ZrO_2$, FG TBCs with any compositional gradation patterns show almost the same and constant for the number of thermal cycles.

3. In case of high creep rates of $ZrO_2$, $ZrO_2$-rich FG TBCs show high maximum equivalent stresses, $\sigma_{max}^{eq}$, while in case of low creep rates of $ZrO_2$, Ti-rich FG TBCs shows high $\sigma_{max}^{eq}$.

4. In comparison with experimental data, the simulation results are reasonable and effective to engineer such FG TBCs subject to cyclic thermal shock loadings. In experiment, in case of high creep rates of $ZrO_2$ ($d_{ZrO_2} = 0.026$ μm), Ti-rich FG TBCs exhibit high resistance to cyclic thermal shock loadings compared to linear and $ZrO_2$-rich FG TBCs. Consequently, the parameters of $\sigma_{max}$, $\sigma_{max}^{eq}$, $\Delta\sigma$ and $R$ are considered to be effective to estimate the thermal cyclic fracture behavior of the FG TBCs in case of high creep rates of $ZrO_2$.

## Nomenclature

| | |
|---|---|
| $\sigma_0^{in}, \sigma_1^{in}$ | In-plane micro-stresses |
| $\sigma_0^{out}, \sigma_1^{out}$ | Out-of-plane micro-stresses |
| $\sigma_0^{eq}, \sigma_1^{eq}$ | Equivalent micro-stresses |
| $\alpha_0, \alpha_1$ | Coefficient of thermal expansion |
| $f_0, f_1$ | Volume fraction |
| $\varepsilon^c$ | Creep strain |
| $\varepsilon^p$ | Plastic deformation with the strain |
| $\varepsilon^d$ | Eigen strain of the particle due to diffusional mass transport along the metal-ceramic interface |
| $\sigma$ | Macro-stress due to balanced bi-axial loadings |
| $\beta_0, \beta_1, \gamma_0, \gamma_1, \beta^*, \gamma^*$ | Micromechanical constants depending on the elastic constants and volume fraction of each phase |
| $N$ | In-plane force |
| $M$ | Bending moment |
| $a, c, n_p$ | Constants in Swift's equation |
| $C$ | Geometric constant |
| $D_{gb}$ | Grain boundary diffusivity |
| $\omega_{gb}$ | Grain boundary width |
| $\Omega$ | Volume of a diffusing atom |
| $k$ | Boltsman's constant |
| $C^{int}$ | Material constant derived from micromechanical considerations |
| $\omega_{int}$ | Interface width for diffusion |

| $D_{int}$ | Interfacial diffusivity |
|---|---|
| $\dot{\sigma}(z,t)$ | Plane stress rate |
| $S^e(z)$ | Overall plane-stress elastic compliance |
| $S^p(z,t)$ | Overall plane-stress plastic compliance |
| $\alpha(z)$ | Overall in-plane thermal expansion coefficient |
| $\dot{\varepsilon}^{p(cd)}(z,t)$ | Overall plastic strain rate due to the difference between creep abilities of each phase and interfacial diffusion |
| $\dot{\varepsilon}^{c\text{-}d}(z,t)$ | Overall creep strain rate |
| $f_m(i), f_c(i)$ | Volume fractions of metal(Ti) and ceramics(ZrO2) phases in the $i$-th sub-layer, respectively |
| $P$ | Total number of sub-layers |
| $n$ | Parameter characterizing the compositional gradation pattern |

## Conflicts of Interest

The author declares no conflict of interest.

## References

1. Xiong, H.P.; Kawasaki, A.; Kang, Y.S.; Watanabe, R. Experimental study on heat insulation performance of functionally graded metal/ceramic coatings and their fracture behavior at high surface temperatures. *Surf. Coat. Tech.* **2005**, *194*, 203–214.

2. Khor, K.A.; Gu, Y.W. Thermal properties of plasma-sprayed functionally graded thermal barrier coatings. *Thin Solid Films* **2000**, *372*, 104–113.

3. Miyamoto, Y.; Kaysser, W.A.; Rabin, B.H.; Kaeasaki, A.; Ford, R.G. *Functionally Graded Materials: Design, Processing and Applications*; Springer: New York, NY, USA, 1999.

4. Tsukamoto, H. Design against fracture of functionally graded thermal barrier coatings using transformation toughening. *Mater. Sci. Eng. A* **2010**, 527, 3217–3226.

5. Tsukamoto, H. Analytical method of inelastic thermal stresses in a functionally graded material plate by a combination of micro- and macromechanical approaches. *Comps. Part B Eng.* **2003**, *34*, 561–568.

6. Cho, J.R.; Ha, D.Y. Averaging and finite element discretization approaches in the numerical analysis of functionally graded materials. *Mater. Sci. Eng. A.* **2001**, *302*, 187–196.

7. Giannakopoulos, A.E.; Suresh, S.; Finot, M.; Olsson, M. Elastoplastic analysis of thermal cycling: layered materials with compositional gradients. *Acta Metall. Mater.* **1995**, *43*, 1335–1354.

8. Grujicic, M.; Zhang, Y. Determination of effective elastic properties of functionally graded materials using voronoi cell Finite Element Method. *Mater. Sci. Eng. A.* **1998**, *251*, 64–76.

9. Aboudi, J.; Pindera, M.J.; Arnold, S.M. Higher-order theory for functionally graded materials. *Comps. Part B Eng.* **1999**, *30*, 777–832.

10. Eshelby, J.D. The determination of the elastic field of inclusion and related problems. *Proc. R. Soc. Lond. A* **1957**, *241*, 376–396.

11. Eshelby, J.D. The Elastic field outside an ellipsoidal inclusion. *Proc. R. Soc. Lond. A* **1959**, *252*, 561–569.

12. Eshelby, J.D. Elastic inclusions and inhomogeneities. *Prog. Solid Mech.* **1961**, *2*, 89–140.

13. Wakashima, K.; Tsukamoto, H. Micromechanical approach to the thermomechanics of ceramic-metal gradient materials. In Proceedings of the 1st International Symposium on Functionally Gradient Materials, Sendai, Japan, 8–9 October 1990; pp. 19–26.

14. Wakashima, K.; Tsukamoto, H.; Ishizuka, T. Numerical approach to the elasticplastic analysis of thermal stresses in a ceramic-metal bi-material plate with graded microstructure. In *Modelling of Plastic Deformation Its Engineering Applications*, Proceedings of the 13th Risø International Symposium on Materials Science, Risø National Laboratory, Roskilde, Denmark, 7–11 September 1992; pp. 503–510.

15. Taya, M.; Lee, J.K.; Mori, T. Dislocation punching from interfaces in functionally graded materials. *Acta Mater.* **1997**, *45*, 2349–2356.

16. Tohgo, K.; Masunari, A.; Yoshida, M. Two-phase composite model taking into account the matricity of microstructure and its application to functionally graded materials. *Compos Part A Appl. S.* **2006**, 37, 1688–1695.

17. Mao, Y.Q.; Ai, S.G.; Fang, D.N.; Fu, Y.M.; Chen, C.P. Elasto-plastic analysis of micro FGM beam basing on mechanism-based strain gradient plasticity theory. *Comps Struct.* **2013**, *101*, 168–179.

18. Nejad, M.Z.; Kashkoli, M.D. Time-dependent thermo-creep analysis of rotating FGM thick-walled cylindrical pressure vessels under heat flux. *Int. J. Eng. Sci.* **2014**, *82*, 222–237.

19. Lidong, T.; Wenchao, L. Residual stress analysis of Ti-ZrO$_2$ thermal barrier graded materials. *Mater. Design.* **2002**, *23*, 627–632.

20. Gravie, R.C.; Hannink, R.H.; Pascoe, R.T. Ceramic steels. *Nature.* **1975**, *258*, 704–730.

21. Kelly, P.M.; Rose, L.R.F. The martensitic transformationin ceramics—Its role in transformation toughening. *Prog. Mater. Sci.* **2002**, *47*, 463–557.

22. Lin, K.L.; Lin, C.C. Reaction between titanium and zirconia powders during sintering at 1500 °C. *J. Am. Ceram. Soc.* **2007**, *90*, 2220–2225.

23. Tsukamoto, H.; Kunimine, T.; Yamada, M.; Sato, H.; Watanabe, Y. Microstructure and mechanical properties of Ti-ZrO$_2$ composite fabricated by spark plasma sintering. *Key Eng. Mater.* **2012**, *520*, 269–275.

# Permissions

The contributors of this book come from diverse backgrounds, making this book a truly international effort. This book will bring forth new frontiers with its revolutionizing research information and detailed analysis of the nascent developments around the world.

We would like to thank all the contributing authors for lending their expertise to make the book truly unique. They have played a crucial role in the development of this book. Without their invaluable contributions this book wouldn't have been possible. They have made vital efforts to compile up to date information on the varied aspects of this subject to make this book a valuable addition to the collection of many professionals and students.

This book was conceptualized with the vision of imparting up-to-date information and advanced data in this field. To ensure the same, a matchless editorial board was set up. Every individual on the board went through rigorous rounds of assessment to prove their worth. After which they invested a large part of their time researching and compiling the most relevant data for our readers.

The editorial board has been involved in producing this book since its inception. They have spent rigorous hours researching and exploring the diverse topics which have resulted in the successful publishing of this book. They have passed on their knowledge of decades through this book. To expedite this challenging task, the publisher supported the team at every step. A small team of assistant editors was also appointed to further simplify the editing procedure and attain best results for the readers.

Apart from the editorial board, the designing team has also invested a significant amount of their time in understanding the subject and creating the most relevant covers. They scrutinized every image to scout for the most suitable representation of the subject and create an appropriate cover for the book.

The publishing team has been an ardent support to the editorial, designing and production team. Their endless efforts to recruit the best for this project, has resulted in the accomplishment of this book. They are a veteran in the field of academics and their pool of knowledge is as vast as their experience in printing. Their expertise and guidance has proved useful at every step. Their uncompromising quality standards have made this book an exceptional effort. Their encouragement from time to time has been an inspiration for everyone.

The publisher and the editorial board hope that this book will prove to be a valuable piece of knowledge for researchers, students, practitioners and scholars across the globe.

# List of Contributors

**Takeo Oku**
Department of Materials Science, The University of Shiga Prefecture, 2500 Hassaka, Hikone, Shiga 522-8533, Japan

**Tetsuya Yamada**
Department of Materials Science, The University of Shiga Prefecture, 2500 Hassaka, Hikone, Shiga 522-8533, Japan

**Kazuya Fujimoto**
Department of Materials Science, The University of Shiga Prefecture, 2500 Hassaka, Hikone, Shiga 522-8533, Japan

**Tsuyoshi Akiyama**
Department of Materials Science, The University of Shiga Prefecture, 2500 Hassaka, Hikone, Shiga 522-8533, Japan

**Parnia Navabpour**
Teer Coatings Ltd., Miba Coating Group, West Stone House, Berry Hill Industrial Estate, Droitwich WR9 9AS, UK

**Soheyla Ostovarpour**
Faculty of Science and Engineering, Manchester Metropolitan University, Chester Street, Manchester M1 5GD, UK

**Carin Tattershall**
Cristal Pigment UK Ltd., P.O. Box 26, Grimsby, North East Lincolnshire, DN41 8DP, UK

**Kevin Cooke**
Teer Coatings Ltd., Miba Coating Group, West Stone House, Berry Hill Industrial Estate, Droitwich WR9 9AS, UK

**Peter Kelly**
Faculty of Science and Engineering, Manchester Metropolitan University, Chester Street, Manchester M1 5GD, UK

**Joanna Verran**
Faculty of Science and Engineering, Manchester Metropolitan University, Chester Street, Manchester M1 5GD, UK

**Kathryn Whitehead**
Faculty of Science and Engineering, Manchester Metropolitan University, Chester Street, Manchester M1 5GD, UK

**Claire Hill**
Cristal Pigment UK Ltd., P.O. Box 26, Grimsby, North East Lincolnshire, DN41 8DP, UK

**Mari Raulio**
VTT Technical Research Centre of Finland, P.O. Box 1000, FI-02044 VTT Espoo, Finland

**Outi Priha**
VTT Technical Research Centre of Finland, P.O. Box 1000, FI-02044 VTT Espoo, Finland

**Sanjay S. Latthe**
Photocatalysis International Research Center, Research Institute for Science & Technology, Tokyo University of Science, Noda, Chiba 278-8510, Japan

**Shanhu Liu**
Photocatalysis International Research Center, Research Institute for Science & Technology, Tokyo University of Science, Noda, Chiba 278-8510, Japan

**Chiaki Terashima**
Photocatalysis International Research Center, Research Institute for Science & Technology, Tokyo University of Science, Noda, Chiba 278-8510, Japan

**Kazuya Nakata**
Photocatalysis International Research Center, Research Institute for Science & Technology, Tokyo University of Science, Noda, Chiba 278-8510, Japan

**Akira Fujishima**
Photocatalysis International Research Center, Research Institute for Science & Technology, Tokyo University of Science, Noda, Chiba 278-8510, Japan

**Christopher O. Phillips**
Welsh Centre for Printing and Coating, College of Engineering, Swansea University, Singleton Park, Swansea SA2 8PP, UK

**David G. Beynon**
Welsh Centre for Printing and Coating, College of Engineering, Swansea University, Singleton Park, Swansea SA2 8PP, UK

**Simon M. Hamblyn**
Welsh Centre for Printing and Coating, College of Engineering, Swansea University, Singleton Park, Swansea SA2 8PP, UK

**Glyn R. Davies**
Welsh Centre for Printing and Coating, College of Engineering, Swansea University, Singleton Park, Swansea SA2 8PP, UK

**David T. Gethin**
Welsh Centre for Printing and Coating, College of Engineering, Swansea University, Singleton Park, Swansea SA2 8PP, UK

**Timothy C. Claypole**
Welsh Centre for Printing and Coating, College of Engineering, Swansea University, Singleton Park, Swansea SA2 8PP, UK

**Roberto Fioretti**
Department of Industrial Engineering and Mathematical Sciences, Università Politecnica delle Marche, Via Brecce Bianche 12, 60131 Ancona, Italy

**Paolo Principi**
Department of Industrial Engineering and Mathematical Sciences, Università Politecnica delle Marche, Via Brecce Bianche 12, 60131 Ancona, Italy

**Peter Zarras**
Naval Air Warfare Center Weapons Division (NAWCWD), Polymer Science & Engineering Branch

**Christopher E. Miller**
Army Research Laboratory, Coatings and Corrosion, Building 4600, ARSRD-ARL-WM-SG
Aberdeen Proving Ground, MD 21005-5069, USA

**Cindy Webber**
Naval Air Warfare Center Weapons Division (NAWCWD), Polymer Science & Engineering Branch

**Nicole Anderson**
Naval Air Warfare Center Weapons Division (NAWCWD), Polymer Science & Engineering Branch

**John D. Stenger-Smith**
Naval Air Warfare Center Weapons Division (NAWCWD), Polymer Science & Engineering Branch

**Nicholas Curry**
Department of Engineering Science, University West, Gustava Melins Gata 2, Trollhattan 461 86, Sweden

**Kent VanEvery**
Progressive Surface, Grand Rapids, MI 49512, USA

**Todd Snyder**
Progressive Surface, Grand Rapids, MI 49512, USA

**Nicolaie Markocsan**
Department of Engineering Science, University West, Gustava Melins Gata 2, Trollhattan 461 86, Sweden

**Moisés Bueno**
Laboratory of Acoustics Applied to Civil Engineering (LA2IC), Universidad de Castilla-La Mancha, Avda. Camilo José Cela s/n, 13071 Ciudad Real, Spain

Road Engineering Laboratory, Empa, Swiss Federal Laboratories for Material Science andTechnology, Ueberladstr. 129, CH-8600 Duebendorf, Switzerland

**Jeanne Luong**
Laboratory of Acoustics Applied to Civil Engineering (LA2IC), Universidad de Castilla-La Mancha, Avda. Camilo José Cela s/n, 13071 Ciudad Real, Spain
Environmental Sciences and Technologies Department, University of Liège, Passage des Déportés 2, 5030 Gembloux, Belgium

**Fernando Terán**
Laboratory of Acoustics Applied to Civil Engineering (LA2IC), Universidad de Castilla-La Mancha, Avda. Camilo José Cela s/n, 13071 Ciudad Real, Spain

**Urbano Viñuela**
Laboratory of Acoustics Applied to Civil Engineering (LA2IC), Universidad de Castilla-La Mancha, Avda. Camilo José Cela s/n, 13071 Ciudad Real, Spain

**Víctor F. Vázquez**
Laboratory of Acoustics Applied to Civil Engineering (LA2IC), Universidad de Castilla-La Mancha, Avda. Camilo José Cela s/n, 13071 Ciudad Real, Spain

**Santiago E. Paje**
Laboratory of Acoustics Applied to Civil Engineering (LA2IC), Universidad de Castilla-La Mancha, Avda. Camilo José Cela s/n, 13071 Ciudad Real, Spain

**Gaetano Licitra**
Department of Lucca, Via A.Vallisneri n.6, Agenzia Regionale per la Protezione Ambientale della Toscana, ARPAT, Lucca I-55100, Italy
Consiglio Nazionale delle Ricerche CNR, IPCF, Via G.Moruzzi n.1, Pisa I-56124, Italy

**Mauro Cerchiai**
Agenzia Regionale per la Protezione Ambientale della Toscana, ARPAT, Area Vasta Costa – Settore Agenti Fisici, Via V.Veneto n.27, Pisa I-56127, Italy

**Luca Teti**
Consiglio Nazionale delle Ricerche CNR, IPCF, Via G.Moruzzi n.1, Pisa I-56124, Italy

**Elena Ascari**
Physics Department, University of Siena, Via Roma n.56, Siena I-53100, Italy
Consiglio Nazionale delle Ricerche CNR, IDASC, Via Fosso del Cavaliere n.100, Roma I-00133, Italy

**Francesco Bianco**
Physics Department, University of Siena, Via Roman.56, Siena I-53100, Italy

**Marco Chetoni**
Consiglio Nazionale delle Ricerche CNR, IPCF, Via
G.Moruzzi n.1, Pisa I-56124, Italy

**Giovanni Di Girolamo**
ENEA, Materials Technology Unit, Casaccia Research
Center, Rome 00123, Italy

**Alida Brentari**
ENEA, Materials Technology Unit, Faenza Research
Center, Faenza 48018, Italy

**Emanuele Serra**
ENEA, Materials Technology Unit, Casaccia Research
Center, Rome 00123, Italy

**Elia Boonen**
Belgian Road Research Center (BRRC), Woluwedal 42,
1200 Brussels, Belgium

**Anne Beeldens**
Belgian Road Research Center (BRRC), Woluwedal 42,
1200 Brussels, Belgium

**Eifion Jewell**
College of Engineering, Swansea University, Swansea
SA2 8PP, UK

**Simon Hamblyn**
College of Engineering, Swansea University, Swansea
SA2 8PP, UK

**Tim Claypole**
College of Engineering, Swansea University, Swansea
SA2 8PP, UK

**David Gethin**
College of Engineering, Swansea University, Swansea
SA2 8PP, UK

**Hideaki Tsukamoto**
Graduate School of Engineering, Nagoya Institute of
Technology, Gokiso-cho, Showa-ku, Nagoya, Aichi, 466-
8555, Japan